© 1998 Algrove Publishing Limited
ALL RIGHTS RESERVED.
No part of this book may be reproduced in any form including photocopying without permission in writing from the publishers, except by a reviewer who may quote brief passages in a magazine or newspaper or on radio or television.

Algrove Publishing Limited
1090 Morrison Drive
Ottawa, Ontario
Canada K2H 1C2

Distributed in the United States by:
Veritas Tools Inc.
12 East River Street
Ogdensburg, New York 13669

Distributed in Canada by:
Lee Valley Tools Ltd.
1090 Morrison Drive
Ottawa, Ontario K2H 1C2

Canadian Cataloguing in Publication Data

Boulton and Paul Limited
    Boulton & Paul, Ltd., 1898 catalogue : Rose Lane Works, Norwich

2nd ed.
(Classic reprint series)
Includes index.
First published as catalogue no. 97: Norwich : Boulton & Paul, 1898.
ISBN 0-921335-21-0

    1. Boulton and Paul Limited--Catalogs.   2. Buildings, Prefabricated--England--Catalogs.   3. Buildings, Prefabricated--Catalogs.   I. Title. II. Series: Classic reprint series (Ottawa, Ont.).

HD9521.9.B6B68 1998   693'.97'0294   C98-900102-4

Printed in Canada
#20499

Publisher's Note

It was just 200 years ago, the fall of 1797, that William Moore left his home on a farm near the village of Warham and moved to Norwich. The move was probably dictated more by fear than hope. The British Navy, caught up in the Napoleonic wars, had press gangs scouring the coasts for young men to man the ships. If you could not amply demonstrate the necessity of your work to the general economy, you could be pressed into a service that Samuel Johnson once described as "…being in a jail with the chance of being drowned." On another occasion Johnson stated that he would prefer jail to the British Navy since "A man in a jail has more room, better food, and commonly better company."

William Moore could well have been motivated by similar views when he found premises on Cocky Lane in Norwich and established himself as an ironmonger. Beginning with fire grates, he expanded into kettles and took on the services of John Barnard, who later became a partner and inherited the business when William Moore died. In 1844, needing someone to help him with the correspondence and reckoning that was required in the small business, John Barnard eventually brought into partnership William Boulton who, in 1853, engaged a 12-year-old boy by the name of Joseph Paul. Paul developed a keen interest in the manufacturing side and became Works Manager and a partner in 1864 at the age of 23.

It was from these modest beginnings that the firm of Boulton & Paul grew. Thirty years later, when this catalog was published, they were not only the main suppliers of greenhouses, aviaries and conservatories of the day, but they were prefabricators of every sort of building. If you happened to be moving to the colonies and were uncertain of the sort of accommodation you would be able to have constructed, you could order a house and a complete set of farm buildings to take on board with you. The empire was still in full flower and missionaries were active, as shown by the various designs of churches, field hospitals and schools that were available.

The catalog unwittingly demonstrates the difficulty Great Britain was having in adjusting to the new industrialized society. The list of patrons in the front of the catalog includes a generous sprinkling of royalty and then some 300 titled gentry. However, a close look at the larger structures that had been installed by Boulton & Paul will reveal that most of them were for the untitled industrialists of the day, the "nouveau riche". Boulton & Paul made their money from the new economy, but were still figuratively tugging their forelocks to the old.

This catalog represents the product line of Boulton & Paul at its zenith. The world continued to change and Boulton & Paul were forced to change with it. Although the firm still exists, it has been through some difficult times and has recently become a small part of a much larger enterprise, Rugby PLC. Today, the firm makes windows, doors and kitchen cupboards for others, not the majestic glass edifices of old.

The great value of this catalog today is not just in the excellent representation of the types of prefabricated structures available in the late Victorian era, but the finely detailed drawings of the numerous small structures that are quite within the capability of a keen amateur builder today. You may not want the small building on page 310B as a chicken roost, but it would make an excellent garden shed.

Leonard G. Lee, Publisher
Ottawa
January, 1998

Boulton & Paul, Ltd.,

Manufacturers,

Rose Lane Works,

Norwich.

ENTERED AT STATIONERS HALL.

<u>All previous Catalogues withdrawn.</u>

No. 97.
Revised Edition.

—÷—

March, 1898.

—÷—

Terms of Payment. NET CASH on Delivery of Goods or completion of Contract.

Prices. Our Prices are fixed to meet the requirements of <u>Cash Buyers only,</u> Customers thereby getting the benefit of Manufacturers' Prices.

Prices are subject to Market fluctuations, and Estimates should always be obtained before ordering.

References. If correspondents unknown to us will kindly send a remittance with their orders, it would save delay.

Delivery of Goods. Carriage Paid on all Orders above 40/- value (except where otherwise stated) to any Railway Station in England and Wales.

Packages. Boxes and Packing charged, and allowed for if returned Carriage Paid. No Credit given for Empties unless advised and returned within seven days.

Breakages. Goods broken in transit, if signed for as such, can be returned and will be replaced Free of Cost and Free of Carriage. No deductions allowed for Breakages or Repairs.

Copyright. This Catalogue is Registered and Copyright secured.

ROSE LANE WORKS, NORWICH.

Contents

A.

	PAGE
Amateur's Garden Engine	216
Angle Posts	195
Annealed Wire	244
Arboretum Shelter	159, 161
,, Seat	227
Arches—Garden	191, 192, 194
Awnings	224
Aviaries	328, 338—345
Aviary for Tropical Birds	342

B.

	PAGE
Ball Room	125
Band Stands	155, 157
Bantam House	315, 316
Barbed Wire Fencing	245
,, Stretcher	245
Bar Fencing	248, 249
Barns	177
Barrows—Water, &c.	202, 203, 216
,, Wheel	218, 219
Bath—Swimming	102
Bathing Chalet	102
Bathing Hut	102
Baths for Pigeons and Aviaries	325
Bay Window with Balcony	155
Bee Hives—Covers for	231
Bee Houses and Hives	230
Benches—Folding	296
Benches for Dog Shows	296
Billiard Room	130, 131
Bins—Dust	219, 220
Bins for Corn	275
Bins—Wine	221
Black Varnish	236, 249
Boat Houses	102, 163
Boilers	62—70
Bottle Racks	221
Boundary Railing	273
Breeding Pens for Pheasantries	333
Bridges	162
Bridge House	166
Bungalows	113—117, 124
Buralls' Poultry Feeder	318

C.

	PAGE
Cabmen's Shelter	155, 159
Calf Crib	274
Carriage Gates	265
Carriage Washing Shed	150
Cart Sheds	145, 177
Carts—Water and Manure	202—207
,, Milk	217
Cast-iron Gratings	79
Cattle Crib	274
Cattle Feeding Troughs	274
Cattle Fittings	151
Cattle Hurdles	253
Cattle Shed and Yard with Keeper's Cottage	176
Cattle Troughs	275
Chaff Cutters	276
Chairs—Garden	225, 226, 228, 229
Chicken House	315, 316
Chicken Hurdles	313
Chicken Rearer	318
Chicken Runs	316
Chicken Nursery	316
Chrysanthemum House	26
Children's Play Room	134
Churches	82—90
Church Fittings	91
Cinder or Gravel Screen	231
Cinder Sifter	218, 219
Cisterns	58, 275
Cisterns—Supply	77
,, Water	77
Cloak Room and Lavatory	161
Club House	164, 165
Club Seat	227
Coach Houses	142, 144
Cockerel Runs	316
Cocks for Boilers	77
Coils and Cases	78
Composite Kennels	286, 288
Combination Aviary and Summer House	328
Combination Cow House, Stable, Piggery, &c.	174
Combination Cote for Pet Birds, Rabbits, Guinea Pigs, &c.	343
Combination Fowl House, Aviary, Dove Cote, and Kennel	306

	PAGE
Combination Fowl House, Aviary, and Pheasantry ...	305
Combination Piggery and Fowl House	175
Combination Pigeon Cote and Fowl House	327
Combination Hut	135
Concert Hall	166
Conservatories	12—31, 35, 38, 39
Conservatory Engine and Pump ...	202, 209, 214, 215
Contractor's Hut	136
"Cookson" Hatching Box	315
"Cookson" Run for Setting Hens	315
Coops	317—319
Coop for Exporting Birds	319
Corn Bins	275, 320
Corn Grinding and Crushing Mills	276
Corrugated Fencing	175
Corrugated Iron Wall for Training Fruit Trees ...	197
Corrugated Sheets	179
Corrugated Skylight	179
Cote for Pets	329
Cottage and Laundry	132
Cottage Hospital	94
Cottage Pump	209
Cottage, Stable, and Coach House	143
Cottages	109, 110, 117—121
Covered Cattle Yards	176, 177
Crestings and Finials	60
Cricket Pavilion	169, 170
Cricket Tents	223
Cucumber Frames	54
Cups—Feeding or Drinking	323

D.

Dairy	152, 153
Deep Well Pumps	211
Diamond Wire Trellis	59, 198
,, Wood Trellis	198
Direction Plates and Posts	233
Dog Benches	296
Dog Kennels	279—282
Dog Travelling Boxes	292
Dog Troughs	296
Door Scrapers	231
Double Kennels	281, 285, 289
Double Leaf Iron Gates	250, 267, 268
Double Park Gates	269
Double Poultry Houses and Runs	305, 308
Dove Cotes	325, 327
Drop Fence	241
Duck Houses	319
Duck Pond	319
Dust Bath	321
Dust Bins	219, 220
Dwelling-houses	121, 122

E.

	PAGE
Eagle House	342
Eel Traps	334
Egg Cabinet	321
Enclosure for Dogs	294
Engines—Garden, &c.	202, 213, 214, 216
Entrance Gates	261, 262, 264, 265—268
Espaliers	194, 195
Esplanade Shelter	160, 161
Exhibition Travelling Boxes for Dogs	292
Explorer's Tent	223
Eyes—Wrought-iron	59, 196

F.

Farm Buildings	174
Fattening Coop	314
Fattening Pens	314
Feeding Cage for Rabbits	298
Feeding Coop	314
Feeding or Drinking Cups	323
Feeding Hoppers for Pigeons and Poultry ...	325, 327
Feeding Troughs	274, 322
Fencing	240—249, 257, 258
Fencing Wire	244
Fencing for Poultry	312, 313
Fencing—Tennis and Cricket	222
Ferret Box	292
Ferret Feeding Pan	292
Ferret Hutch	292
Ferret Kennels	292
Field Gates	251
Field Hurdles	253
Finials and Crestings	60
Fire Bucket	214
Fire Engines	214
Fishing Temple and Boat House	163
Fittings for Cattle	151
Fittings for Stables	151, 181—188
Folding Day and Night Bench for Dogs ...	296
Flower Court	14
Folding Hurdles—Sheep	254
Folding Steps	231
Foot Bridges	162
Footpaths—Wood Batten	58
Footpath Gate	250
Foot Rests	225, 229
Force Pumps	208, 209, 210, 215
Forcing Houses	42, 43
Forcing Pit	45, 51
Fountains	232
Fountains for Poultry	322
Fowl Houses	308, 309, 310, 311, 340
,, On Wheels	311

ROSE LANE WORKS, NORWICH.

	PAGE
Frames—Garden	50—55
Fruit Protector	197
Furnace Fittings	76
,, Fronts	76
,, Tools	69—76

G.

Galvanized Raidisseurs	59, 195, 196
Galvanized Wire	195, 244
Galvanized Wire Netting	232—240
Galvanized Wire Strawberry Protectors	199
Game Larder	135, 152
Game Proof Hurdles	235, 313
Gamekeeper's Cottage	111
Gamekeeper's Hut	135, 136
Garden Arches	191, 192, 194
Garden Barrows	202, 203, 216
Garden Bordering	232—234
Garden Chairs and Seats	224, 226—229
Garden Engines	202, 213, 214, 216
Garden Fountains	232
Garden Frames	50—55
Garden Hose	212
Garden Houses	136, 154, 156, 157
Garden and Park Hurdles	258
Garden Pumps	208—211, 213, 215
Garden etc., Rollers	200
Garden Scrapers	231
Garden Stakes	199
Garden Swing	224
Garden Tent	223
Garden Truck	231
Garden Vases	232
Garden—Winter	12, 23, 27
Gates	250, 251, 256, 257, 259—265, 268
Goat House	325
Golf Houses	167—169
Golf Locker	168
Grand Stand	160
Gratings—Iron	79
Gravel Screen	231
Greasing Jack	218
Greenhouses	39, 40, 44, 46—48
Guards—Tree	235, 252
Gun Room	152
Gun Room Fittings	153
Gymnasium	99

H.

Handlights	57
Harness Room	150
Harness Room Fittings	181, 182
Harness Room Stove	120, 150
Hatching Boxes	315
Hay Racks	240
Heads for Stone Posts	273
Heating Apparatus	61—80
Hives—Bee	230
,, Covers for	231
Holdfasts	196
Hooks	239
Hose	212
Hose Reels	212
Hospitals	93—95
Hot-water Pipes	72
Hound Kennel	279, 280, 280A
House for St. Bernards or Mastiffs	291
Houses—Portable	82, 177
House for Bears and Kangaroos	281
House for Guinea Pigs and Rabbits	316
Hunting Establishment	141, 146, 281A
Hunting Lodge	114
Hurdles	253—257
Hurdles for Poultry	312, 313
Hutches for Ferrets and Rabbits	292, 297, 298
Huts	135, 136

I.

Intermediate Standards	244
Iron Heads for Stone Posts	273
Iron Roofs	178, 179
Iron Railing	270—273
Iron Staples	242
Iron Tables	226, 229

K.

Keeper's Dog Kennel	291
Keeper's Coop	317
Keeper's Watch Hut	136
Kennel Fencing	296
Kennel Railing	294—296
,, For Puppies	293
Kennels	279—292
Kitchen Range	126
Kiosks	157—171

L.

Lambing Hut	135
Larders	135, 152, 323
Large Poultry House	304
Lattice Hurdle	255
Laundries	132
Laundry Stoves	132

Lawn Conservatory 50	Pascall's Chick Feeding Run 314
Lawn Mowers 201	Pavilions ... ... 164, 166—172
Lawn Watering Machines 213, 214, 216	Payne-Gallwey Coop 317
Lean-to Fowl House 307	Peach Houses ... ... 35—37
Lean-to Porch 127	Pea Fowl House 340
Lights—For Pits, &c. 57	Pea Guards 199
Linen-Posts and Socket 231	Pea Trainer 199
Liquid Manure Barrows and Carts ... 202—205, 207	Pens for Pheasant Breeding 333
Liquid Manure Pumps 209	Perches for Pigeon Lofts 325
Lodges 115, 118	Pheasant Feeder 318, 332
	Pheasantries 335, 341, 342
M.	Pheasantry, Poultry, or Pigeon House ... 336
	Pheasant Trap 334
Manure Barrows and Carts ... ... 202—205, 207	Pigeon Bath 325
Materials for Roof ... ... 178—180	Pigeon Cotes ... ... 325—330, 341, 343, 344
Materials required for Wiring Garden Walls 56, 59, 195, 196	Pigeon Feeding Hoppers 325
Meat Safes 323	Pigeon House ... ... 329—330
Method of Wiring in Fruit Gardens 194	Pigeon Loft with Flight 330
Melon Frames 54	Pigeon Nest and Nesting Boxes 325
Milk Carts and Carriages 217	Pigeon Shooting Trap 327
Mission Room 82, 88, 90, 95	Piggeries 175
Mission Room Fittings 101	Piggery, Cow House, Stable, and Fowl House ... 174
Model Poultry Establishment 301	Pig Trough 274
Monkey House or Aviary 345	Pipe Fittings and Connections ... ... 73—75
Morant Hutch 297	Pipes—Hot-water 72
Movable Fowl House ... ... 308—311	Pipe Rests 80
Music Room 125	Pit Lights 57
	Plant Carrier 231
N.	Plant House 17, 35, 38, 42
	Plant Preserver 50, 51
Nesting House on Wheels 315	Plant Protector 197
Nest Boxes for Poultry 315	Plates and Eyes 56, 59, 196
Nesting Boxes for Pigeons 325	Pond for Ducks 319
Netting ... ... 232—240	Porches—Glazed 32, 33
Netting Stakes 238, 239	Porches 127, 128
	Portable Buildings ... ... 82—177
O.	Portable Cellarets 221
	Portable Dust Bin 219, 220
Offal Pans 220	Portable Fowl House ... ... 308—311
Orchard House 17, 31, 36	Portable Iron Buildings 134, 135
Orchid Houses, &c. 34	Portable Wood Buildings 134, 135
Ornamental Cottage 112, 115	Portable Iron Studio 96
Ornamental Cast-Iron Grating 79	Portable Pumps 208, 209
Ornamental Railing 270, 273	Posts 231
Out-houses—Portable 136	Poultry Establishments 301, 302
	Potting House 136
P.	Poultry and Pheasant Fencing 312, 313
	Poultry and Pheasant Feeder 318
Paints—Mixed 236	Poultry Drinking Fountains 322
Palisading 259	Poultry Houses and Runs ... 300, 303—305, 307
Pans—Offal 220	Poultry Hurdles 312, 313
Parish Room 97	Poultry on Farms 311
Park Fencing 257, 258	Poultry Pens 312, 313
Parochial Hall 88	Poultry Shelter 321

ROSE LANE WORKS, NORWICH.

	PAGE
Poultry Show Pens	323
Poultry Troughs	321, 322
Propagating Glasses	57
Protectors—Wall Fruit Tree	56
Pumps	208—211, 213, 215
Puppy House	293
Puppy Railing	293

R.

	PAGE
Rabbit Fencing	240, 241
Rabbit Hutches	297, 298
Racks—Bottle	221
,, Hay	240
Raidisseurs—Galvanized	59, 195, 196
Range of Covered Yards for Cattle	176, 177
Range of Dog Houses and Yards	283
Range of Kennels	280, 282, 284, 285, 287, 289
Range of Poultry Houses and Yards	300, 303, 304
Railing	270—273, 294, 295
Railing for Puppies	293
Raspberry Trainer	199
Registered Dog Kennel	290, 291
Residences	101—108, 122, 123
Rick Covers	180
Roofing Materials	178, 179
Roof Principals	178
Rope—Strand	231
Roseries	189—193
Rustic Table	226

S.

	PAGE
Sack Barrow	218
Sanitary Barrow	220
Sanitary Bin	220
Sanitary Pans	220
Sanitary Truck	220
School Desks and Seats	100, 101
Schools	96—98
Scorer's Tent	223
Scrapers	231
Screen—Cinder or Gravel	231
Screw Jack	218
Scullery	153
Seats—Garden	224—229
Seed Protector	197
Setting Boxes, with runs	315
Sheds	145, 150, 177
Shed to Wash Carriages under	150
Sheep Fold Hurdles	254
Sheep Netting	139
Sheep Shelters	274
Sheep Troughs	274

	PAGE
Shelter for Watering Places, Parks, Schools, &c.	159
Shelters	155, 158—160, 274
Shelves for Greenhouses	58
Shepherd's Hut	135
Shooting Boxes	108, 109, 114
Side Stays	243
Sifter—Cinder	218, 219
Sinks for Cottage	126
Skylight	179
Smoke Room	131
Soot Boxes	76
Sportsman's Tent	223
Stables	137, 139—141, 144, 147—149
Stable and Coach House	138, 143
Stable and Shed	145
Stable, Coach House, and Cottage	138, 143
Stable, Cow House, Piggery, and Fowl House	174
Stable Fittings	151, 181—188
Stable with Loose Boxes, Harness Room, &c.	142
Stable for Two Loose Boxes	144
Stages in Greenhouses	58
Stakes—Garden	199
Stakes—Netting	238, 239
Stand for Storing Fruit	197
Standards	244
Staples	242
Step Ladder	231, 247, 250
Stock Feeding Machinery	276
Stove for Harness Room	126, 150
Stoves for Iron Buildings	126
Straight Wire Lattice	198
Straining Bolt	196
Straining Posts	243
Strained Wire Fences	246, 247
Strainers	243
Strand Rope	231
Strawberry Protector	199
Street Watering Cart	205, 207
Studios	96
Study	131
Summer Houses	136, 154, 156, 157
Summering Boxes for Hunters	142
Swift on Wheels	245
Swing Water Barrows	202—204
Swing Water Carts	203, 204

T.

	PAGE
Tables—Iron	226, 229
Tablets	233
Tandy's Chicken Rearer	318
Tanks—Water	57, 77, 206, 215, 275
Tennis Chair	229
Tennis Fencing	222

	PAGE
Tennis Pavilion	170
Tents	223
Terminal Posts	195
Tomb Railing	273
Tool House	145
Tools for Erecting Fencing	242
Trainers	195, 196, 199
Training Hooks and Eyes	196
Traps—For Shooting Pigeons and Birds ...	327
,, For Eels	334
,, For Pheasants	334
,, For Catching Weasels, Stoats, or Rats ...	334
Travelling Boxes for Dogs	292
Tree Guards	235, 252
Tree Planter	242
Tree Protectors—Wall Fruit	56
Trellis—Wire and Wood	59, 198
Troughs	240, 321
Truck for Garden or House	231

U.

Umbrella Stand	231
Unclimbable Hurdles, Railway Gates and Palisading, and Fencing	256, 257
Unions—Brass	212

V.

Valves, &c.	71
Vapour Troughs	72
Varnish	236, 249
Vases	232
Verandahs	32, 33, 128, 129
Villa Poultry House	307
Vineries	35, 38, 40, 41

W.

	PAGE
Wall Fruit-tree Protector	56
Wall Trainers	195
Waiting Room	159
Warehouses	133, 144
Warren Fencing	240, 241
Water Barrows	202, 203, 216
Water Carts	203—207
Watering Machine	213
Watchman's Hut	136
Water Director	212
Water Supply Pipes	77
Water Tanks	58, 77, 206, 215, 275
Weighing Machine Room	159
Wheelbarrows	218, 219
Wicket Gates	250, 260, 263, 264
Wigwam Fowl House and Run	305
Winter Garden	12, 23, 27
Wire—Barbed	245
Wire Cage Trap	334
Wire—Galvanized	59, 195, 244
Wire Fences	246, 247
Wire for Fencing	244
Wire Holders for Preserving Grapes in Pickle Bottles	197
Wire Netting	232—240
Wire Netting—Tree and Plant Guards ...	235
Wire Strainers	243
Wiring Walls	56, 59, 195, 196
Wood Batten Floors for Dog Kennels ...	296
Wood Batten Paths for Greenhouses ...	58
Wood Kennels for Hounds	275
Wood and Wire Trellis	198
Workshops	96, 133
Wrought-iron Pipes	77
,, Fittings	77
Wrought-iron Tank on Wheels	206, 215, 219
Wrought-iron Railing	270, 273

Under the Distinguished Patronage of
HER MOST GRACIOUS MAJESTY THE QUEEN.

H.R.H. THE PRINCE OF WALES, K.G.
H.R.H. THE DUKE OF EDINBURGH, K.G.
H.R.H. THE DUKE OF CONNAUGHT, K.G.
H.R.H. THE DUKE OF YORK, K.G.
H.I.M. THE SULTAN OF TURKEY.

HER MAJESTY'S BOARD OF WORKS, THE ADMIRALTY, THE WAR DEPARTMENT, THE ROYAL GARDENS AT KEW, THE ROYAL HORTICULTURAL SOCIETY, ETC., ETC.

His Grace the Archbishop of York
His Grace the Duke of Abercorn
His Grace the Duke of Bedford
His Grace the Duke of Devonshire
His Grace the Duke of Grafton
His Grace the Duke of Hamilton
His Grace the Duke of Leinster
His Grace the Duke of Marlborough
His Grace the Duke of Northumberland
His Grace the Duke of Norfolk
His Grace the Duke of Richmond
His Grace the Duke of Sutherland
His Grace the Duke of Westminster
Her Grace the Duchess of Abercorn
Her Grace the Duchess of Montrose
Her Grace the Dow. Duch. of St. Albans
The Mt. Hon. the Mar. of Abergavenny
The Mt. Hon. the Marquis of Bath
The Mt. Hon. the Marquis of Bute
The Mt. Hon. Lord Breadalbane
The Mt. Hon. the Mar. of Cholmondeley
The Mt. Hon. the Marquis of Ely
The Mt. Hon. the Marquis of Exeter
The Mt. Hon. the Marquis of Hertford
The Mt. Hon. the Marquis of Townshend
The Mt. Hon. the Marchioness of Ailsa
The Mt. Hon. the Marchioness of Anglesey
The Mt. Hon. the Marchioness of Camden
The Mt. Hon. the March. of Conyngham
The Mt. Hon. the March. of Downshire
The Mt. Hon. the Marchioness of Hastings
The Mt. Hon. the March. of Headfort
The Mt. Hon. the Marchioness of Huntly
The Mt. Hon. the Marchioness of Lothian
The Right Hon. the Earl of Aberdeen
The Right Hon. the Earl of Bathurst
The Right Hon. the Earl of Bective
The Right Hon. the Earl of Brownlow
The Right Hon. the Earl Cadogan
The Right Hon. the Earl of Carnarvon
The Right Hon. the Earl of Carysfort
The Right Hon. the Earl of Cork
The Right Hon. the Earl of Darnley
The Right Hon. the Earl of Dartmouth
The Right Hon. the Earl Delawarr
The Right Hon. the Earl of Derby
The Right Hon. the Earl of Desart
The Right Hon. the Earl of Drogheda
The Right Hon. the Earl of Dysart
The Right Hon. the Earl of Effingham
The Right Hon. the Earl of Egmont
The Right Hon. the Earl of Ellesmere
The Right Hon. the Earl of Eldon
The Right Hon. the Earl of Essex
The Right Hon. the Earl of Gainsborough
The Right Hon. the Earl of Granard
The Right Hon. the Earl of Howe
The Right Hon. the Earl of Kilmorey
The Right Hon. the Earl of Kimberley
The Right Hon. the Earl of Leitrim
The Right Hon. the Earl of Leven
The Right Hon. the Earl of Lindsey
The Right Hon. the Earl of Lisburn
The Right Hon. the Earl of Listowel
The Right Hon. the Earl of Lonsdale

The Right Hon. the Earl of Lovelace
The Right Hon. the Earl of Lytton
The Right Hon. the Earl of Malmesbury
The Right Hon. the Earl of Manvers
The Right Hon. the Earl Nelson
The Right Hon. the Earl of Normanton
The Right Hon. the Earl of Northesk
The Right Hon. the Earl of Northbrook
The Right Hon. the Earl of Onslow
The Right Hon. the Earl of Pembroke
The Right Hon. the Earl of Radnor
The Right Hon. the Earl of Roden
The Right Hon. the Earl of Romney
The Right Hon. the Earl of Rosebery
The Right Hon. the Earl of Stanhope
The Right Hon. the Earl of Sefton
The Right Hon. the Earl Sondes
The Right Hon. the Earl Spencer
The Right Hon. the Earl of Tankerville
The Right Hon. the Earl of Warwick
The Right Hon. the Earl of Wharncliffe
The Right Hon. the Earl of Westmoreland
The Right Hon. the Earl of Wicklow
The Right Hon. the Earl of Winterton
The Right Hon. the Earl of Yarborough
The Rt. Hon. the Dow. C'tess of Aylesford
The Rt. Hon. the Dow. C'tess of Craven
The Rt. Hon. the Dow. C'tess of Ellesmere
The Rt. Hon. the Countess of Aylesford
The Rt. Hon. the Countess of Bandon
The Right Hon. the Countess of Bective
The Rt. Hon. the Countess Cairns
The Rt. Hon. the Countess of Cowley
The Rt. Hon. the C'tess of Ashburnham
The Rt. Hon. the Countess of Guildford
The Rt. Hon. the Countess of Hardwick
The Rt. Hon. the Countess of Kingston
The Rt. Hon. the C'tess of Lanesborough
The Rt. Hon. the Countess of Leven
The Rt. Hon. the Countess of Lonsdale
The Rt. Hon. the Countess of Lovelace
The Rt. Hon. the Countess of Mayo
The Rt. Hon. the Countess of Stradbroke
The Right Hon. the Viscount Camden
The Right Hon. the Viscount Cole
The Right Hon. the Viscount Dangan
The Right Hon. the Viscount Downe
The Right Hon. the Viscount Falmouth
The Right Hon. the Viscount Galway
The Right Hon. the Viscount Hill
The Right Hon. the Viscount Massereene
The Right Hon. the Viscount Pollington
The Right Hon. the Viscount Sidmouth
The Right Hon. the Viscount Templeton
The Right Hon. the Viscount Tyre
The Right Hon. the Viscount Valentia
The Rt. Hon. the Viscountess Chewton
The Rt. Hon. the Viscountess Dangan
The Rt. Hon. the Viscountess Downe
The Rt. Hon. the Viscountess Harberton
The Rt. Hon. the Viscountess Helmsley
The Rt. Hon. the Viscountess Lascelles
The Rt. Hon. the Dow. Viscountess Gort
*The Vice-Chancellor of Ireland
The Right Hon. Lord Aberdour

The Right Hon. Lord Acton
The Right Hon. Lord Annaly
The Right Hon. Lord Ashburton
The Right Hon. Lord Battersea
The Right Hon. Lord Belper
The Right Hon. Lord Blackford
The Right Hon. Lord Bingham
The Right Hon. Lord Brougham
The Right Hon. Lord Camoys
Major-Gen. Lord Clarina
The Right Hon. Lord Carew
The Right Hon. Lord Carberry
The Right Hon. Lord Carrick
The Right Hon. Lord A. Cecil
The Right Hon. Lord E. Cecil
The Right Hon. Lord T. H. P. Cecil
The Right Hon. Lord Clinton
The Rt. Hon. Lord Hope Pelham Clinton
The Right Hon. Lord Randolph Churchill
The Rt. Hon. Lord de l'Isle and Dudley
The Right Hon. Lord de Tabley
The Right Hon. Lord Dunsany
The Right Hon. Lord Dunboyne
The Right Hon. Lord Dynevor
The Right Hon. Lord Otho Fitzgerald
The Right Hon. Lord A. Fitzroy
The Right Hon. Lord Esher
The Right Hon. Lord Esme Gordon
The Right Hon. Lord Fitzhardinge
The Right Hon. Lord Gerard
The Right Hon. Lord Halifax
The Right Hon. Lord Harris
The Right Hon. Lord Hastings
The Right Hon. Lord Headley
The Right Hon. Lord Henniker
The Right Hon. Lord Heytesbury
The Right Hon. Lord Hillingdon
The Right Hon. Lord Huntingfield
The Right Hon. Lord Hylton
The Right Hon. Lord Kensington
The Right Hon. Lord Kesteven
The Right Hon. Lord Kinnaird
The Right Hon. Lord Lawrence
The Right Hon. Lord Londesborough
The Right Hon. Lord Lurgan
The Right Hon. Lord Lyvedon
The Right Hon. Lord Manners
The Right Hon. Lord Methuen
The Right Hon. Lord Morton
The Rt. Hon. Lord Mowbray and Stourton
The Right Hon. Lord Geo. Neville
The Right Hon. Lord Ormathwaite
The Right Hon. Lord Berkeley Paget
The Right Hon. Lord Penzance
The Right Hon. Lord Poltimore
The Right Hon. Lord Powerscourt
The Right Hon. Lord Raglan
The Right Hon. Lord Rayleigh
The Right Hon. Lord Romilly
The Right Hon. Lord Rothschild
The Right Hon. Lord Russell
The Right Hon. Lord Straithnairn
The Right Hon. Lord Suffield
The Right Hon. Lord Tennyson
The Right Hon. Lord Tredegar

Lieut.-Gen. Lord Templeton
The Right Hon. Lord Templemore
The Right Hon. Lord Charles Thynne
The Right Hon. Lord Walsingham
The Right Hon. Lord Winmarleigh
The Right Hon. Lord Windsor
The Right Hon. Lord Wolverton
The Right Hon. Lord Wodehouse
The Right Hon. the Master of the Rolls
The Right Hon. Baron Huddlestone
Lord Justice Bowen
Lord Justice Cotton
Lord Justice Fitzgibbon
Lord Justice Lopes
Lord Justice Wills
Baron Ferdinand de Rothschild
Baron William Schroeder
Baron de Barreto
The Dowager Lady Belper
Lady Blomfield
Lady Borthwick
Lady Mary Boscawen
Lady Emily Bury
The Dowager Lady Buxton
Lady Gordon Cathcart
Lady de Clifford
The Dowager Lady de Clifford
Lady Barbara Coventry
Lady Daly
Lady Anne Daly
Lady Dawnay
The Dowager Lady Dyer
Lady Egerton
Lady Elphinstone
Lady Fermoy
Lady Ffrench
Lady Filmer
Lady Fitzgerald
Lady Fitzpatrick
Lady Emily Foley
Lady Forester
Lady Gaskell
Lady Graves
Lady Gray
The Dowager Lady Gresley
Lady H. Grosvenor
Lady Gordon
Lady Margaret Gore
Lady Hampton
Lady Harlech
Lady Heathcote
Lady Beatrice Herbert
Lady Emily Howard Bury
Lady Huntingfield
Lady Katherine Hutton
Lady Charles Innes-Kerr
Lady Lefevre
Lady Lyttleton
Lady Maitland
Lady Sophia Melville
The Dowager Lady Middleton
Lady Milton
Lady Sybil Montgomerie
Lady Mostyn
Lady Murray
Lady Neave
Lady Nelson
Lady Neville
Lady Paget
Lady Reddington
Lady Riddell
Lady Sebright

Lady Jane Swinburne
The Dowager Lady St. John
Lady Shuckburgh
Lady Smyth
Lady Van-de-Weyer
Lady Charles Wellesley
Lady Elena Wickham
Lady Ethel Wickham
Lady Williams Bulkeley
The Dowager Lady Williams Wynn
Lady Chamberlain
Lady Hume Campbell
Lady Jenkinson
Lady Prescott
Lady Ramsden
The Hon. Mrs. Adams
The Hon. H. W. E. Agar
The Hon. Mrs. W. H. Allsopp
The Hon. Fred W. Anson
The Hon. Mrs. Arundell
The Right Hon. A. E. M. Ashley
The Hon. Ashburnham
The Hon. Sir R. Baggallay, Bart.
The Hon. C. W. Bamfylde
The Hon. Mrs. C. Brand
The Hon. A. H. Baring
The Hon. Mrs. Barrington
The Hon. C. B. Bellew
The Hon. Bethell
The Hon. Mrs. Bethell
The Hon. and Rev. J. T. Boscawen
The Hon. and Rev. H. Boscawen
The Hon. Miss Boscawen
The Hon. Mrs. Brassey
The Hon. Reginald B. Brett
The Hon. Mrs. Brooke
The Hon. H. Burke
The Hon. H. Butler
The Hon. Fred Cadogan
The Hon. Mrs. Meynell Ingram
The Hon. C. Irby
The Hon. Mrs. Cecil Ives
The Hon. Lieut.-General F. Keane, C.B.
The Hon. Henrietta Kenyon
The Hon. Mrs. G. Kenzon
The Hon. and Rev. E. S. Keppel
The Hon. Mrs. Ker
The Hon. Fred Lascelles
The Hon. Colonel Le Poer Trench
The Hon. Lieutenant H. W. Mansfield
The Hon. F. Majoribanks
The Hon. and Rev. Marsham
The Hon. Lieut.-Colonel L. Massey
The Hon. C. Maude
The Hon. Stewart Menzies
The Hon. Mrs. Stewart Menzies
The Hon. Miss Melles
The Hon. Mrs. Molyneaux
The Hon. F. C. Morgan
The Hon. and Rev. Nelson
The Hon. and Rev. L. Neville
The Hon. F. S. O'Grady
The Hon. W. T. Orde Powlett
The Hon. Edward Palk
The Hon. Mrs. Palk
The Hon. A. R. Pelham
The Hon. Mrs. Dudley Pelham
The Hon. E. Pert
The Hon. E. Pierrepont
The Hon. Leopold W. H. Powys
The Hon. J. W. Plunkett
The Hon. Maud Russell

The Hon. Rushout
The Hon. Mrs. Stirling
The Hon. A. Stourton
The Hon. S. Calthorpe
The Hon. A. Campbell
The Hon. Ivan Campbell
The Hon. Miss Canning
The Hon. Mrs. Capell
The Hon. Colonel Charles Crighton
The Hon. Mrs. A. Wynne Corrie
The Hon. E. H. Dawnay
The Hon. Guy C. Dawnay
The Hon. A. Dawson
The Hon. A. H. T. de Montmorency
The Hon. Mrs. H. Devereux
The Hon. Admiral Douglas
The Hon. A. Egerton
The Hon. and Rev. Ellis
The Hon. A. W. Erskine
The Hon. Ailwyn Fellowes
The Hon. H. W. Fitzwilliam
The Hon. T. Fitzwilliam
The Hon. C. W. Fitzwilliam
The Hon. Fitzwilliam
The Hon. J. T. Fitzmaurice
The Hon. Fulke Greville
The Hon. and Rev. J. Gifford
The Hon. and Rev. Hanbury
The Hon. Paulyn Hastings
The Hon. E. Henniker
The Hon. Mrs. Herbert
The Hon. R. C. Herbert
The Hon. Mrs. Noel Hill
The Hon. R. G. C. Hill
The Hon. Mrs. Holland
The Hon. Miss Hughes
The Hon. and Rev. W. C. Talbot
The Hon. A. Talbot
The Hon. R. H. Temple
The Hon. and Rev. Vernon
The Hon. Eustace Vesey
The Hon. Mrs. Vesey
The Hon. William Vernon
The Hon. Mrs. Vereker
The Hon. E. Willoughby
The Hon. C. H. Winn
The Hon. H. W. T. W. Woodhouse
The Hon. C. W. Wynn
The Hon. Fred Wynn
The Hon. Captain John Manners Yorke
Sir Alfred W. Bagge, Bart.
Sir David Baird, Bart.
Sir T. R. Bailey, Bart.
Sir Walter Barttelot, Bart.
Major-General Sir H. P. de Bathe, Bart.
Sir T. H. Bathurst, Bart.
Sir W. Baynes, Bart.
Sir Reginald Beauchamp, Bart.
Sir A. Bellington, Bart.
Sir H. H. Berney, Bart.
Sir P. Blake, Bart.
Sir Edward Birkbeck, Bart.
Sir Thomas Birklebank, Bart.
Sir Edward Buckley, Bart.
Sir Thomas Butler, Bart.
Sir Charles Butt, Bart.
Sir T. F. Buxton, Bart.
Sir H. Campbell, Bart.
Rev. Sir C. Clarke, Bart.
Sir W. Gibson Carmichael, Bart.
Sir Milles Cave Brown Cave, Bart.

And upwards of 12,000 of the Leading Families in the United Kingdom.

Horticultural Buildings

COMPRISING

Winter Gardens, Conservatories

GLAZED PORCHES AND VERANDAHS,

Orchid Houses, Plant Houses,

VINERIES AND PEACH HOUSES,

Garden Frames,

HEATING APPARATUS,

ETC., ETC.

BOULTON & PAUL, ROSE LANE WORKS, NORWICH.

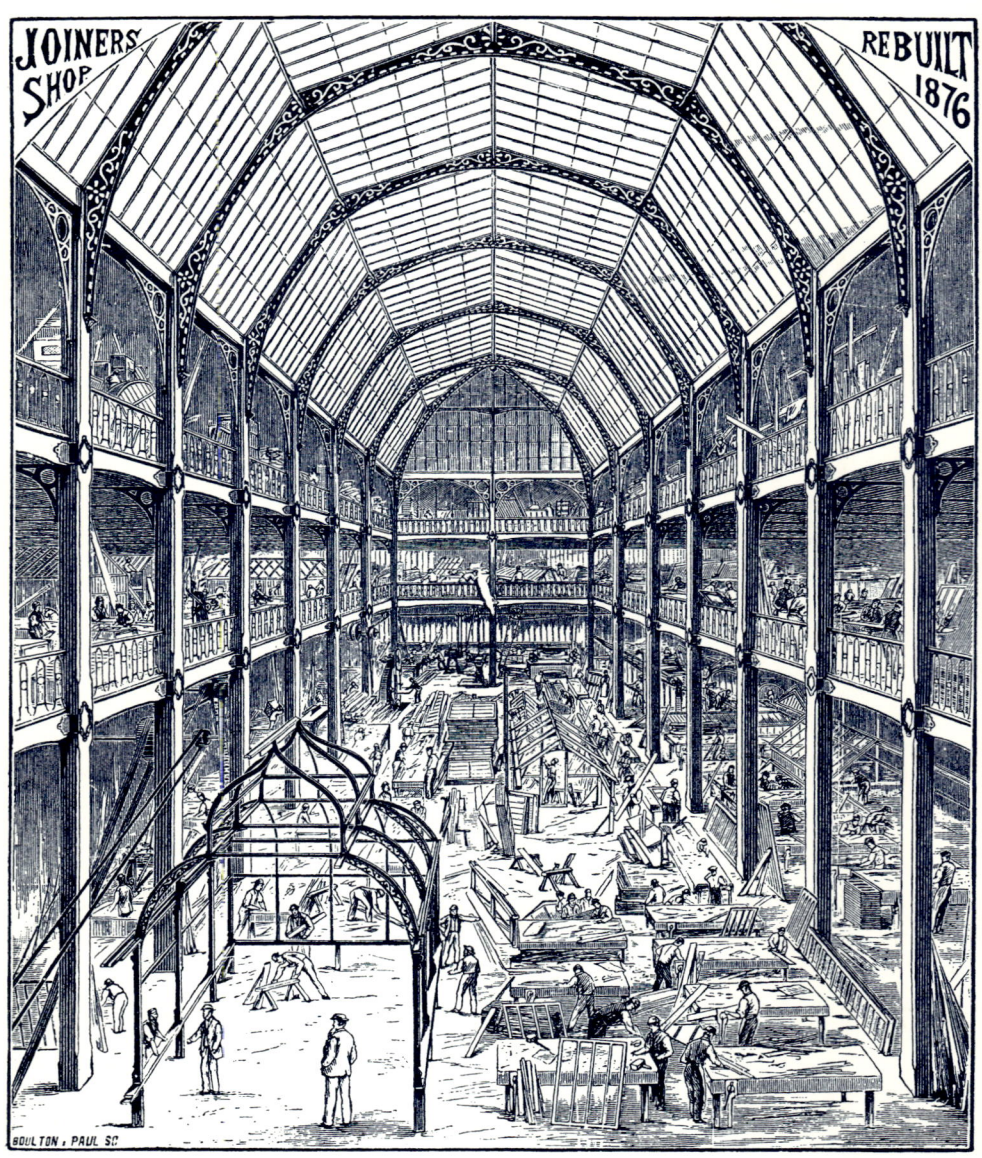

FLOOR AREA OF GREENHOUSE BUILDING DEPARTMENT, 40,000 FEET.

Conservatories in all Styles

Vineries, Peach Houses, Orchid Houses, Garden Frames,

and every description of Horticultural Buildings and Appliances.

Conservatory erected at Derry Ormond, Cardiganshire

Surveys made in any part of the country. Ladies and Gentlemen waited upon by appointment.

BOULTON & PAUL. MANUFACTURERS.

WINTER GARDEN.

Erected for J. Corbett, Esq., M.P., Impney, Droitwich.

The above structure is heated by one of our celebrated No. 4 Cheek End Boilers, the boiler house being 500 ft. away from the building.

CONSERVATORY AND PALM HOUSES.

CONSERVATORY.

Erected for T. Pim, Esq., Bexley Heath, Kent.
Connected with this Conservatory is a very fine Range of Three-Quarter Span Houses, comprising Fernery, Vineries, and Peach Houses, 250 ft. in length.

CONSERVATORIES.

WINTER GARDENS, CONSERVATORIES, FERNERIES, &c., DESIGNED TO SUIT ANY SITUATION.

Surveys made in any part of the Country.

Ladies and Gentlemen waited upon by special appointment

Erected at Alresford, Hants.

TESTIMONIAL.

GUILDFORD.

GENTLEMEN,—I am pleased to inform you that the Baron & Baroness de Worms are highly pleased with the House erected. Yours respectfully,

A. J. ELPHINSTONE

CONSERVATORIES.

TESTIMONIAL.

Moore Lane,
Milton Street,
London, E.C.

Dear Sirs,—I enclose cheque in settlement of your account. Your men have been most careful with their work, and I am really pleased with the result. There is not a prettier Conservatory in Forest Hill.

Yours truly,
C. H. HINTON.

Interior of Lantern Roof Conservatory.

CONSERVATORY AND CORRIDOR.

Erected at Scio House, Putney Heath, S.W.

CONSERVATORY.

This design is well adapted for a Palm House.

CONSERVATORY.

Erected at Hatley St. George, Cambridgeshire.

RANGE OF PLANT AND ORCHARD HOUSES.

The above shows an adaptation of "Working" Houses to rising ground.

CONSERVATORY.

This design is admirably adapted for placing in the angle of a building, and could be entered either from billiard or reception room.

CONSERVATORY AND FERNERY.

Erected for Mrs. Monins, Ringwould, Dover.

CONSERVATORY.

This design is most suitable for a small Conservatory or Glazed Entrance, the latter making a charming extension to the hall space.

CONSERVATORY.

The above represents a large Central House, well adapted for Palms, with Side Houses for smaller plants.

CONSERVATORIES.

TESTIMONIAL.

THE PALACE, ELY.

GENTLEMEN,—I am quite satisfied with the work here and the workmen also. Their conduct throughout was very excellent.

(*Signed*)
ALWYN, ELY.

Erected at Bidborough, Sussex, for MRS. ATKINSON.

TESTIMONIALS.

AVISFORD, ARUNDEL.

DEAR SIRS,—I am extremely pleased with the Conservatory and Houses, and the way in which the work has been done.

(*Signed*)
C. P. HENTY.

THE RYELANDS, LEOMINSTER.

MISS WOODS encloses a cheque, and is much pleased with the Greenhouse erected for her by Messrs. Boulton & Paul.

The above Conservatory was erected at Briton Ferry, for MRS. LLEWELLYN.

CONSERVATORIES.

Erected at Cleckheaton, Yorks, for A. LAW, ESQ.

TESTIMONIAL.

HAZELSHAW, KENLEY, SURREY.

DEAR SIRS,—I am much pleased with the way you have built the Greenhouse, it is very complete, and a pretty house.

Yours truly,
W. C. STRAKER.

Erected at Clacton-on-Sea, for H. GRANT, ESQ.

TESTIMONIAL.

CARMARTHEN.

DEAR SIRS,—I am thoroughly satisfied with my Greenhouse, and with your men.

Yours faithfully,
JOHN FRANCIS.

BOULTON & PAUL, MANUFACTURERS.

CONSERVATORIES DESIGNED TO SUIT SPECIAL SITUATIONS.

Conservatory and Ante-Room erected at Dorchester for A. Pope, Esq.

This Conservatory was designed to attach to Billiard Room with doors into Verandah to connect the whole with Reception Rooms.

HEADLEY PARK, HANTS.

DEAR SIRS,—I enclose cheque for Greenhouse which is very satisfactory. Your men were in every way civil and trustworthy.

(Signed)
R. S. WRIGHT.

ROSE LANE WORKS, NORWICH.

WINTER GARDENS IN ALL STYLES.

Conservatory erected at Oaklands, Oxshott, Surrey, for the late SIR CHARLES BUTT.

BOULTON & PAUL, MANUFACTURERS.

SEMI-OCTAGON CONSERVATORY.

This is an inexpensive form of Conservatory, and has been erected for S. J. BROWN, ESQ., at Naas, County Kildare, and at many other places.

CONSERVATORIES, &c.

This Block of Buildings consists of Conservatories, Plant and Forcing Houses, with Garden House beneath, as shown. Erected for B. J. BRIDGEWATER, ESQ., at Tufnell Park. The upper Conservatory is entered from the Drawing Room, and connected with the lower Houses by a flight of stone steps; the whole forming a most complete arrangement.

CONSERVATORY.

Erected for J. Jacques, Esq., Banstead, Surrey.

CONSERVATORY.

This Conservatory was erected over a Billiard Room for D'Arcy Reeve, Esq., Beechcombe, Basildon, Berks

CONSERVATORY.

TESTIMONIAL.

BARROW-IN-FURNESS.

GENTLEMEN,
Mr. Butler is very pleased with the workmanlike way in which your men have finished the Conservatory. It has been admired by every one.
(*Signed*)
A. B. BUTLER.

TESTIMONIAL.

MILTON STREET,
LONDON, E.C.

DEAR SIRS,—I enclose cheque in settlement of your account. Your men have been most careful with their work, and I am really pleased with the result. There is not a prettier Conservatory in Forest Hill.
(*Signed*)
C. H. HINTON.

INTERIOR OF CHRYSANTHEMUM HOUSE.

TESTIMONIAL.

CRESCENT WOOD ROAD, LONDON.
The New Span Greenhouse is a very fine one; please send account. (*Signed*) J. PRINCE.

ROSE LANE WORKS, NORWICH. 26A

CONSERVATORY.

TESTIMONIAL.

Montem,
Slough.

Gentlemen,—I am thoroughly pleased with the addition made to the Conservatory you put up last year.

Yours faithfully,
(*Signed*)
W. CREAK.

TESTIMONIAL.

South-Eastern
Railway.

Gentlemen,—I am very pleased with the Conservatory, and it is greatly admired by my friends.

Yours truly,
(*Signed*)
C. A. PRIKLER.

Erected at Montem, Slough, for W. CREAK, ESQ.

CONSERVATORY.

TESTIMONIAL.

GODFREY HOUSE, CHELTENHAM.

MRS. THORP is very much pleased with the Conservatory, which all her friends greatly admire.

BOULTON & PAUL, MANUFACTURERS,

CONSERVATORY.

TESTIMONIAL.

THE GROVE
    GARDENS,
        WATFORD.

DEAR SIRS,—I am pleased to say that Lord Clarendon has expressed his satisfaction with the work done.
Yours truly,
(*Signed*)
    J. MYERS.

TESTIMONIAL.

EAST
    FINCHLEY.

The Conservatory & Verandah have turned out very well, and are generally much admired.
(*Signed*)
HY. COLLIER.

Erected at Blackburn for E. HEYWORTH, ESQ.

CONSERVATORY.

Erected at Hampstead for J. W. SMITH, ESQ.

CONSERVATORY WITH GABLE ENTRANCE.

The above Illustration represents a Conservatory, the interior of which is planned with a view to the effective display of Show Plants, etc.

WINTER GARDEN.

Designed with a Palm House in centre, and Orchid Houses at each end. Suitable for Botanic Gardens or any large establishment.

CONSERVATORY.

TESTIMONIAL.

CASTLEMANS,
TWYFORD.

GENTLEMEN,—It is a pleasure to inform you that the Range of Glass, new apparatus, and extra piping give every satisfaction. I shall be pleased at any time to recommend your firm.

Yours truly,

C. PARRY.

TESTIMONIAL.

HERONS GHYLL,
UCKFIELD.

DEAR SIRS,—The Range of Vineries put up here I consider are substantial and well built. The hot-water apparatus works to perfection, and the whole is a credit to the men, who behaved well.

(*Signed*)

G. GRIMSELL.

CONSERVATORY AND GREENHOUSE RANGE.

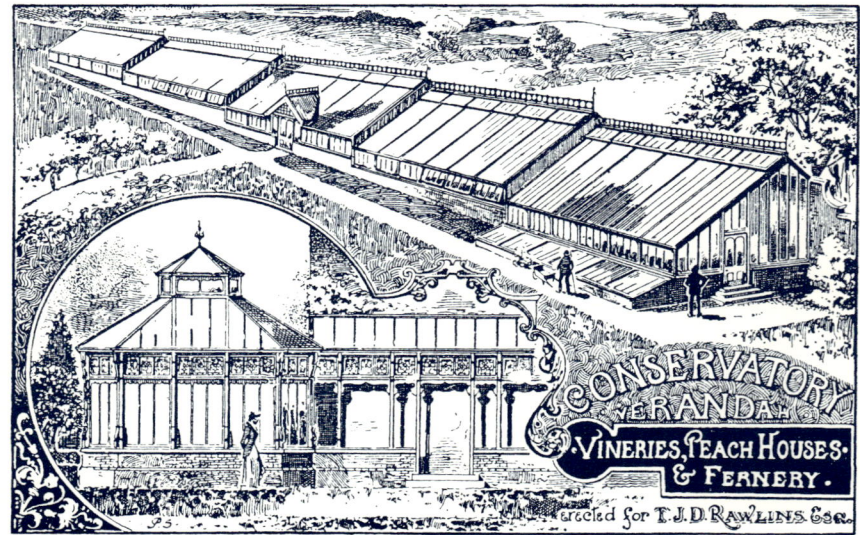

ROSE LANE WORKS, NORWICH. 28A

SPAN-ROOF PLANT HOUSE.

TESTIMONIAL.

WILLSBORO',
LONDONDERRY.

I beg to enclose cheque for Greenhouse. We have the house now complete and it is very satisfactory.

Yours truly,
(*Signed*)
W. E. SCOTT.

TESTIMONIAL.

LLANISHEN,
near CARDIFF.

I am greatly pleased with the Greenhouse and Conservatory supplied by you last autumn.

(*Signed*)
W. SAUNDERS.

Erected at Keighley for PRINCE SMITH, ESQ.

CONSERVATORY.

Erected at North Creake for THE BISHOP OF THETFORD.

CONSERVATORY OR GLAZED PORCH.

LANTERN-ROOF CONSERVATORY.

For Interior of a similar structure see page 15.

ROSE LANE WORKS, NORWICH. 29

CONSERVATORIES & WINTER GARDENS

Designed to suit Special Situations, and to meet all Requirements.

Winter Garden erected at Moatlands, Paddock Wood, for EDWARD BEANES, ESQ.

Horticultural Buildings, Garden Frames, Boilers, Pipes, Valves, Tanks, & every description of Greenhouse Appliances.

SURVEYS MADE IN ANY PART OF THE COUNTRY.

BOULTON & PAUL, MANUFACTURERS.

CONSERVATORIES TO SUIT ANY SITUATION.

Erected at Drayton, Norfolk, for J. J. Winter, Esq.

CONSERVATORY.

CONSERVATORY.

Recently erected at Carrow House, Norwich.

CONSERVATORY.

Recently erected at Woodford, Essex.

BOULTON & PAUL, MANUFACTURERS,

RANGES OF HORTICULTURAL BUILDINGS.

Erected and fitted complete with Stages, Trainers, Heating Apparatus, Footpaths, Tanks, &c.

Range of Horticultural Buildings recently erected at Frensham Place, Farnham, for C. Arthur Pearson, Esq.

Surveys made in any part of the kingdom.

SPECIAL DESIGNS PREPARED AND ESTIMATES SUBMITTED.

FORCING HOUSES, PITS, FRAMES, &c., of every description.

SPAN-ROOF CONSERVATORY.

SPAN-ROOF ORCHARD HOUSE.

The above House can be made to any length and width.

BOULTON & PAUL, MANUFACTURERS.

No. 1. GLAZED VERANDAH.

TESTIMONIAL.

LIMPSFIELD, SURREY

SIRS,—I wish to say that as far as I can judge the work has been done very satisfactorily, and that your men have given no trouble, but have conducted themselves well, and I think worked honestly for you.

Yours truly,
D. G LANDALE

TESTIMONIAL.

DRAYTON, NORFOLK.

DEAR SIRS,—I desire to express my acknowledgement of the admirable way in which your representatives and workmen have carried out the Conservatory here. The workmen deserve a special word of praise for the quiet manner in which they carried out their respective parts.

(*Signed*)
J. J. WINTER

THE Verandah above illustrated can be made to any length or width, and makes an important feature when carried round the main fronts of the dwelling. The specification would be the same as for the porches below.

GLAZED PORCHES.

No. 2.

No. 3.

THESE Porches can be arranged either for stone or wood foundation; the latter being tenants' fixtures, removable at the expiration of tenancy or lease. All the wood is of selected red deal, painted four times with good oil colour; and the glazing done with 21-oz. sheet glass. If desired, the roofs can be glazed with rolled plate glass, and tinted glass used for the transoms or upper sashes of side framing.

Estimates on application.

ROSE LANE WORKS, NORWICH. 33

No. 4. GLAZED VERANDAH.

TESTIMONIAL.

Hazelshaw,
Henley

Dear Sirs,—I am much pleased with the way you have built the Greenhouse. It is very complete, and a pretty house.

Yours truly,
W C STRAKER

TESTIMONIAL

Kenley Park,
Guildford.

Gentlemen,—I am pleased to inform you that the Baron and Baroness de Worms are highly pleased with the House erected.

Yours respectfully,
A J. ELPHINSTONE

This design can be adapted to almost any situation. If desired, the front and ends can be entirely closed with glazed framing. The specification would be the same as for the porches below.

GLAZED PORCHES.

No. 5.

No. 6.

These porches can be arranged either for stone or wood foundation; the latter being tenants' fixtures, removable at the expiration of tenancy or lease. All the wood is of selected red deal, painted four times with good oil colour; and the glazing done with 21-oz. sheet glass. If desired, the roofs can be glazed with rolled plate glass, and tinted glass used for the transoms or upper sashes of side framing.

Estimates on application.

BOULTON & PAUL, MANUFACTURERS.

Plant Houses, Vineries, &c.

ORCHID HOUSES.

Of every description erected and fitted complete with Heating Apparatus, Tanks, Stages, Trays, and Shelvings.

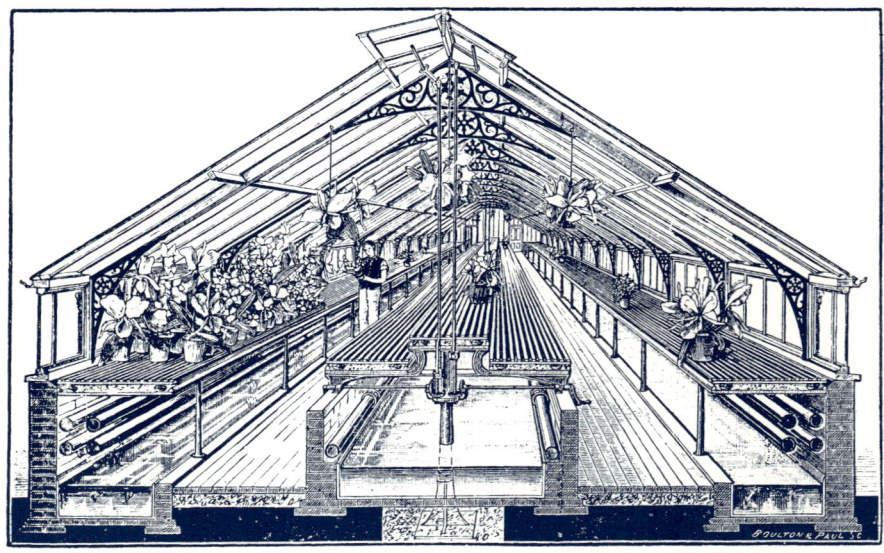

Interior View of Span-roof Orchid House at Downside, Leatherhead.

The well known Orchid Houses of

W. LEE, Esq., Downside, Leatherhead, Surrey.
C. W. WALKER, Esq., Milnthorpe, Westmoreland.
W. SPINDLER, Esq., Old Park, Isle of Wight.
W. VANNER, Esq., Chislehurst, Kent.

SIR W. HUTT, Apley Towers, Ryde, Isle of Wight.
CHARLES YOUNG, Esq., The Thorns, Sevenoaks.
G. LE DOUX, Esq., Langton House, East Moulsey.
GENERAL HUTCHINSON, Owthorpe, Bournemouth,

and many others in the kingdom have been erected by us.

TESTIMONIAL.

CORYPTON PARK, AXMINSTER.

GENTLEMEN,—I enclose you a cheque in settlement of account. I have much pleasure in telling you that the Houses are satisfactory, and that your workmen throughout did their work excellently and in a workmanlike manner.

Yours truly, W. G. KING.

ROSE LANE WORKS, NORWICH

HORTICULTURAL BUILDINGS DESIGNED TO SUIT ANY SITUATION.

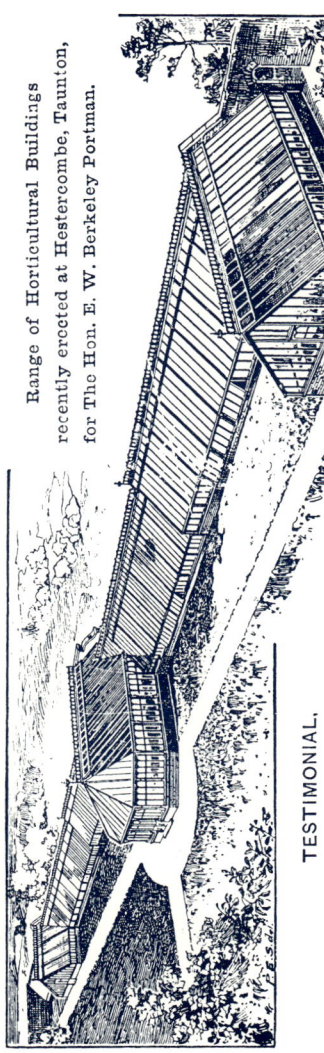

Range of Horticultural Buildings recently erected at Hestercombe, Taunton, for The Hon. E. W. Berkeley Portman.

TESTIMONIAL.

HESTERCOMBE, TAUNTON.

GENTLEMEN,—I am very pleased indeed with the Range, and the way your workmen have done their work.

Yours faithfully, *(Signed)* A. J. KEEN.

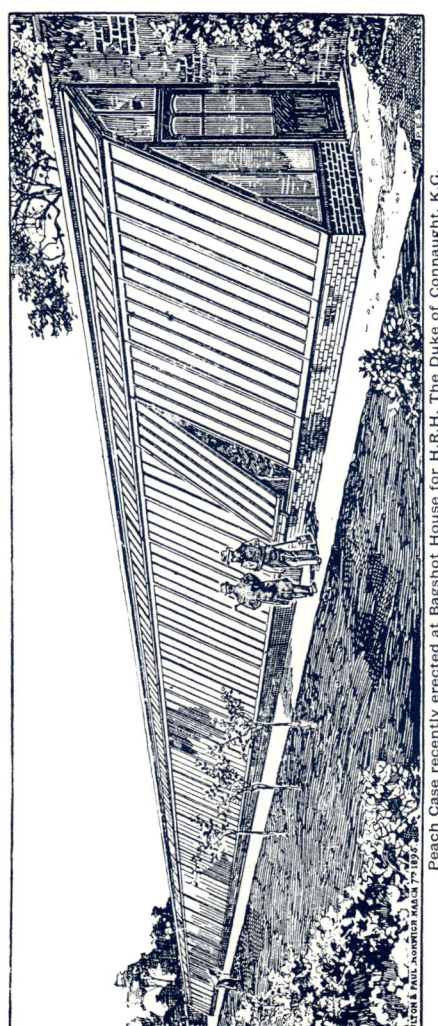

Peach Case recently erected at Bagshot House for H.R.H. The Duke of Connaught, K.G.

BOULTON & PAUL, MANUFACTURERS.

HORTICULTURAL BUILDINGS ERECTED IN ANY PART OF THE KINGDOM.

Range of Horticultural Buildings recently completed at Chorley Wood for Lady Ela Russell.

SURVEYS MADE. LADIES AND GENTLEMEN WAITED UPON.

PLANT HOUSES, VINERIES, AND PEACH HOUSES.

Erected for HENRY TUBB, ESQ., Chesterton Lodge, Bicester.

CONSERVATORY AND VINERIES.

Erected for THE RIGHT HONOURABLE W. E. GLADSTONE, at Hawarden Castle.

INTERIOR OF A SPAN-ROOF PEACH HOUSE.

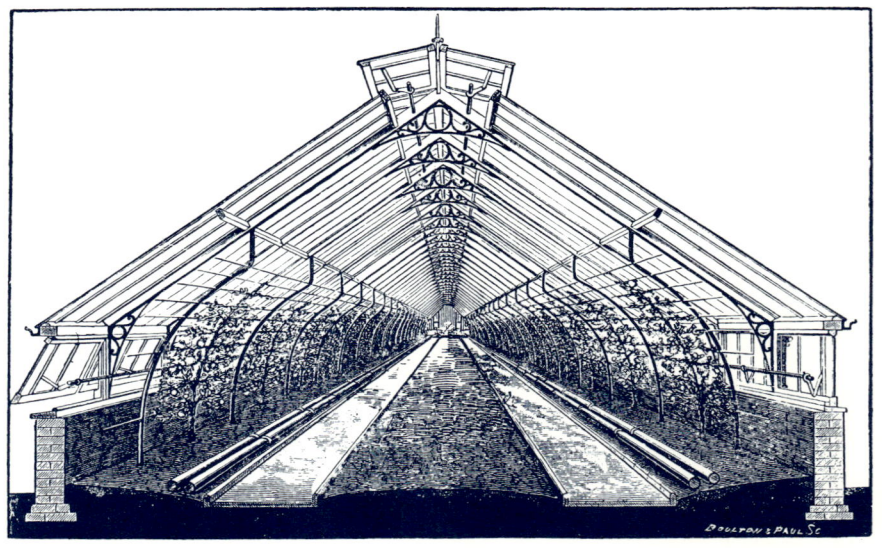

SPAN-ROOF ORCHARD HOUSE.

RANGE OF LEAN-TO PEACH HOUSES.

THREE-QUARTER SPAN-ROOF PEACH HOUSES.

We have erected several very fine ranges of Peach Houses to this design.

RANGE OF PLANT HOUSES AND VINERIES.

This is a good arrangement for a moderate-sized range of houses, and can be adapted to meet the wants of any establishment; divided to form Plant House, Vineries, Peach House, Forcing House, etc. Erected for R. Daws, Esq., Ealing; A. Law, Esq., Cleckheaton, Yorks; C. D. Harrod, Esq., Morebath, Devon; and at many other places.

TESTIMONIALS.

MEDWAY VIEW, BORSTAL, ROCHESTER.
Dear Sirs,—I have great pleasure in reporting that the Conservatory is completed entirely to my satisfaction.
(*Signed*) GEO. LEE WHARTON.

The Hawthorne, Castle Bar Park, Ealing, W.
Gentlemen,—I bought one of your Three-quarter Span Greenhouses about seven years ago which has given great satisfaction.
(*Signed*) G. PRICE.

CONSERVATORIES AND VINERIES.

Erected for the late Sir W. Edward Hanmer, Bart., Bettisfield Park, Whitchurch, Salop.

CONSERVATORY.

TESTIMONIAL.

Epping House,
Hertford.

Dear Sirs,
 The work has been most satisfactorily completed in every detail, and I am exceedingly pleased with it.

(Signed)
H. F. BARCLAY.

TESTIMONIAL.

Bury
St. Edmunds.

Gentlemen,
 The Conservatory is in every way satisfactory, and has been greatly admired by several ladies who have seen it.

(Signed)
GEORGE LAST.

Erected at Maidenhead.

RANGES OF HORTICULTURAL BUILDINGS

Erected complete in any part of the kingdom.

RANGE OF PALM & PLANT HOUSES.

Recently erected at Newmarket.

CONSERVATORY.

Recently erected at Chepstow, Mon.

CONSERVATORY.

Recently erected at Epsom.

No. 35. LEAN-TO CONSERVATORY.

The above Illustration shows a very effective Lean-to House, suitable for attaching to a Villa Residence; it is one of those designs which much improve the appearance of the house and gardens without excessive outlay. All woodwork would be of selected red deal, painted four times with good oil colour, and the glazing done with 21-oz. sheet glass.

Estimates for House, Stages, and Heating Apparatus will be sent on receipt of particulars of requirements.

No. 37. SPAN-ROOF GREENHOUSE.

The materials for this House would be the same as specified above.

BOULTON & PAUL. MANUFACTURERS.

No. 38. SPAN-ROOF GREENHOUSE.

ALL woodwork of selected red deal, painted four times, and the glazing done with 21-oz. sheet glass, door fitted with brass mortise-lock; gutters and down pipes provided, and iron cresting and terminals fixed to the ridge.

Estimates for House, Stages, and Heating Apparatus will be sent on receipt of particulars of requirements.

No. 40. LEAN-TO GREENHOUSE OR VINERY.

The materials for this House would be the same as specified above.

No. 41. THREE-QUARTER SPAN-ROOF GREENHOUSE OR VINERY.

This is a most useful form of House where the height of the back wall is limited. All the woodwork is of selected red deal, painted four times with good oil colour, and the glazing done with 21-oz. sheet glass. The door is fitted with brass mortise-lock; gutters and down-pipes provided; and cresting and terminals fixed to the ridge. The front sashes open by lever gear, and the roof ventilators by our screw gearing.

These Houses are made to any size.

Estimates on application.

No. 42. SPAN-ROOF GREENHOUSE.

All the materials for this House would be the same as specified above.

PLANT HOUSE.

ERECTED for C. R. GILMAN, ESQ., Norwich. Similar Houses have been erected for F. H. KAY, ESQ., Holmbrook, Frant; C. CRAWSHAY, ESQ., Hingham, Norfolk; and at many other places.

LEAN-TO RANGE OF FORCING HOUSES.

THIS Illustration shows a range of Lean-to Houses suitable for Propagating or Forcing, for Melons or Stove Plants. This is a smaller size of our No. 40 class.

All woodwork of selected red deal, painted four times with good oil colour, and the glazing done with 21-oz. sheet glass.

FORCING AND GROWING HOUSES.

The sections on this page show the various forms of useful Houses, which are adaptable to any situation.

The Three-quarter Span could be arranged with front framing if desired, and Lean-to Pits could, of course, be fixed either to the Span or Lean-to.

Interior fittings are given as suggestions only, they would vary with individual requirements.

Estimates on application.

Section of a Lean-to Forcing House.

TESTIMONIAL.

AYLSHAM, NORFOLK.

GENTLEMEN,—I enclose cheque for the Greenhouse you put up for me. I am thoroughly satisfied with it in every part.

Yours truly,

W. R. BANSALL.

Section of a Span-Roof Forcing House

TESTIMONIAL.

HAZELSHAW, KENBY.

DEAR SIRS,—You will be glad to hear the Glass Houses erected here are finished off well, and the Heating Apparatus works capitally, I am very pleased with it.

Yours truly,

JOHN JOHNSON.

Section of a Three-quarter Span-Roof Forcing House, with Lean-to Pits in front.

BOULTON & PAUL, MANUFACTURERS,

PLANT HOUSE.

TESTIMONIAL.

KILLINGWORTH VICARAGE, NEWCASTLE-ON-TYNE.

The Greenhouse is an excellent piece of work, and so very simple to put together.
(*Signed*)
REV. G. REED.

TESTIMONIAL.

THE BEECHES, KINGSTON HILL.

DEAR SIRS,--The new House seems all right, and the work very satisfactorily done.
Yours truly,
(*Signed*)
DAVID WILLIS.

GLAZED COVERED WAY.

ROSE LANE WORKS, NORWICH.

GREENHOUSES FOR AMATEURS.

No. 47a. LEAN-TO GREENHOUSE.

TESTIMONIAL.

UTRECHT,
HOLLAND.
I have the pleasure to announce the little Greenhouse arrived in perfect order, and is to the entire satisfaction of my daughters.
(Signed)
A. W. VAN BEECK CALKOEN.

10 ft. by 7 ft., Price, Carriage Paid, £8 10 0

THESE Houses are specially designed to meet the requirements of Amateurs, and those having only a limited space at command. They are a cheaper type of our well known Tenant's Fixture Houses, are thoroughly well made, accurately fitted and numbered, ready for easy re-erection.

No. 49a. SPAN-ROOF GREENHOUSE.

TESTIMONIAL.

FERRER'S RECTORY,
CHELMSFORD.
DEAR SIRS,—I am obliged to you for sending the very nice little Greenhouse, which I consider is well worth the money.
Yours faithfully,
CHAS. P. PLUMPTRE.

10 ft. by 8 ft., Price, Carriage Paid, £10 10 0

THESE structures are properly **painted and glazed** at our Works with **21-oz. glass**, all **carefully packed and Carriage Paid** to nearest Goods Station; and not as sent out by some of the so called cheap makers, who put on rail, at purchaser's risk, a bundle of cills and bars, and a box of odd-sized Foreign glass.

Small Greenhouses

Suitable for General Purposes.

This series of Greenhouses, Nos. 44 to 52, embraces those useful for small requirements. In every case the work is fitted, painted, and glazed all ready for putting together, and not, as usually advertised, with glass cut up and packed, etc., leaving half or more of the fitting to be done on delivery. From the many testimonials we have received, it will be seen that every satisfaction is given, that trouble is avoided, and the houses themselves give pleasure at once, as in a few hours they are quite ready to receive the plants.

No. 44. LEAN-TO FORCING PIT.

LEAN-TO PIT suitable for Propagating House, Cucumbers, Melons, &c. Made of selected red deal, painted three times, and glazed with 21-oz. sheet glass. All sashes are 2 in. thick. The roof ventilator is made the full length, hinged and opened by lever from inside. Good lock with brass handles to door. All parts carefully fitted and numbered, ready for re-erection by any handy man.

						8 ft. wide.	10 ft. wide.	12 ft. wide.
Height to eaves	...	...	...	...	...	4 ft. 6 in.	4 ft. 6 in.	4 ft. 6 in.
Height at back	...	..	...	...	...	8 ft. 6 in.	9 ft. 6 in.	10 ft. 6 in.

This Pit is made to any length.   Plan for Brickwork supplied free on receipt of order.

ESTIMATES including Ventilating Boxes, Stages, Footpaths, and Heating Apparatus on application.

No. 45. THREE-QUARTER SPAN FORCING PIT.

Three-quarter Span Pit suitable for Propagating House, Cucumbers, Melons, &c. Made of selected red deal, painted three times, and glazed with 21-oz. sheet glass. All sashes are 2 in. thick. The roof ventilator is made the full length, hinged and opened by lever from inside. Good lock, with brass handles, to door. All parts carefully fitted and numbered, ready for re-erection by any handy man.

	8 ft. wide.	10 ft. wide.	12 ft. wide.
Height to eaves	4 ft. 6 in.	4 ft. 6 in.	4 ft. 6 in.
Height of back wall (about)	6 ft. 6 in.	7 ft. 3 in.	8 ft. 0 in.

No. 46. SPAN-ROOF FORCING PIT.

All the materials for No. 46 Forcing Pit would be the same as specified for No. 45.

	8 ft. wide.	10 ft. wide.	12 ft. wide.
Height to eaves	4 ft. 6 in.	4 ft. 6 in.	4 ft. 6 in.
Height to ridge	7 ft. 3 in.	8 ft. 0 in.	8 ft. 9 in.

These Pits are made to any length.   Plans for Brickwork supplied free on receipt of order.

ESTIMATES including Ventilating Boxes, Stages, Footpaths, and Heating Apparatus on application.

No. 47. LEAN-TO GREENHOUSE.

LEAN-TO HOUSE suitable for Plants, Cucumbers, Melons, &c., made of selected red deal, painted three times, and glazed with 21-oz sheet glass. All the sashes are 2 in. thick The roof ventilator is made full length, and arranged to open by lever and cord. The front sashes open by set-opes. Good lock and brass handles to door, and iron gutters and down-pipes provided. All carefully fitted and numbered ready for re-erection on arrival.

	8 ft. wide.	10 ft. wide.	12 ft. wide
Height to eaves, including 2 ft. 9 in. of brickwork	5 ft. 3 in.	5 ft. 3 in.	5 ft. 3 in.
Height at back	9 ft. 6 in.	10 ft. 6 in.	11 ft. 6 in.

No. 48. THREE-QUARTER SPAN-ROOF GREENHOUSE.

All the materials for No. 48 Greenhouse would be the same as specified for No. 47.

	8 ft. wide.	10 ft. wide.	12 ft. wide
Height to eaves, including 2 ft. 9 in. of brickwork	5 ft. 3 in.	5 ft. 3 in.	5 ft. 3 in.
Height of back wall, about	7 ft. 0 in.	7 ft. 6 in.	8 ft. 0 in.

These Houses are made to any length as Tenants' Fixtures or Permanent Buildings Plans for Brickwork supplied free on receipt of order.

ESTIMATES for Houses, Stages, Footpaths, and Heating Apparatus on application.

No. 49. SPAN-ROOF GREENHOUSE.

SPAN-ROOF. HOUSE for Cucumbers, Melons, Tomatoes, and general Plant House. Made of selected red deal, painted three times, and glazed with 21-oz. sheet glass. The roof is strengthened with cast-iron spandril brackets. All the sashes are 2 in. thick. The roof ventilator is made full length one side of ridge, and arranged to open by lever and cord. All the side sashes open by set-opes, good lock and brass handles to door, and iron gutters and down-pipes provided. The whole carefully fitted and numbered ready for re-erection by any handy man.

	8 ft. wide.	10 ft. wide.	12 ft. wide.
Height to eaves, including 2 ft. 9 in. of brickwork	5 ft. 3 in.	5 ft. 3 in.	5 ft. 3 in.
Height to ridge	7 ft. 10 in.	8 ft. 6 in.	9 ft. 2 in.

No. 50. LEAN-TO GREENHOUSE.

All the materials for No. 50 Greenhouse would be the same as specified for No. 49, the roof ventilator being made to open by screw-gearing instead of lever and cord.

	8 ft. wide.	10 ft. wide.	12 ft. wide.
Height to eaves, including 2 ft. 6 in. of brickwork	5 ft. 6 in.	5 ft. 6 in.	5 ft. 6 in.
Height to top of capping	9 ft. 9 in.	10 ft. 9 in.	11 ft. 8 in.

These Houses are made to any length as Tenants' Fixtures or Permanent Buildings. Plans for Brickwork supplied free on receipt of order.

ESTIMATES for Houses, Stages, Footpaths, and Heating Apparatus on application.

BOULTON & PAUL, MANUFACTURERS.

No. 51. THREE-QUARTER SPAN-ROOF GREENHOUSE.

THREE-QUARTER SPAN HOUSE constructed of selected red deal, thoroughly seasoned, and strengthened with iron work. Lights 2 in. thick. Roof ventilators made to open the full length on one side of ridge by our improved screw-gearing, front sashes by set-opes. Iron gutters fitted to the eaves, and down-pipes provided for conveying the rain-water to the ground level. Ornamental cresting with end terminals for fixing to the ridge. Door fitted with strong hinges, good rim lock, and brass handles. Glazed with 21-oz. sheet glass, and painted three coats of good oil colour. Every part carefully fitted and numbered before leaving our Works, ready for fixing by any handy man.

	8 ft. wide.	10 ft. wide.	12 ft. wide.
Height to eaves, including 2 ft. 6 in. of brickwork	5 ft. 6 in.	5 ft. 6 in.	5 ft. 6 in.
Height of back wall (about)	7 ft. 3 in.	7 ft. 9 in.	8 ft. 3 in.

No. 52. SPAN-ROOF VILLA GREENHOUSE.

All the materials for No. 52 Greenhouse would be the same as specified for No. 51.

	8 ft. wide.	10 ft. wide.	12 ft. wide
Height to eaves, including 2 ft. 6 in. of brickwork	5 ft. 6 in.	5 ft. 6 in.	5 ft. 6 in.
Height to ridge	8 ft. 3 in.	9 ft. 0 in.	9 ft. 9 in.

These Houses are made to any length as Tenants' Fixtures or Permanent Buildings. Plans for Brickwork supplied free on receipt of order.

ESTIMATES for Houses, Stages, Footpaths, and Heating Apparatus on application.

MELON AND CUCUMBER FRAMES

PLANT PRESERVERS, FORCING PITS, WALL FRUIT-TREE PROTECTORS,

HANDLIGHTS, PROPAGATING GLASSES,
TRAINING APPLIANCES, &c.

ADAPTED FOR EVERY PURPOSE FOR WHICH THESE GOODS ARE APPLICABLE.

Special Estimates for Long Ranges of Pits on receipt of full particulars.

No. 60. PORTABLE UNIVERSAL PLANT PRESERVER.

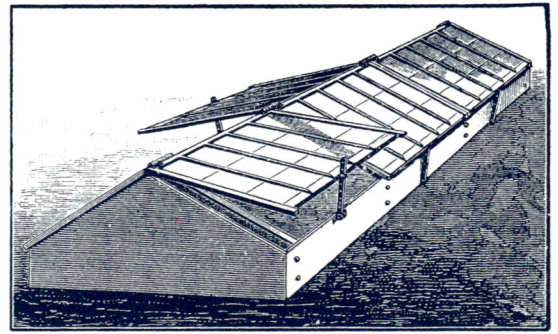

TESTIMONIAL.

CHESTERFIELD.

DEAR SIRS,—The eight large Lights arrived quite safely in capital condition, and I am very pleased with them.

Yours very truly,
J. BORE.

TESTIMONIAL.

HEREFORD.

SIRS,—It is ten years since I got four Garden Frames from you, and they are as good now as when I bought them.

Yours truly,
J. T.

THIS Plant Preserver is made of the undermentioned sizes, of good red deal, painted three times, and glazed with 21-oz. sheet glass. The lights are made to turn quite over, and are provided with set-opes and stays, giving free access to the plants. Every part is carefully fitted and marked.

Length.	Width.	Height to ridge.	Cash Prices. £ s. d.	Length.	Width.	Height to ridge.	Cash Prices £ s. d.
6 ft.	3 ft.	1 ft. 8 in.	1 10 0	12 ft.	4 ft.	1 ft. 11 in.	3 8 0
12 ft.	3 ft.	1 ft. 8 in.	2 17 6	12 ft.	5 ft.	2 ft. 6 in.	4 7 0
6 ft.	4 ft.	1 ft. 11 in.	2 0 0				

No. 61. PORTABLE LAWN CONSERVATORY.

TESTIMONIAL.

THROGMORTON HOUSE, E.C.

GENTLEMEN,—I must say that I am exceedingly pleased with the Frame. It is the first thing I have bought from you, but I can assure you it will not be the last. I never saw anything turned out so thoroughly, with an attention to details which is astonishing.

Yours faithfully,
T. CHARLES.

Breakages seldom occur, but should any glass be broken in transit, we will send sufficient to replace it, carriage free.

Goods received broken should be signed for as such, and returned by the same Railway Company.

All the materials for this Frame are the same as specified for No. 60.

Length.	Width.	Height of sides.	Height to ridge.	Cash Prices, Carriage Paid. £ s. d.	Length.	Width.	Height of sides.	Height to ridge.	Cash Prices, Carriage Paid. £ s. d.
6 ft.	4 ft.	18 in.	2 ft. 6 in.	3 10 0	12 ft.	5 ft.	18 in.	3 ft. 0 in.	6 15 0
12 ft.	4 ft.	18 in.	2 ft. 6 in.	5 10 0	12 ft.	6 ft.	18 in.	3 ft. 6 in.	8 5 0

Packing Cases charged extra, and allowed for if returned Carriage Paid.

No. 62. LEAN-TO FRAME OR FORCING PIT.

TESTIMONIAL.
STANLEY LODGE,
CADOGAN ROAD,
SURBITON,
SURREY.
GENTLEMEN,—I am very pleased with the Frame I had of you this Spring; it being thoroughly well-made.
Yours truly,
THOS. COLLIN

TESTIMONIAL.
FRENSHAM HALL,
SHOTTERMILL.
DEAR SIRS,—The Frames all came in first-rate order, not a pane of glass cracked. They are well made, the wood is well seasoned, and they give entire satisfaction.
Yours truly,
S. KEVAN

THIS frame is prepared to set on brick foundation against a garden wall, or at the side of a greenhouse. All the woodwork is of good red deal, painted three times, and glazed with 21-oz. sheet glass. The sashes are provided with set-opes and stays, and the joints arranged that each light may be easily lifted off.

	Length.	Width.	Height at back above wall.	Cash Prices, Carriage Paid.	Length.	Width.	Height at back above wall	Cash Prices, Carriage Paid.	
Made up to any length.	10 ft.	3 ft.	1 ft. 9 in.	£ s. d. 2 4 0	10 ft.	4 ft.	2 ft. 0 in.	£ s. d. 3 0 0	*Other widths at proportionate prices.*
	15 ft.	3 ft.	1 ft. 9 in.	3 3 0	15 ft.	4 ft.	2 ft. 0 in.	4 0 0	
	20 ft.	3 ft.	1 ft. 9 in.	4 2 0	10 ft.	5 ft.	2 ft. 3 in.	3 15 0	
					15 ft.	5 ft.	2 ft. 3 in.	4 17 0	

No. 64. PATENT PLANT PRESERVER.

TESTIMONIAL.
TEKLEY-IN WHARFDALE.
The Wall Fruit tree Protector and the Lights for Forcing Pit, supplied to me last Autumn, have given me great satisfaction.
(*Signed*)
F. T. TREMEL.

Made up to any length at proportionate prices.

Packing Cases charged extra and allowed for if returned Carriage Paid.

All materials same as specified for No. 62 Frame, the lights being made to turn quite over.

Length.	Width.	Cash Prices, Carriage Paid.	Length.	Width.	Cash Prices, Carriage Paid.	Length.	Width.	Cash Prices, Carriage Paid.
		£ s. d.			£ s. d.			£ s. d.
12 ft.	5 ft.	4 15 0	12 ft.	6 ft.	5 14 0	12 ft.	7 ft.	7 10 0
18 ft.	5 ft.	6 13 0	18 ft.	6 ft.	8 0 0	18 ft.	7 ft.	9 13 0
24 ft.	5 ft.	8 11 0	24 ft.	6 ft.	10 6 0	24 ft.	7 ft.	11 16 0

Packing Cases charged extra and allowed for if returned Carriage Paid

BOULTON & PAUL, MANUFACTURERS.

No. 70. SPAN-ROOF GARDEN FRAME.

THESE Frames are provided with lifting ridge ventilators in addition to the sashes which open on both sides. If set on brickwork and a path sunk, they make very useful houses, and may be heated or not as required. The sides are 14 in. in height. All the woodwork is of good red deal, painted three times; and the glazing done with 21-oz. sheet glass.

	Length.	Width.	Height to ridge.	Cash Prices, Carriage Paid.	
If arranged with door and frame, lock and inside lever for ventilators and prepared for brick walls 35/- extra.	10 ft.	8 ft.	4 ft.	£ s. d. 9 6 0	Packing Cases charged extra, and allowed for if returned Carriage Paid.
	15 ft.	8 ft.	4 ft.	12 8 0	
	20 ft.	8 ft.	4 ft.	15 10 0	
	25 ft.	8 ft.	4 ft.	18 12 0	

No. 70. ARRANGED AS A SMALL GREENHOUSE.

This Frame can be set on brick walls, when it fulfils the purposes of a useful Growing House.

Glazed sides can be arranged as shown in illustration, and ventilating boxes for side walls.

Estimates given for Stages & Heating Apparatus as suggested in Illustration.

Prices on application.

TESTIMONIAL.

BELGROVE,
QUEENSTOWN,
Oct., 1893.
GENTLEMEN,—I beg to inform you Sashes, Bearers, and Wallplates arrived here in excellent order and sustained no damage in transit; owing I must say to careful packing. They are a perfect fit, and I had them erected in half an hour after unpacking.

Yours truly,
M. LEARY.

Packing Cases charged extra, and allowed for if returned Carriage Paid.

No. 72. THREE-QUARTER SPAN-ROOF FORCING FRAME.

These Frames are arranged for building on brickwork, as shown. All the woodwork is of good red deal, painted three times, and glazed with 21-oz. sheet glass. The sashes are provided with set-opes and stays, and are arranged to turn quite over.

Plans and sections of brickwork sent on receipt of order.

Made in any other widths or lengths.

Estimates given for heating any length by our Portable Heating Apparatus.

Cash Prices, Carriage Paid.

Length.	Width.	Cash Prices, Carriage Paid. £ s. d.	Packing Cases Returnable. s. d.	Length.	Width.	Cash Prices, Carriage Paid. £ s. d.	Packing Cases Returnable. s. d.
10 ft.	6 ft.	4 17 0	5 0	10 ft.	7 ft.	5 16 0	6 0
15 ft.	6 ft.	6 10 0	6 0	15 ft.	7 ft.	6 14 0	7 0
20 ft.	6 ft.	8 6 0	7 0	20 ft.	7 ft.	9 15 0	8 0

No. 73. NEW SPAN-ROOF GARDEN FRAME.

This is a very strong and handy Frame, 14 in. high at the sides, and 27 in. at the ridge. The lights turn quite over, are fitted with set-opes and stays, and are glazed with 21-oz. sheet glass, and painted three times. Carefully fitted. Easily put together. Made up to any length.

Packing Cases charged extra, and allowed for if returned Carriage Paid.

No.	Length.	Width.	Cash Prices, Carriage Paid. £ s. d.	Packing Cases Returnable. s. d.	No.	Length.	Width.	Cash Prices, Carriage Paid. £ s. d.	Packing Cases Returnable. s. d.
No. 1 size	4 ft.	6 ft.	2 9 0	4 0	No. 4 size	16 ft.	6 ft.	6 18 0	5 6
No. 2 ,,	8 ft.	6 ft.	3 18 0	4 6	No. 5 ,,	20 ft.	6 ft.	8 9 0	6 0
No. 3 ,,	12 ft.	6 ft.	5 7 0	5 0	No. 6 ,,	24 ft.	6 ft.	9 19 0	6 6

BOULTON & PAUL, MANUFACTURERS.

No. 74. THREE-QUARTER SPAN-ROOF GARDEN FRAME.

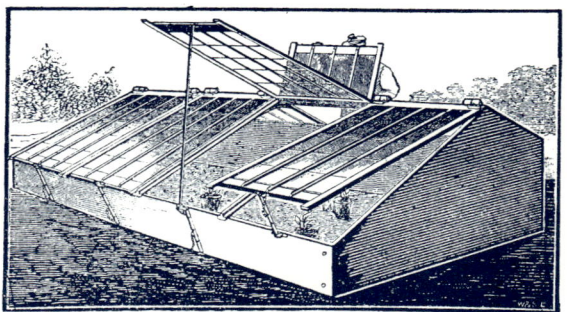

TESTIMONIAL.
HAYWARDS HEATH.
GENTLEMEN,—I have great pleasure in saying the Sixteen Lights arrived in capital condition, not a pane of glass broken, and scarcely a scratch on the paint, thanks to your system of packing. A better lot of Lights I never saw.
H. SAWYER.

TESTIMONIAL.
WHITCHURCH.
GENTLEMEN,—I have received Frame, which I am glad to say has sustained no damage in transit owing to careful packing. I consider it a marvel of cheapness, good material and workmanship. Thanking you for your prompt attention to my order,
Yours faithfully,
A. B. G.

THE most useful of all Frames that are made, owing to the extra height and convenience for attention. The fronts are 11 in. high, and the backs 22 in. high, and 32 in. at the ridge, bolted at the corners, easily taken to pieces if required. The lights are fitted with set-opes, and arranged to turn over, back and front, for ventilation. They are glazed with 21-oz. sheet glass, and painted three coats.

No.	Size.	Cash Prices, Carriage Paid.			Packing Cases Returnable.		No.	Size.	Cash Prices, Carriage Paid.			Packing Cases Returnable.	
		£	s.	d.	s.	d.			£	s.	d.	s.	d.
No. 1 size	4 ft. by 6 ft.	2	10	0	4	0	No. 4 size	16 ft. by 6 ft.	7	2	0	5	6
No. 2 ,,	8 ft. ,, 6 ft.	4	0	0	4	6	No. 5 ,,	20 ft. ,, 6 ft.	8	14	0	6	0
No. 3 ,,	12 ft. ,, 6 ft.	5	10	0	5	0	No. 6 ,,	24 ft. ,, 6 ft.	10	6	0	6	6

No. 75. MELON AND CUCUMBER FRAME.

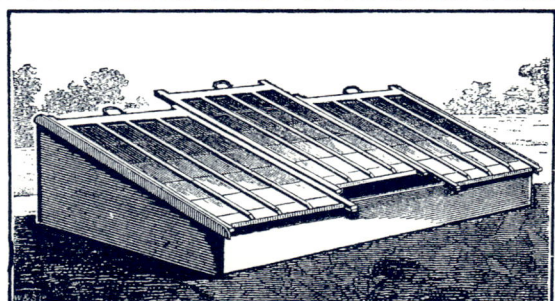

GODALMING,
Jan., 1894.
GENTLEMEN,—I enclose cheque in payment of your account. I am much pleased with the Greenhouse, which looks thoroughly satisfactory.
Yours faithfully,
A K LEIGHTON

Angle-iron corner standards are supplied for the No. 75 Frames at an extra cost of 11/- the set of four. These keep the Frame a few inches off the ground.

A LARGE stock of these Frames always ready. The fronts are 11 in. high, and the backs 22 in. high. All materials would be the same as for No. 74 Frame.

Lights only, 6 ft. by 4 ft., unglazed and unpainted, 5/- each.
Glazed with 21-oz. sheet glass, and painted three coats, 10/6 each.

No.	Size.	Cash Prices.			Packing Cases Returnable.	
		£	s.	d.	s.	d.
1-Light Frame	4 ft. by 6 ft.	1	15	0	4	0
2 ,, ,,	8 ft. ,, 6 ft.	2	15	0	4	6
3 ,, ,,	12 ft. ,, 6 ft.	3	15	0	5	0
4 ,, ,,	16 ft. ,, 6 ft.	4	15	0	5	6
5 ,, ,,	20 ft. ,, 6 ft.	5	15	0	6	0
6 ,, ,,	24 ft. ,, 6 ft.	6	15	0	6	6

TESTIMONIAL.
CHISWICK, W.
GENTLEMEN,—Enclosed I beg to hand you cheque in payment of your account. I am very much pleased with the appearance of the four 12 by 6 No. 75 Melon Frames.
(Signed) M. F. MAY.

Packing Case: charged extra, and allowed for if returned Carriage Paid.

No. 76. FORCING PIT WITH SLIDING LIGHTS.

Prepared for brickwork.

Made up to any length.

THE sills are prepared for brickwork, with a bearer and parting piece between each Light, painted three coats of good oil colour, and glazed with 21-oz. sheet glass. Iron handle at top, and strengthening bar across centre. A large stock of these always ready. Can be easily fixed by any handy man. Plans for brickwork sent on receipt of order. Estimates given for heating any length with our Independent Boilers.

Cash Prices, Carriage Paid.

	Length.	Width—6 ft.	Packing Cases returnable.	Width 7 ft. 6 in.	Packing Cases returnable.
		£ s. d.	s. d.	£ s. d.	s. d.
Sills and 2 Lights	8 ft.	2 8 0	3 6	2 17 6	4 0
,, 3 ,,	12 ft.	3 7 0	4 0	4 0 0	4 6
,, 4 ,,	16 ft.	4 6 0	4 6	5 2 6	5 0
,, 5 ,,	20 ft.	5 5 0	5 0	6 5 0	5 6
,, 6 ,,	24 ft.	6 4 0	5 6	7 7 6	6 0
,, 7 ,,	28 ft.	7 3 0	6 0	8 10 0	6 6
,, 8 ,,	32 ft.	8 0 0	6 6	9 12 6	7 0

No. 77. VIOLET OR BORDER FRAMES.

With Sliding Lights.

With Hinged Lights and Set-opes.

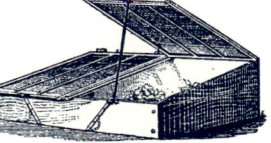

Front 8 in.
Back 16 in.

These Frames are made 4 ft. wide, painted three times, and glazed with 21-oz. sheet glass.

EMSWORTH, HANTS.

GENTLEMEN,—This afternoon the Frame has been delivered safely, and I am satisfied with it. I am obliged for your attention to my small order, and hope to trouble you again.

FRED LETT.

No.	Length.	Width.	With Sliding Lights.	With Hinged Lights & Set-opes.
			£ s. d.	£ s. d.
No. 1 size	6 ft.	4 ft.	1 10 0	1 15 0
No. 2 ,,	9 ft.	4 ft.	2 3 0	2 10 6
No. 3 ,,	12 ft.	4 ft.	2 15 0	3 5 0

THE CEDAR GARDENS, GT. MALVERN.

DEAR SIRS,—The Ten Pit Lights you supplied me with in March, have given me great satisfaction,
Yours truly,
W. T. ACOCK.

Estimates given for heating long ranges of Pits or Frames with our Portable Heating Apparatus.
Packing Cases charged extra, and allowed for if returned Carriage Paid.

BOULTON & PAUL, MANUFACTURERS,

No. 65. WALL FRUIT-TREE PROTECTOR.

This Illustration shows our Protectors for wall fruit-trees, supported on iron brackets, secured to the wall by bolts. The lights slide in grooves formed in the brackets, and are firmly held in position by a clamp; but can easily be removed in a few minutes. This is often a great advantage during a shower, as the rain saves syringing. Bolts are sent for a 9-in. wall unless ordered for a special thickness. The lights can be used during the summer on beds of salads, etc. Made up to any lengths.

Hooks and Rods for Blinds extra if required.

Length	Width	Cash Prices	Length	Width	Cash Prices
		£ s. d.			£ s. d.
12 ft	2 ft	1 8 0	12 ft	2 ft 6 in	1 12 0
18 ft	2 ft	2 1 0	18 ft.	2 ft. 6 in.	2 7 0
24 ft	2 ft	2 14 0	24 ft	2 ft 6 in.	3 2 0
30 ft	2 ft	3 7 0	30 ft	2 ft 6 in.	3 17 0

No. 65a. WALL FRUIT-TREE PROTECTOR.

This is a cheaper form of Protector, made without sashes. The glass is let into a grooved top rail and secured by lead clips at the bottom, as shown, on the system of glazing without putty, which answers very well for this purpose. Made in any length, complete with bolts for wall Brackets 6 ft. apart.

Cash Prices.

24 ft length, 2 ft projection	..	£2 0 0
24 ft ,, 2 ft 6 in ,,	..	2 10 0

MATERIALS FOR WIRING ROOFS AND WALLS.

GALVANIZED HANGERS.
No. 1   No. 2

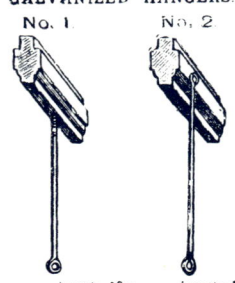

	Length—12-in	Length—15-in
No. 1	2/- per doz	2/4 per doz
No. 2	1/- ,,	1/2 ,,

BEST QUALITY. GALVANIZED WIRE.

No. 11	..	..	3/- per 100 yards
No. 12	..	..	2/6 ,,
No. 13	..	..	2/- ,,
No. 14	..	..	1/6 ,,

GALVANIZED STRAIGHT RODS.

⅜-in. diameter 1d per foot

Carriage Paid on Orders of 40/- value and upwards

PLATES AND EYES.

For Straining Wires in Vineries, and on Walls generally

No. 3.

Made with right and left-hand screws.

Cash Price 3/6 per doz

Packing Cases charged extra, and allowed for if returned Carriage Paid.

PROPAGATING GLASSES.   CAST-IRON HANDLIGHTS.

Most useful and indispensable for raising seeds and cuttings, and for covering tender plants.

THE Handlights are made with strong cast-iron frames, the tops arranged that they can be raised without removing the whole of the Light, for easy access to the plants. Painted three coats, and glazed with 21-oz. sheet glass.

Propagating Glasses sold only in sets of twelve.

8 in. diameter	Four of each size	
10 in. ,,	sent Carriage Free	
12 in. ,,	for 21/-	
10 in. ,,	Four of each size	
12 in. ,,	sent Carriage Free	
14 in. ,,	for 29/-	

14 in. by 14 in., glazed and painted	£0	10	0	each.
16 in. ,, 16 in. ,, ,,	0	11	6	,,
18 in. ,, 18 in. ,, ,,	0	13	0	,,
20 in. ,, 20 in. ,, ,,	0	14	6	,,
22 in. ,, 22 in. ,, ,,	0	17	0	,,
24 in. ,, 24 in. ,, ,,	0	19	6	,,

Packing Cases charged extra and allowed in full if returned Carriage Paid.

PIT LIGHTS GLAZED.

Always in Stock.

Cash Price.

Lights, 6 ft. by 4 ft., painted and glazed, **14/-** each.

Lights with Four Bars made to order.

Sash Bars, Rafters, Sills, and Plates quoted for on receipt of full particulars.

TESTIMONIAL.

BARROW HILL, CHESTERFIELD.
GENTLEMEN,—The eight large Lights arrived quite safely, in capital condition, and I am very pleased with them.   J. B.

PIT LIGHTS UNGLAZED.

Always in Stock.

Cash Price.

Lights, 6 ft. by 4 ft., unpainted and unglazed, **5/3** each.

Lights with Four Bars made to order

Pit Lights made to any size; also Frames to suit odd-sized Lights.

TESTIMONIAL.

THE CEDARS, GT. MALVERN.
DEAR SIRS,—The ten Pit Lights you supplied me with last year have given me great satisfaction.   W T

Packing Cases charged extra and allowed in full if returned Carriage Paid.

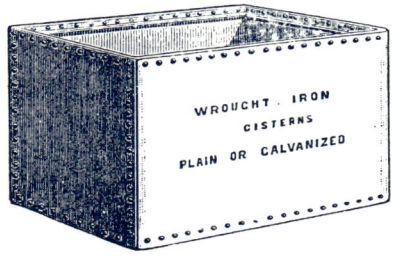

WROUGHT-IRON TANKS OR CISTERNS.

Galls. about.	Length	Width	Depth.	Prices		
25	2 ft. 0 in.	1 ft. 9 in.	1 ft. 3 in.	£0	18	0
30	2 ft. 0 in.	2 ft. 0 in.	1 ft. 3 in.	1	0	0
50	2 ft. 7 in.	2 ft. 0 in.	1 ft. 7 in.	1	6	0
100	3 ft. 0 in.	2 ft. 6 in.	2 ft. 2 in.	2	1	6
150	3 ft. 7 in.	2 ft. 10 in.	2 ft. 5 in.	2	14	6
200	4 ft. 0 in.	3 ft. 0 in.	2 ft. 8 in.	3	9	0
300	6 ft. 0 in.	3 ft. 0 in.	2 ft. 8 in.	4	5	6

Loose Covers for Cisterns to fit on angle-iron frame **1d.** per gall. extra.

Prices for Wrought-iron Pipes and Fittings on application.

BOULTON & PAUL. MANUFACTURERS.

GREENHOUSE SHELVES.

No. 1.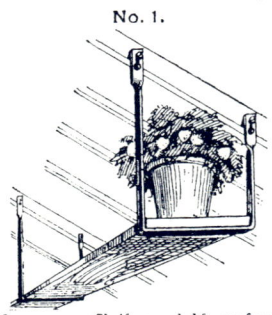

No. 2.

No. 3.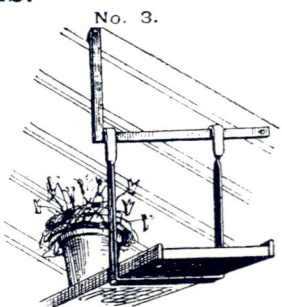

STRAWBERRY Shelf suspended from rafters, with single hangers. Hangers slotted for easy removal. Placed 5 ft. apart.
Cash Price.
20 ft. long by 8 in. wide, 16/-

SHELF to attach to back wall; made with one or two boards according to width. Brackets placed 5 ft. apart.
Cash Price.
20 ft. long by 12 in. wide, £1 4 0 including bolts.

SHELF suspended on double hangers, the upper iron screwed to rafter, with lower hanger made to hook on. Placed 5 ft. apart.
Cash Price.
20 ft. long by 9 in. wide, £1 1 0

WOOD BATTEN FOOT-PATHS FOR GREENHOUSES.

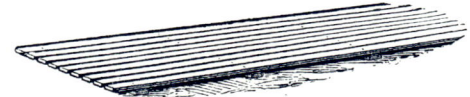

Made of good red deal, painted three times. Battens 3 in. by 1 in., placed ½-in. apart.
Cash Price.
Path, 20 ft. long by 3 ft. wide £1 15 0
For each additional foot run add 1/9

GREENHOUSE SIDE STAGES.

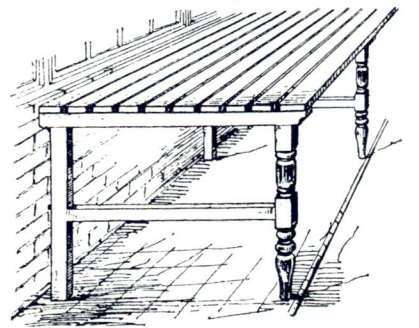

Wood Batten Stage suitable for general Greenhouse purposes.

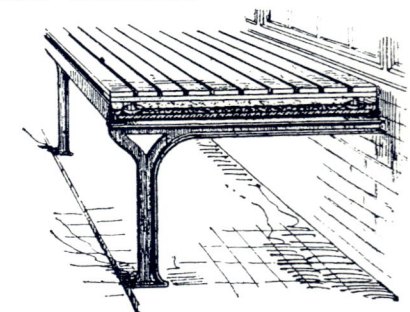

Slate and Iron Stages, with Movable Batten on top, suitable for Stove or Orchid Houses.

Special quotations given upon receipt of requirements.

Carriage Paid on all Orders above 40/- value to the principal Railway Stations in England and Wales, also to Dublin, Belfast, Cork, Londonderry, Edinburgh, or Glasgow.

ROSE LANE WORKS, NORWICH.

MATERIALS FOR WIRING ROOFS AND WALLS OF CONSERVATORIES, ORCHARD HOUSES, &c.

GALVANIZED HANGERS.

No. 1. No. 2.

		Length—12 in.	Length—15 in.
No. 1	..	2/- per doz.,	2/4 per doz.
No. 2	...	1/- ,,	1/2 ,,

BEST QUALITY GALVANIZED WIRE.

No. 11	...	.	3/- per 100 yards.
No. 12	..	...	2/6 ,,
No. 13	..	..	2/- ,,
No. 14	...	...	1/6 ,,

GALVANIZED STRAIGHT RODS.

¾-in. diameter ... 1d. per foot.

PLATES AND EYES.

For Straining Wires in Vineries and on Walls generally.

No. 3.

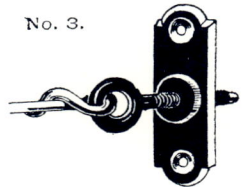

Made with right and left-hand screws.

Cash Price 3/6 per doz.

DIAMOND WIRE TRELLIS.

For Training Climbing Plants, &c.

Cash Prices.

Stock sizes—5-in. Mesh, Light Quality.

| 6 ft. by 3 ft. | ... each 2/6 | 6 ft. by 5 ft. | ... each 4/6 |
| 6 ft. ,, 4 ft. | ... ,, 3/6 | 6 ft. ,, 6 ft. | ... ,, 5/6 |

Made any size to order at the following Prices per square foot.

Mesh.	ADAPTED FOR	Medium Quality. Gauge.	
1½ in.	Protecting Windows	14	7d.
2 in.		13	5¼d.
4 in.	Plants in Conservatories, on House Walls, &c.	11	4d.
5 in.		11	4d.
6 in.		11	3½d.

Prices of other Meshes on application.

No. 108. RAIDISSEURS.

For Straining Wires.

One required for each.

Price 3/- per doz.   Keys for ditto, 4d. each.

WROUGHT-IRON EYES.

For Guiding Wires & Fixing Trellis on Walls.

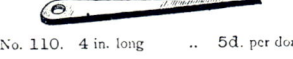

No. 110. 4 in. long .. 5d. per doz.

No. 112. 2 in. long ... 3d. per doz.
         3½ ,,   ... 9d. ,,

PEACH STANDARDS.

WROUGHT-IRON Standards, with necessary Stays, Wires, &c., for Peach Houses, supplied and fixed, to suit all requirements.

Carriage Paid on all Orders above 40/- value to the principal Railway Stations in England and Wales, also to Dublin, Belfast, Cork, Londonderry, Edinburgh, or Glasgow.

BOULTON & PAUL, MANUFACTURERS.

FINIALS AND CRESTINGS.

No. 1. Cast-iron Cresting.
6¼ in. in height.

Prices and full Particulars on Application.

No. 4. Cast-iron 2-way Terminal.

No. 6. Wrought-iron Terminal.

No. 5. Cast-iron 4-way Terminal.

No. 2. Cast-iron Cresting.
Made in three sizes, 5½ in., 7½ in., and 10 in.

No. 3. Cast-iron Cresting.
9¾ in. high.

HEATING APPARATUS.

BOILERS, HOT-WATER PIPES, VALVES COILS, COIL-CASES, TANKS, &c.,

OF EVERY DESCRIPTION FOR

HORTICULTURAL AND OTHER BUILDINGS.

DEFECTIVE OR WORN-OUT BOILERS replaced by our Improved Boilers at the shortest notice. arranged either to work by one Boiler, or by our improved system of duplicate Boilers.

ESTIMATES ON APPLICATION.

No. 4. CHECK-END SADDLE BOILER.

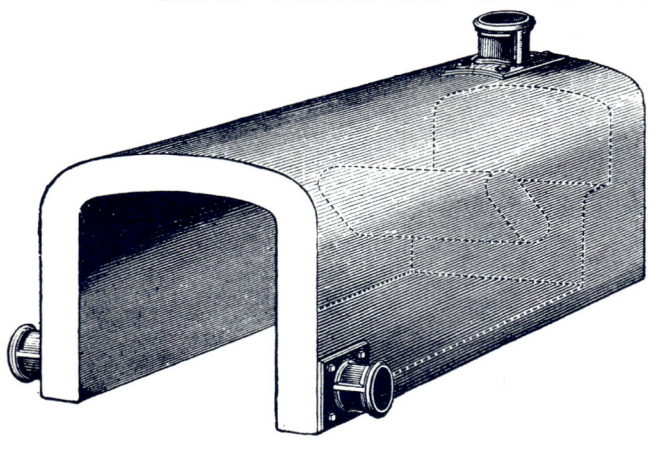

TESTIMONIAL.

CULVER, NEAR EXETER,
May 16th, 1893.

GENTLEMEN,—After giving the Boiler a fair trial, I find it does its work admirably. It is quicker in action, does its work with less amount of stoking and less consumption of fuel, than any other Boiler I have yet had to do with.

Yours truly,
G. CAMP.

AFTER the most careful and extended trials, we can confidently recommend this Boiler as the one best suited for general horticultural purposes.

It maintains a steady heat, requires very little attention; will burn any common fuel, and will commend itself to all requiring a Boiler to heat a large amount of piping with the smallest consumption of fuel.

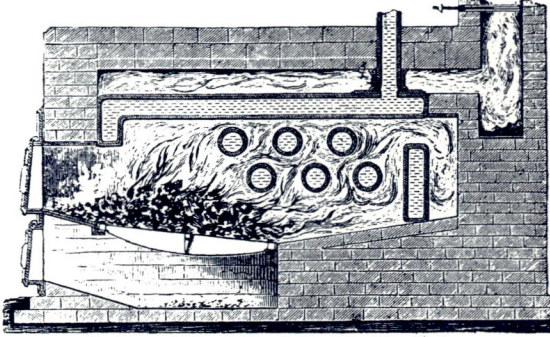

Section of No. 4 Check-End Saddle Boiler

No.	Outside Length.	Outside.		Heating Power, 4-in. Piping.	Cross Tubes.
		Width.	Height.		
1	30 in.	21 in.	18 in.	300 ft.	1
2	36 in.	21 in.	18 in.	500 ft.	1
3	42 in.	22 in.	18 in.	750 ft.	1
4	48 in.	24 in.	20 in.	1000 ft.	1
6	54 in.	27 in.	22 in.	1500 ft.	2
8	60 in.	33 in.	24 in.	2000 ft.	4
10	72 in.	36 in.	26 in.	3000 ft.	6

For Prices see List.

The guaranteed power given with each Boiler is the number of lineal feet of 4-in. piping the Boilers will effectually and economically heat.

No. 5. TERMINAL-END SADDLE BOILER,
WITH DOUBLE FLUES AND CROSS TUBES.

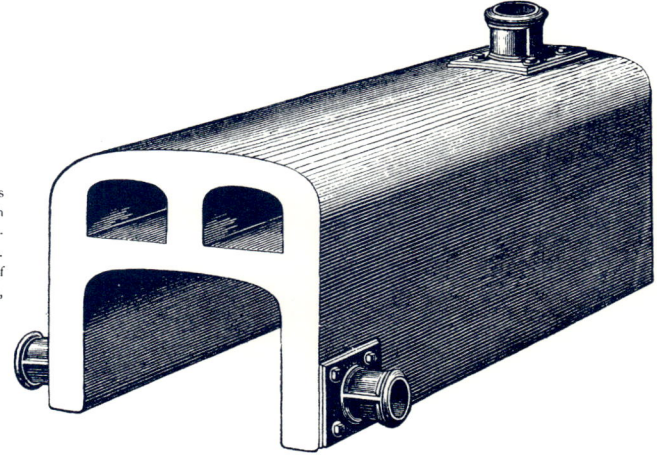

TESTIMONIAL

Rose Bank,
  Marden Park,
    Sept. 20th, 1893.

Dear Sirs,—I wish to express my satisfaction at the way in which your workmen have done the connection of the two Boilers here. We have tested the working of them both together and separately, and they work admirably.

Yours respectfully,
  J. M. TUCKER.

This Boiler is similar to our No. 4 Check-End, but made with top flues, arranged to heat in the most economical manner.

The following sizes are recommended for heating effectually the quantity of piping stated.

We recommend our No. 4 Check-End Boiler for heating from 300 to 3,000 ft. of 4-in. pipes.

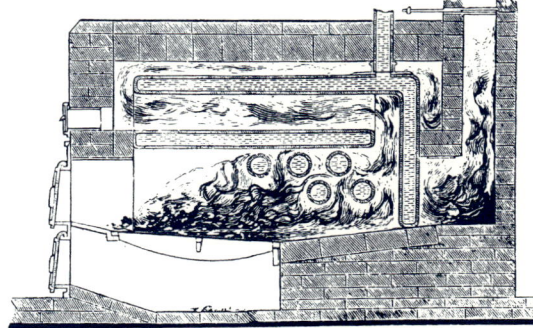

Section of No. 5 Terminal-End Saddle Boiler.

No.	Length.	Width.	Height.	No. of Cross Tubes.	Heating Power, 4-in. Piping.
1	48 in.	33 in.	36 in.	1	**1700** ft.
2	60 in.	36 in.	38 in.	2	**2300** ft.
3	72 in.	39 in.	42 in.	3	**3000** ft.
4	78 in.	42 in.	45 in.	4	**4000** ft.
5	84 in.	42 in.	48 in.	5	**5000** ft.

For Price see List.

Boilers arranged to work singly or in duplicate on our improved system.

BOULTON & PAUL, MANUFACTURERS.

No. 6 TERMINAL-END SADDLE BOILER,

WITH TOP FLUE.

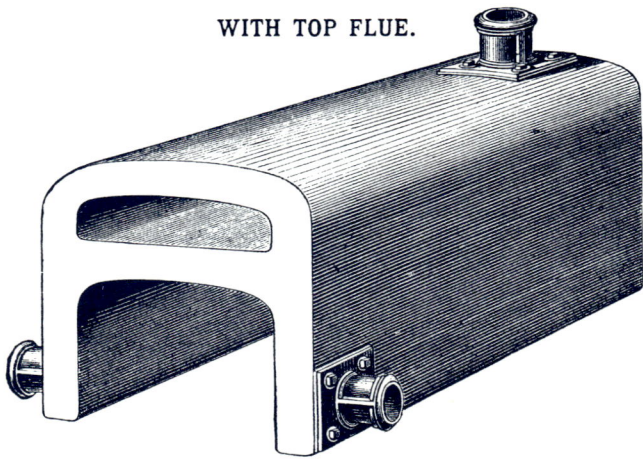

TESTIMONIAL

MOTCOMBE HOUSE,
DORSET.
Jan. 11th, 1894.

GENTLEMEN,—With reference to the Check-end Boilers No. 4 and Hotwater Pipes, I can say that, after forty years' experience with other boilers, that I prefer yours, and that during the late severe weather with twenty-five degrees of frost, we easily kept up the required heat.

Yours sincerely,
W. DENNIS.

THIS form of Boiler is much in use. The sizes mentioned below are quite capable of heating the quantity of 4-inch piping we guarantee them to do. This Boiler, however, requires more draught and consumes more fuel in proportion to work done than our No. 4 on page 62.

Section of No. 6 Terminal-End Saddle Boiler.

Total Length.	Size of Boiler.				Heating Power, 4-in. Piping.
	Inside the Arch.		Outside Measure.		
	Width.	Height.	Width.	Height.	
30 in.	16 in.	16 in.	22 in.	27 in.	500 ft.
36 in.	16 in.	16 in.	22 in.	27 in.	700 ft.
42 in.	18 in.	16 in.	24 in.	27 in.	900 ft.
48 in.	21 in.	18 in.	27 in.	30 in.	1100 ft.
54 in.	24 in.	18 in.	31 in.	30 in.	1300 ft.
60 in.	24 in.	18 in.	31 in.	30 in.	1500 ft.
72 in.	24 in.	21 in.	31 in.	33 in.	2000 ft.

For Prices see List.

Estimates given for Erecting Heating Apparatus on receipt of particulars, or from measurement taken by us.

No. 7. PLAIN SADDLE BOILER.

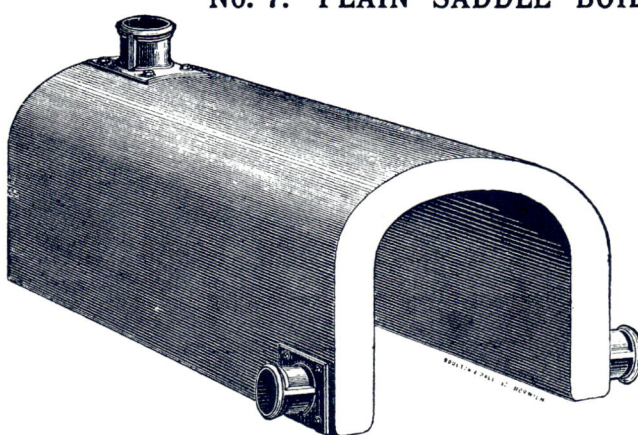

TESTIMONIAL.

Hardwick,
Bury St. Edmunds,
Jan. 10th, 1893.

Dear Sirs,—I am pleased to inform you that we have the Boiler (No. 4 Check-End) fixed and at work. It is doing its work in a very satisfactory manner; in fact, it quite surpasses my expectations, and I am more than satisfied with it.

Yours truly,

B. MARKS.

Plain Saddle Boilers are in general use for heating from 100 feet to 300 feet of 4-inch piping; they are readily fixed, easily fired, and very durable. If more than the quantity of 4-inch piping mentioned is required to be heated, it is more economical to use our Terminal Check-End Saddle Boiler, as one of these, 3 feet long, will heat double the quantity of piping with the same quantity of fuel as a 3-feet Plain Saddle.

Plain Saddle Boilers are made in any size up to 6 feet in length.

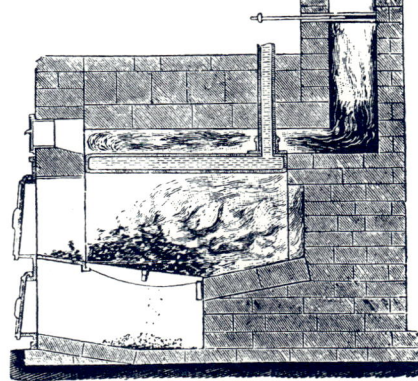

Section of No. 7 Plain Saddle Boiler.

Size of Boiler inside Arch.			Heating Power, 4-in. Piping.
Length.	Width.	Height.	
18 in.	12 in.	10 in.	100 ft.
21 in.	12 in.	10 in.	125 ft.
24 in.	12 in.	12 in.	150 ft.
27 in.	14 in.	14 in.	200 ft.
30 in.	14 in.	14 in.	250 ft.
36 in.	16 in.	16 in.	300 ft.

For Prices see List.

Defective or Worn-out Boilers replaced by our improved Boilers at the shortest notice.

No. 8. INDEPENDENT SLOW COMBUSTION STOVE.

TESTIMONIAL.

STOKE-ON-TRENT.

GENTLEMEN,—I received the Boiler (Amateur's Independent) safely, have fixed it, and it gives great satisfaction. I am very pleased with it, and shall certainly recommend your Boilers in this neighbourhood.

Yours, etc.,

A. HEELING.

TESTIMONIAL.

THEOBALD'S PARK,
WALTHAM CROSS.

GENTLEMEN,—I am very pleased to say your Hot-Water Pipes and Boiler (No. 4 Check-end Saddle) are working capitally, and I am very pleased with the Houses.

Yours truly,

R. GIBBS.

THIS Boiler is designed for heating large quantities of piping. It is perfectly self-contained, requires no brickwork, and is easily removed when required. It will be found most convenient for working where a small space only is available.

It should be understood that any Boiler not set in brickwork will consume more fuel than a Saddle Boiler of the same heating power.

The smoke flue is placed at back of Boiler, and the flow-pipe on the top; the return sockets can be placed as required.

No.	Height.	Diameter.	Heating Power. 4-in. Piping.
1	30 in.	18 in.	300 ft.
2	36 in.	18 in.	400 ft.
3	38 in.	21 in.	475 ft.
4	38 in.	25 in.	550 ft.
5	44 in.	30 in.	750 ft.
6	50 in.	30 in.	950 ft.
7	56 in.	30 in.	1200 ft.

Any other size made to order.

For Prices see List.

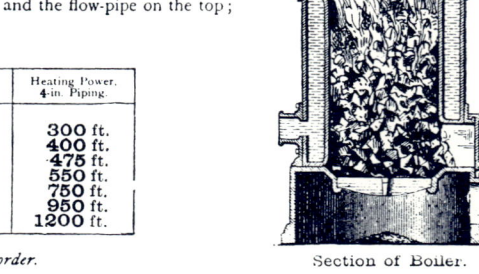

Section of Boiler.

No. 10. PHŒNIX SLOW COMBUSTION BOILER.

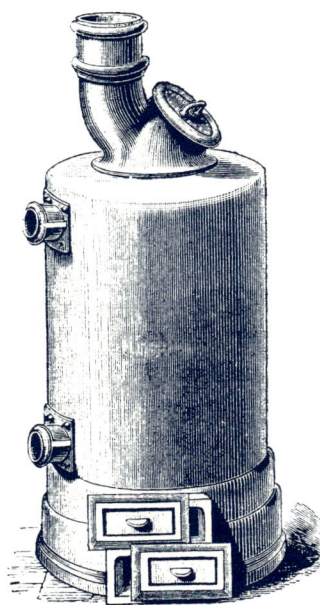

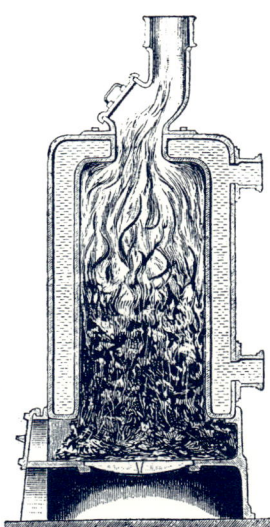

Section B, showing No. 4 size, 2 ft. 6 in. high.

TESTIMONIALS.

GLAIS HOUSE,
SWANSEA VALLEY,
Oct. 9th, 1893.

DEAR SIRS,—Your Phœnix Boiler works splendidly. It will burn for eighteen hours without putting any more fuel on.

Yours faithfully,
JOHN ETTLE.

BOOKHAM LODGE,
COBHAM.

GENTLEMEN,—I have had two of these Boilers (Phœnix) in use nearly ten years, and they give every satisfaction.

Yours faithfully,
A. SANDERS.

THIS Illustration shows our Phœnix Upright Slow Combustion Boiler, for which we claim the following good qualities, viz. It is made of the best materials; has no parts that are liable to failure; is provided with a flue that cannot be choked with fuel, will heat effectually the quantity of pipes stated with a small amount of fuel; and will hold sufficient to keep up the proper heat for twelve hours; requires no brickwork, and takes up the smallest space of any boiler of its power.

No.	Height without Feed Hole.	Diameter of Boiler.	Heating Power, 4-in. Piping.
1	26 in.	16 in.	150 ft.
2	29 in.	16 in.	200 ft.
3	32 in.	16 in.	250 ft.
4	32 in.	18 in.	300 ft.
5	38 in.	18 in.	400 ft.

For Prices see List.

No. 11. UPRIGHT SLOW COMBUSTION BOILER.

Same as our No. 8, but built in Brickwork.   An improvement on the well-known Excelsior Boiler.

Elevation of Boiler.   Vertical Section of Boiler.

Fig **A** shows the upper doors for feeding, **B** the lower doors for raking out, **C** ash pit, **D** flow pipe, and **E** flue to chimney

THESE Boilers are especially suitable where a large amount of heat is required, and where the space is limited. They will work freely from twelve to eighteen hours without attention.   The entire surface is exposed to the heating power before the draught goes up the chimney, and every provision is made for easy attention.   The dimensions given are the sizes of the boilers without foundations.

Complete sets of furnace fittings are supplied, which include upper and lower furnace doors, sockets, dead plates, bearing bars, fire bars, damper, and draw-off tap.

No	Height.	Diameter.	Heating Power, 4-in Piping.
1	36 in.	30 in	1000 ft.
2	42 in.	36 in.	1500 ft.
3	48 in.	42 in.	2000 ft.
4	54 in.	54 in.	3000 ft.
5	60 in.	60 in.	4000 ft.

For Prices see List.

ROSE LANE WORKS, NORWICH. 69

No. 12.
THE AMATEUR'S INDEPENDENT

SLOW COMBUSTION BOILER.

TESTIMONIAL.

Hockley Heath,
Feb., 1894.

Gentlemen,—I am very pleased with the Amateur's Boiler we had from you last December; it is easily fixed, does its work well, and requires very little attention; for small houses or a range of pits it would be very difficult to beat.

Yours faithfully,
H. STARK.

TESTIMONIAL.

Hingham.

Gentlemen,—I am glad to say the Amateur's Boiler I had of you last year works beautifully and most economically.

Yours truly,
CHAS. LEE.

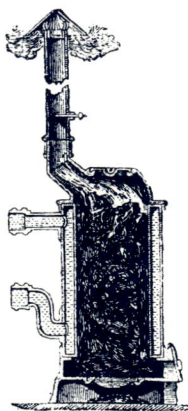

Section of Boiler.

This Boiler is well adapted for small Greenhouses; can be used either for quick heating or slow combustion. It is one of the most durable boilers made; quite portable; will heat properly for twelve hours; and is guaranteed to heat satisfactorily the length of piping named.

Only one-tenth the cost of heating by gas, and much more effectual. Made with strong wrought-iron cylinder, set in a cast-iron base. Fitted with two 2-inch cast sockets, fire bars, sliding doors, top door for feeding, and socket for smoke flue.

It is fitted with dome top, which prevents the fuel from choking the chimney. The section shows the working of the Boiler and the mode of firing.

No.	Total Height.	Diameter of Boiler.	Heating Power, 4-in. Piping.
1	28 in.	14 in.	40 ft.
2	32 in.	14 in.	70 ft.
3	36 in.	14 in.	100 ft.

Set of Furnace Tools for Independent Boilers, consisting of Poker, Rake, Shovel, and Flue Brush.

For Prices see List.

COMPLETE HEATING APPARATUS FOR LEAN-TO AND SPAN-ROOF GREENHOUSES.

THE Apparatus consists of one of our No. 12 Amateur's Independent Slow Combustion Boilers, fitted with 8 ft. of chimney, with damper and cap, and includes connecting pipes through the wall into Greenhouse, either from the back, as shown, or from the end, connected to the pipes as shown on section below. A supply cistern is fitted, together with draw-off tap. The pipes are all fitted, and the joints made, as far as possible, at our Works, rubber rings being supplied for making the remainder. The parts are all numbered, and a plan sent, which will enable any handy man to put it together.

Any other plan can be carried out to suit existing greenhouses, and the Boiler placed in almost any position.

 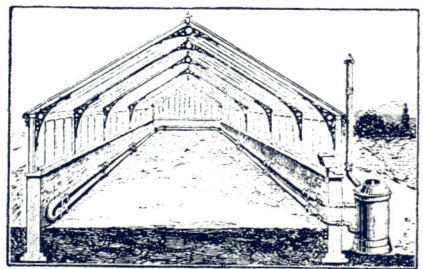

It will save time and correspondence if in all enquiries for Heating Apparatus, our customers will kindly give the particulars mentioned below.

A plan of the house to be heated, showing whether a lean-to or span-roof.

Exact measurements inside.

Position of door, or doors.

The most convenient position for Boiler, together with thickness of back wall.

Particulars of any stages, beds. etc., that are likely to interfere with the course of the piping.

Plan of Lean-to House.   Plan of Span-roof House.

Estimates on Application.

IMPROVED VALVES FOR HOT-WATER APPARATUS.

New and improved designs, warranted to stop the circulation in the pipes. Any valve can be inspected or repaired without removing it from the pipes.

No. 1. Valve with Socket & Spigot Ends

Made for 2-in., 3-in., and 4-in. Pipes.

No. 2. Valve with Double Sockets.

Made for 2-in., 3-in., and 4-in. Pipes.

No. 3. Elbow Valve.

Will save a common Elbow, and give full water-way. The most useful valve that is made. Can be used in most places.

Made for 2-in., 3-in., and 4-in. Pipes.

No. 4. New H-Piece Screw Valve.

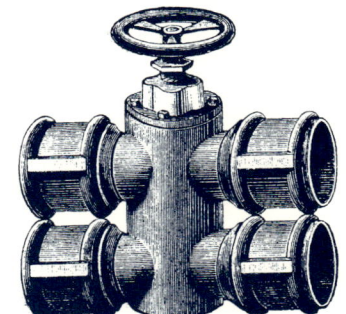

This Valve is intended to take the place of an H-Piece with three Valves.

Made for 3-in., and 4-in. Pipes.

No. 5. Improved Throttle Valve, with Socket and Spigot Ends.

With loose valve-piece, which can be removed to inspect or repair the Valve.

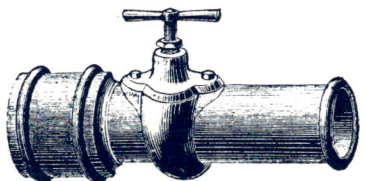

Made for 2-in., 3-in., and 4-in. Pipes.

No. 6. Improved Throttle Valve, with Double Sockets.

Made for 2-in., 3-in., and 4-in. Pipes.

For Prices see List.

HOT-WATER PIPES AND CONNECTIONS.

No. 1. HOT-WATER PIPES

Size	2" Pipes.	3" Pipes.	4" Pipes.
	In 6 ft. lengths.	In 6 ft. or 9 ft. lengths.	In 6 ft. or 9 ft. lengths.

No. 2. LOOSE TROUGH.

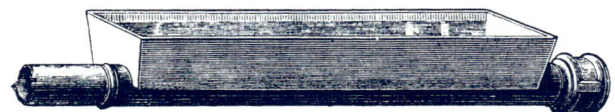 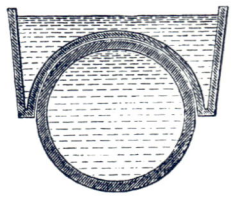

Made in 2 ft. 6 in. lengths only.

No 3. IMPROVED VAPOUR TROUGHS.

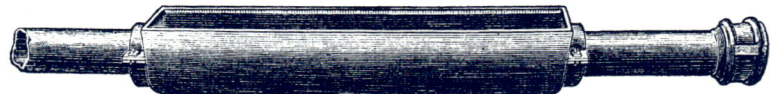

3 ft. Vapour Trough, as fixed on a 9 ft. Length of Piping.

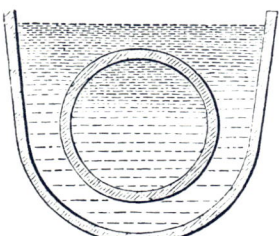

The Inside Ring shows a 4-in. Pipe.

MADE to bolt on to pipes already fixed; these Troughs being made so that the water entirely surrounds the Pipe, give off vapour at a lower temperature than any other form of Trough, and can be easily fixed on at any time in a few minutes.

Made in 3 ft. and 6 ft. Lengths, complete, with bolts for fixing on 3-in. or 4-in. Piping.

For Prices see List.

HOT-WATER PIPE CONNECTIONS.

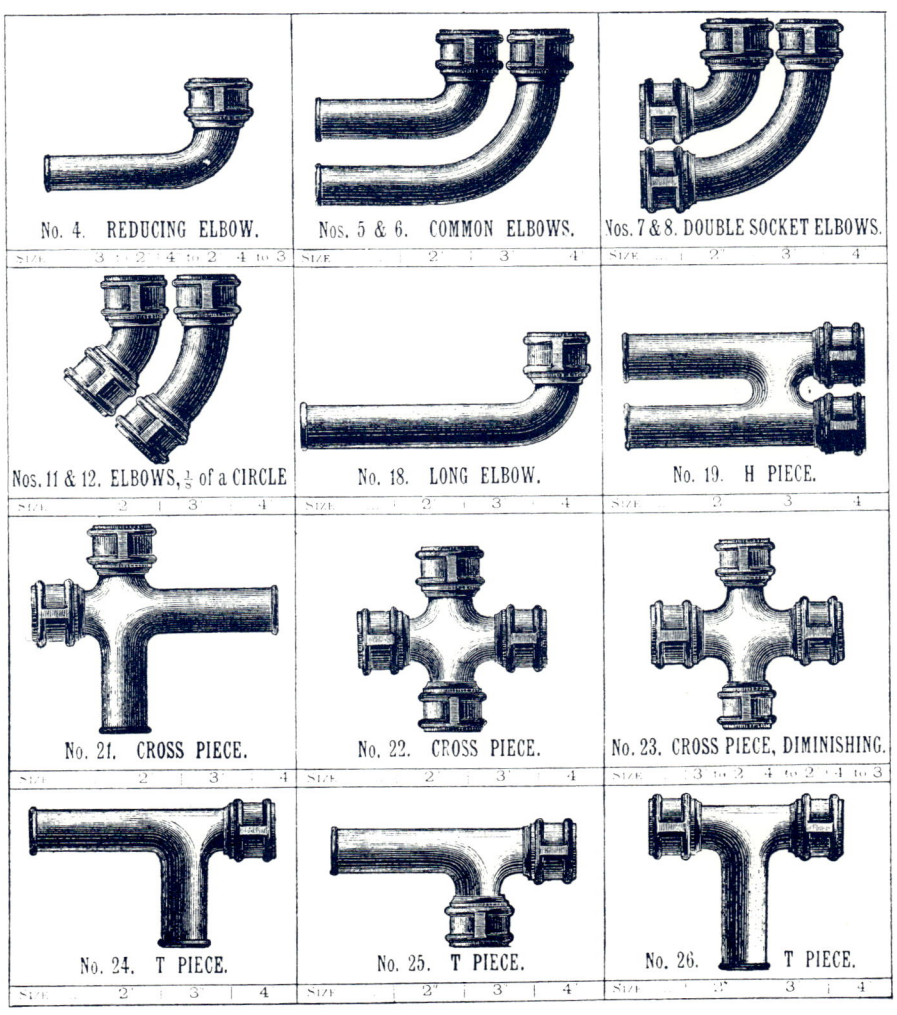

For Prices see List.

HOT-WATER PIPE CONNECTIONS.

No. 27. T PIECE.	No. 28. T PIECE, DIMINISHING.	No. 29. T PIECE, DIMINISHING.
Size ... 2 3 4	Size 3 to 2 4 to 2 4 to 3	Size 3 to 2 4 to 2 4 to 3
No. 30. T PIECE, DIMINISHING.	No. 31. CLOSE SYPHON.	No. 32. OPEN SYPHON.
Size 3 to 2 4 to 2 4 to 3	Size ... 2 3 4	Size ... 2 3 4
No. 33. 3-WAY SYPHON.	No. 34. 2-WAY ELBOW SYPHON.	No. 35. 3-WAY ELBOW SYPHON.
Size ... 2 3 4	Size ... 2 3 4	Size ... 2 3 4
No. 36. 2-WAY SYPHON, SOCKET OUTLET.	No. 37. 2-WAY SYPHON, SPIGOT OUTLET.	No. 38. 3-WAY SYPHON, SOCKET OUTLET.
Size ... 2 3 4	Size ... 2 3 4	Size ... 2 3 4

For Prices see List.

HOT-WATER PIPE CONNECTIONS.

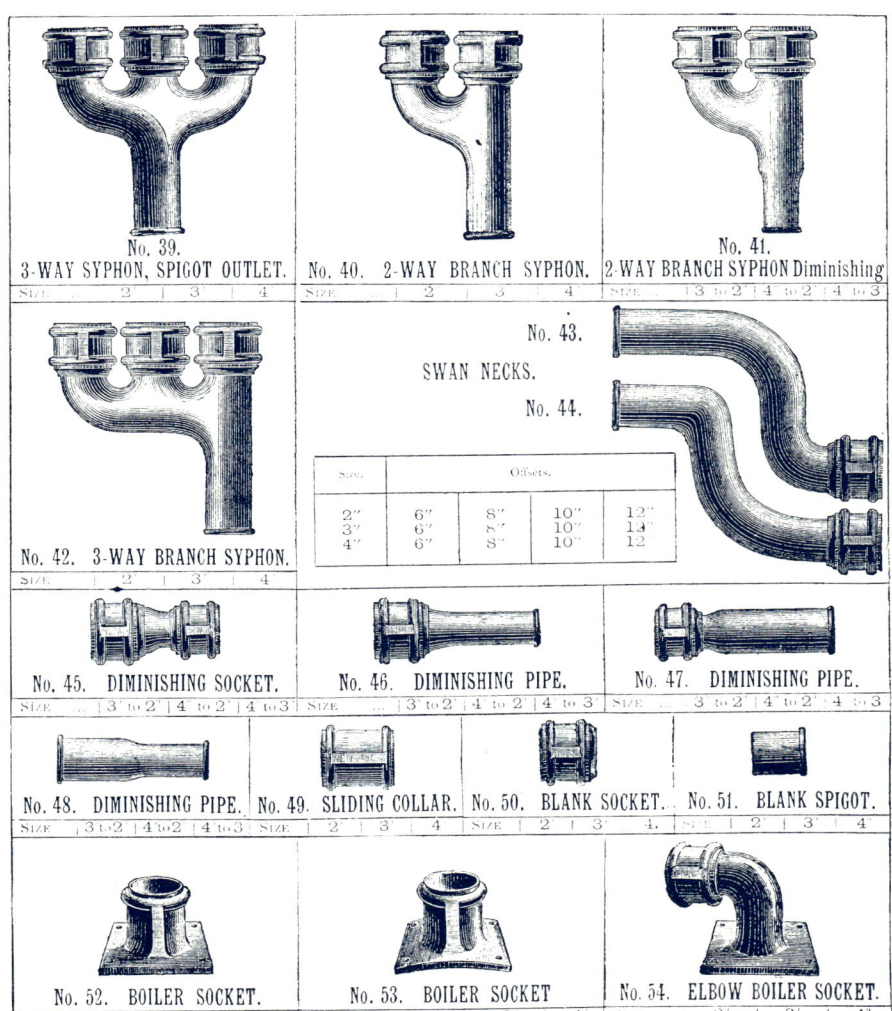

BOULTON & PAUL, MANUFACTURERS.

FURNACE FRONTS, FIRE BARS, DEAD PLATES, &c.

FURNACE FRONTS.

FURNACE Fronts with improved sliding doors. The doors are made large, giving plenty of room for easy stoking. Made in four sizes to suit the Boilers illustrated in our Catalogue.

FURNACE FRONTS.

No.	Height.		Width.		Size of Boiler.	
	ft.	in.	ft.	in.	in.	in.
1	2	4	1	11	12 by 12
2	2	8	2	0	14 ,, 16
3	2	9	2	3	18 ,, 20
4	3	0	2	8	22 ,, 24

FIRE BARS.

Heavy make, to resist a sharp fire.
The 12, 15, and 18 in. Bars are 1⅜ in. wide.
The 21, 24, 27, 30, and 36 in. Bars are 1½ in. wide.

SOOT BOXES WITH SLIDING DOORS.

4 in. by 6½ in., and 6 in. by 10 in.
Upright Boxes same size.

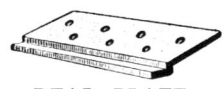

DEAD PLATE.

Made 16, 18, 22, 26, 30, and 36 in. long by 9 in. wide.

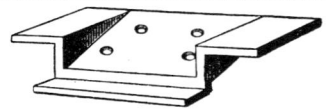

NEW PATTERN DEAD PLATE.

For large Boilers where Cleaning Tubes are used.

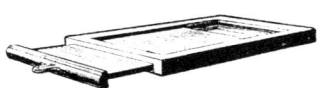

DAMPER AND FRAME.

14 in. by 9 in., and 18 in. by 14 in.

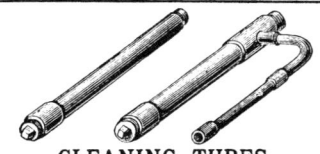

CLEANING TUBES.

To admit of sediment being removed from Boilers.

BEARING BAR.

18, 22, 26, 30, and 36 in. long.

CRAMPS.

For No. 1, 2, 3, 4 Fronts.

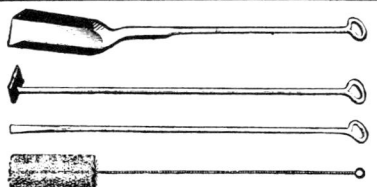

FURNACE TOOLS FOR LARGE BOILERS.

SHOVEL, POKER, HOE, AND FLUE BRUSH.

No.	Length of Boiler
1	30 in.
2	42 in.
3	48 in. and larger

For Prices see List.

ROSE LANE WORKS, NORWICH

No. 1. CAST-IRON PLAIN SUPPLY CISTERN.

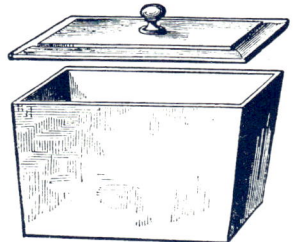

No.	Length.	Width.	Depth.	Approximate number of Gallons.
1	15 in.	8 in	12 in.	5
2	20 in.	12 in.	15 in.	13
3	24 in.	12 in.	18 in.	18

BOILER COCKS.

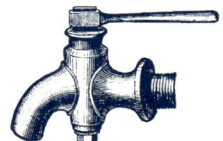

No. 2. WROUGHT-IRON WATER TANKS OR SUPPLY CISTERNS.

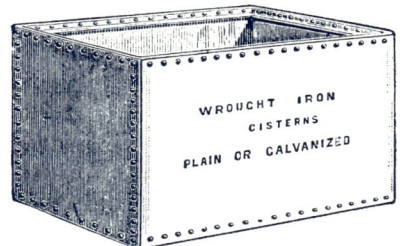

Galls.	Length.		Width.		Depth.	
about	ft.	in.	ft.	in.	ft.	in.
25	2	0	1	8	1	3
30	2	0	2	0	1	3
40	2	0	2	0	1	7
50	2	7	2	0	1	7
80	3	0	2	2	2	0
100	3	0	2	6	2	2
150	3	7	2	10	2	5
200	4	0	3	0	2	8
250	5	0	3	0	2	8
300	6	0	3	0	2	8

Loose Covers for Cisterns, to fit on Angle-iron Frame, 1d. per gallon extra.

WROUGHT-IRON PIPES FOR WATER SUPPLY TO GARDENS & GREENHOUSES.
Unpainted, Painted, or Galvanized.

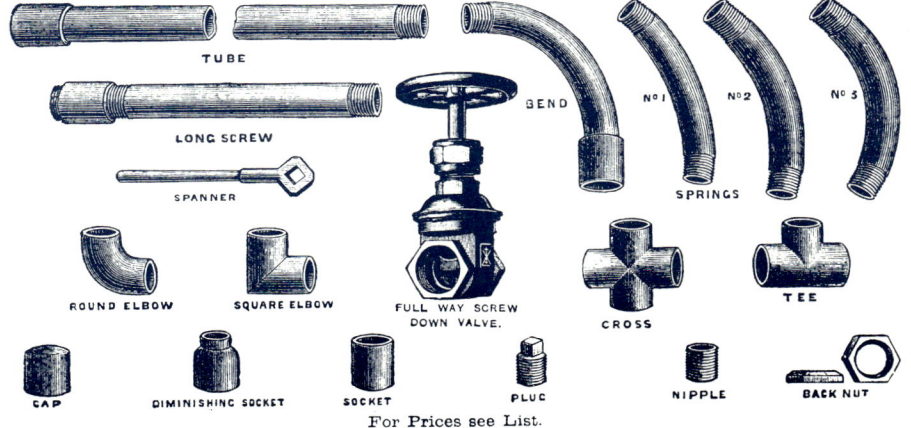

For Prices see List.

COILS AND COIL CASES.

ROUND COIL CASE.

3 ft. high, 2 ft. 0½ in. diameter.

SQUARE COIL CASE.

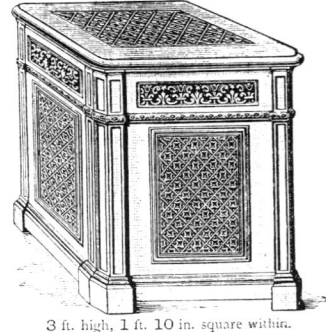

3 ft. high, 1 ft. 10 in. square within.

FLAT COIL CASE, TO COVER COILS Nos. 1 and 2.

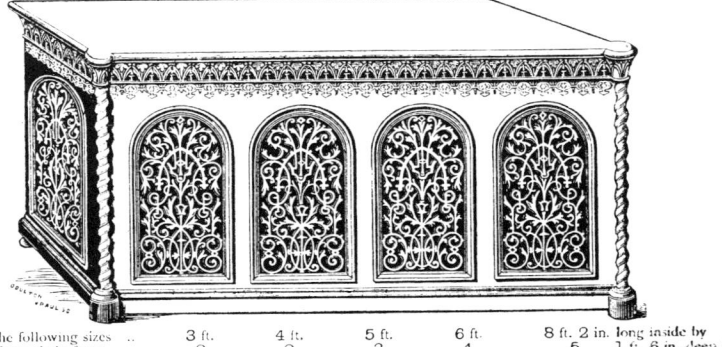

Made of the following sizes	3 ft.	4 ft.	5 ft.	6 ft.	8 ft. 2 in. long inside by 1 ft. 6 in. deep inside.
Number of panels in front	2	2	3	4	5

No. 1.

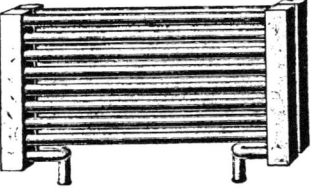

No. 2.

Coils Nos. 1 and 2 for 2-in. and 3-in. pipes, single and double, made to any length and number of pipes.

For Prices see List.

ORNAMENTAL CAST-IRON GRATINGS.

No. 1. PATTERN.

Light.—Made up to 3 ft. lengths, 9½ in. and 15½ in. wide only.

No. 2. PATTERN.

Medium.—Made up to 3 ft. lengths, 12 in. and 18 in. wide only.

No. 3. PATTERN.

Strong.—Made up to 4 ft. lengths, 12, 18, 24, and 30 in. wide.

No. 4. PATTERN.

Medium weight, but very strong.—Made up to 4 ft. lengths, 9, 12, 15, 18, 24, and 30 in. wide.

Recommended for peach houses and vinery paths.

NOS. 2, 3, & 4 MADE CONTINUOUS IN PATTERN WITHOUT END BORDERS.

The patterns are made without End Borders unless they are specially ordered.

Other patterns in stock.

Moulding or Cutting Mitres to Angles are charged for any of the above patterns as follows :—12 in. and under, 1/- ; 18 in. and under, 1/6 ; 24 in. and under, 2/- ; 30 in. and under, 2/6 each.

WHEN ORDERING, A PLAN SHOULD BE SENT WITH ALL DIMENSIONS CLEARLY SHOWN.

For Prices see List.

BOULTON & PAUL, MANUFACTURERS.

ALL DESCRIPTIONS AND SIZES OF BOILERS MADE ON THE SHORTEST NOTICE,

INCLUDING THE

'GOLD MEDAL,' 'TRENTHAM,' 'EXCELSIOR,' 'FINSBURY,' &c.

CAST AND WROUGHT-IRON PIPE RESTS.

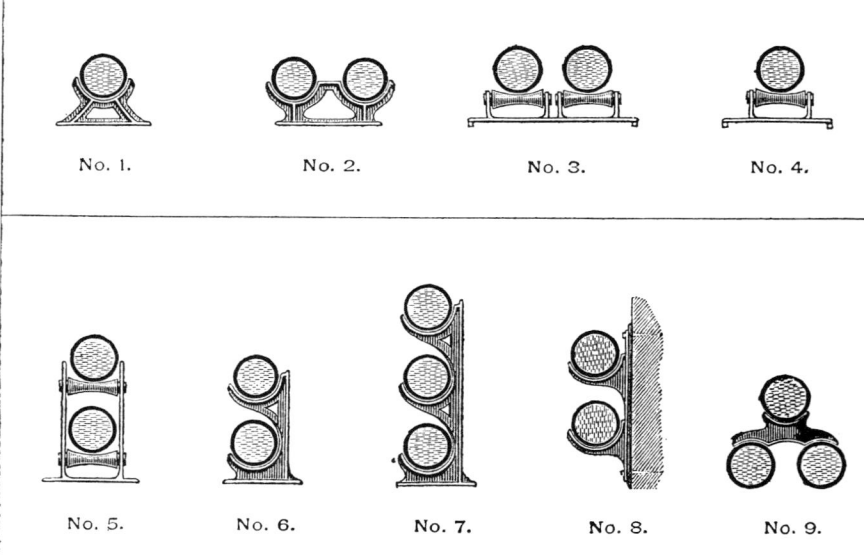

These Rests are made for 2, 3, and 4-in. pipes.

Patterns Nos. 5 and 7 can be constructed for a larger number of pipes if required.

For Prices see List.

PORTABLE BUILDINGS

FOR LEASEHOLD PROPERTY,

OF THE MOST ARTISTIC DESIGNS IN

WOOD AND IRON.

GENTLEMEN WAITED UPON & SITES SURVEYED IN ANY PART OF THE COUNTRY.

The Prices in this Catalogue are not binding, being subject to alteration without notice.

BOULTON & PAUL, MANUFACTURERS.

Wood & Iron Churches, and Chapels.

Wood & Iron Mission Rooms, and Schools.

Interior of Church erected at Norwich, 1893.

SPECIAL DESIGNS PREPARED TO SUIT ANY SITUATION OR REQUIREMENTS. SURVEYS MADE AND ESTIMATES GIVEN FREE OF CHARGE.

ROSE LANE WORKS, NORWICH. 83

No. 600. WOOD CHURCH.

As erected at Norwich, 1893.

Constructed according to the Specification on page 86.

For Fittings, see page 91.

PLAN.

TESTIMONIAL

From THE REV. M. S. JACKSON,
Vicar of St. Paul's, Norwich.
July 4th, 1893.

I have much pleasure in stating that Messrs. Boulton and Paul have built a new Mission Church of wood for this parish.

The building has been planned, designed, and carried out entirely by them, and the material and workmanship appear to be thoroughly sound, strong, and good, and promise to give complete satisfaction.

I have heard only expressions of pleasure and admiration in reference to the style and appearance of the building. It is certainly, both inside and out, a charming little church.

No. 600a. WOOD CHURCH.

As erected at Norwich, 1893, for the Rev. G. S. Barrett.

Constructed according to the Specification on page 86.

For Fittings, see page 91.

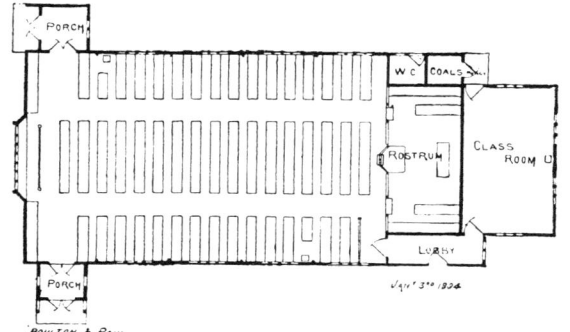

PLAN

TESTIMONIAL.

From the Rev. G. S. Barrett,
Bracondale.

I have much pleasure in testifying to the entire satisfaction which Messrs. Boulton and Paul's work has given to the Committee.

The Church is well built, tasteful in design, and commodiously furnished, and has been erected at a very reasonable cost.

ROSE LANE WORKS, NORWICH. 87

No. 300. IRON CHURCH, CHAPEL-OF-EASE.

TESTIMONIALS.

From Miss PALIN, Tallwylan.
Miss Palin is much satisfied with the Iron Building. Should she ever require such a building again, will employ them instead of local carpenters.

From
ANDREW LOW, Esq., Bradfield.
Mr. Low has to inform Messrs. Boulton and Paul that their workman has this morning finished the erection of the Iron Building; he wishes to express his satisfaction with the work and the workman, whom he has found most civil and obliging.

No. 301. HOSPITAL, SCHOOL, OR MISSION ROOM, &c.

PRICES for materials, framing morticed and tenoned, Carriage Paid to most Stations in England, ready for fitting and erection by purchaser, including galvanized corrugated iron sheets, flooring, matchboard, and felt lining; all necessary bolts and screws, gutters, and down-pipes; doors and windows packed complete; the latter glazed with 21-oz. sheet glass, and all supplied with requisite fittings:

Approximate Prices:

	£	s.	d.
20 ft. by 12 ft., 11 ft. 0 in. high to ridge,	36	10	0
30 ft. ,, 15 ft., 12 ft. 6 in. ,,	61	0	0
40 ft. ,, 20 ft., 15 ft. 0 in. ,,	99	0	0

No. 402. GALVANIZED IRON INFECTIOUS HOSPITAL.

REGISTERED DESIGN, No. 5954.

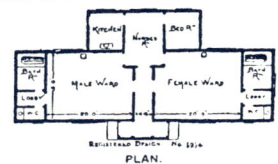

PLAN.

Size of main building 60 ft. by 17 ft.
Three nurses' rooms 26 ft. by 10 ft.

Cash Price, £250.

Constructed according to the Specification on page 92, but with walls covered outside with iron instead of wood; thoroughly well ventilated.

No. 462. WOOD CHURCH OR CHAPEL-OF-EASE.

Erected by us at Lancaster for the County Asylum.

Constructed according to the Specification on page 86.

TESTIMONIAL.

From DAVID M. CASSIDY, Esq., M.D., Medical Superintendent, The County Asylum, Lancaster.

Sept. 10th, 1892.

I beg to inform you that the building of our new Church has now been completed by your workmen. The whole has been done to the satisfaction of the Visiting Committee and myself, and we are all well pleased with the appearance the Church now presents.

REGISTERED COPYRIGHT.

No. 403. GALVANIZED IRON MISSION ROOM OR PAROCHIAL HALL.

Approximate Cash Prices.

Length outside.	Width outside.	Height at eaves.	Height at ridge.	Price.
40 ft.	18 ft.	8 ft.	14 ft.	£114
40 ft.	20 ft.	8 ft.	15 ft.	125
45 ft.	20 ft.	8 ft.	15 ft.	137
50 ft.	25 ft.	9 ft.	18 ft.	180
58 ft.	26 ft.	9 ft.	18 ft.	206
60 ft.	28 ft.	10 ft.	20 ft.	246
70 ft.	30 ft.	10 ft.	20 ft.	279

Moulded Wood String, 4d. per foot run extra.

Carriage Paid to most Stations in England.

REGISTERED COPYRIGHT.

Constructed according to the Specification on page 86, but with iron outside walls in lieu of wood.

If brickwork is not desired, we will estimate for supplying Iron Standards as shown above, the cost being about the same as for brickwork. The advantages of Iron Standards are, no harbour for vermin, no possibility of damp rising.

Prices.

1 ft. high, 5/-; 1 ft. 6 in. high, 6/-; 2 ft. high, 7/6.

For Seats, Desks, and Fittings, see page 91. Estimates on application.

REGISTERED COPYRIGHT.

REGISTERED DESIGN, No. 5947.

ROSE LANE WORKS, NORWICH.   89

No. 453. CHURCH.

N° 453
ELEVATION.

PLAN.

CONSTRUCTED according to the Specification on page 86, but with iron outside walls in lieu of wood.

25 ft. by 15 ft., 15 ft 6 in. high to ridge.

Cash Price, £92.

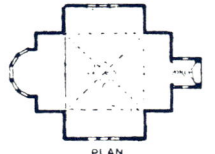
N° 461
ELEVATION

No. 461. PRIVATE CHAPEL.

PLAN

Estimates on application

No. 454. CHAPEL.

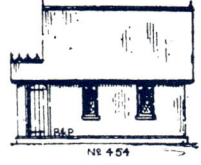
N° 454
ELEVATION

PLAN.

CONSTRUCTED according to the Specification on page 86, but with iron outside walls in lieu of wood.

20 ft. by 12 ft., 13 ft. high to ridge.

Cash Price, £62 10 0

No. 456. CHURCH.

N° 456
ELEVATION

30 ft. by 20 ft., 16 ft. high to ridge.

Cash Price, £159.

BOTH these Churches constructed according to the Specification on page 86, but with iron outside walls in lieu of wood.

Carriage Paid to most Railway Stations in England

PLAN

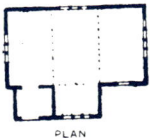
PLAN

No. 465. CHURCH.

N° 465
ELEVATION.

30 ft. by 20 ft., 16 ft. high to ridge.

Cash Price, £121 15 0

BOULTON & PAUL, MANUFACTURERS.

No. 460. CHURCH.

50 ft. by 25 ft., 20 ft. high to ridge.

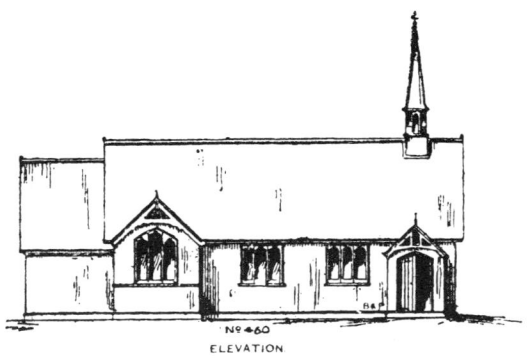
ELEVATION

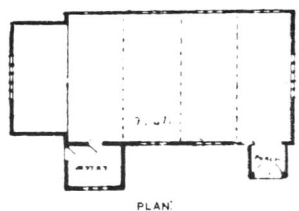
PLAN.

No. 459. MISSION ROOM.

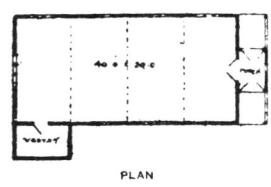
PLAN

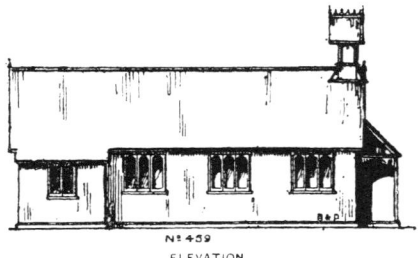
ELEVATION.

No. 458. MISSION ROOM.

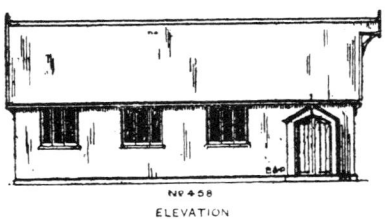
ELEVATION.

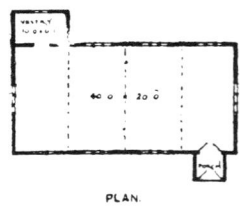
PLAN.

ESTIMATES AND CATALOGUES FREE ON APPLICATION.

CHURCH FITTINGS.

No. 507. LECTERN.

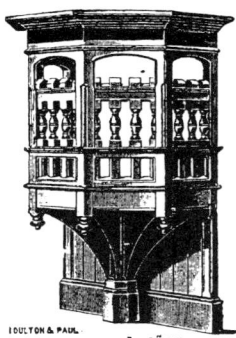
No. 10. PULPIT.

No. 507a. LECTERN.

Our Church Furniture is constructed of Oak or Pitch Pine.

Special Estimates given and Designs prepared to meet any requirements.

GLASTONBURY
CHANCEL CHAIR.

ALTAR TABLE.

NAVE CHAIR
With Book Rail.

No. 506a. MISSION ROOM SEAT.
Pitch pine, varnished, with wrought-iron standards.

No. 506. MISSION ROOM SEAT.
Pitch pine, varnished, with cast-iron standards.

SPECIFICATION

OF

IMPROVED PORTABLE WOOD HOUSES.

Walls. — Constructed of strong unplaned deal framing, morticed and tenoned, and covered outside part with matchboarding, and part with rustic joint weather-boarding, ⅝-in. thick.

Lining. — External walls lined with inodorous sheet felt and matchboarding, and partitions matchboarded on both sides.

Doors. — Six-panelled square-framed doors, having ovolo moulded stops, fitted with strong hinges, good locks and handles.

Windows. — Ovolo moulded casements about 18 in. wide, divided by mullions with moulded stops, glazed with 21-oz. sheet glass. Those casements which open for ventilation are fitted with fastener and set-ope.

Flooring. — Red deal sleeper plates, joists, and ⅞-in. planed floor boards.

Ceilings. — Matchboarded, and when beneath roof felted as well.

Roof. — Consisting of strong deal principals, purlins, and best galvanized corrugated iron, 24 gauge, coated with pure Silesian spelter. Felt for laps of same. Capping to ridge. Gutters to eaves, and down-pipes to the ground. Projecting barge boards to gables.

Painting and Staining. — External woodwork stained with our preparation. Inside matchboarding sized. Door and window frames, and overlay pieces picked out in different colour. Doors, casements, barge boards, gutters, and down-pipes painted three coats.

Carriage. — Carriage paid to any principal goods station in England, or to docks for shipment. Carting from railway station to site of erection is not included in our estimates. Packing will be charged for, but the amount will be allowed in full if cases are returned here carriage paid, purchaser carting to station.

Erection. — Erection by our men on purchaser's foundation, purchaser providing assistant labour.

Prices. — Our prices are put as low as possible compatible with best material and workmanship.

Terms of Payment. — Half on delivery and the remainder on completion. Until fully paid for the title and ownership of any building shall remain the property of Boulton and Paul, and the same shall not be subject to any mortgage of the land on which it may be fixed

Foundation. — Brickwork is recommended as the best foundation, and we strongly advise brick chimneys, the plan for which we will supply free on receipt of order, but we do not include any brickwork or drains in our estimates. This part of the work can usually be done by a local bricklayer, to the purchaser's advantage.

Notice to Local Authorities. — All necessary notices to Local Authorities as to the erection of our buildings must be given by purchasers, as we cannot hold ourselves in any way responsible for the same.

BOULTON & PAUL, MANUFACTURERS.

No. 606. WOOD AND IRON SCHOOL BUILDINGS.

Erected by us at Whitby for the MARQUIS OF NORMANBY.
Constructed according to the Specification on page 92.

TESTIMONIAL
From HERBERT KIRKLEY, ESQ., Whitby.   21st Jan., 1893.

I have this morning looked over the new School you have erected for Lord Normanby, and have pleasure in stating the work has been completed to the satisfaction of his lordship.

WOOD STUDIOS OR WORKSHOPS.

WALLS covered outside with our rustic joint boarding, stained; galvanized corrugated iron roof, and matchboard roof lining, casement windows glazed with 21-oz. sheet glass, panelled doors, fitted with hinges, lock, and handles; cast-iron gutters and down-pipes to eaves; capping to ridge. In sections ready for easy erection by purchaser on his foundation.

No. 607.

Cash Prices of No. 608.
12 ft. by 8 ft., 6 ft. 6 in. high at back, 8 ft. 6 in. in front, £14; floor 34/-
15 ft. ,, 10 ft., 7 ft. 0 in.   9 ft. 6 in.   ,,   20; ,, 55/-
18 ft. ,, 12 ft., 7 ft. 0 in.   10 ft. 0 in. ,, £25 10/-,, 80/-

Varnished, matchboard and felt lining to walls extra, £4 7 6.
£5 15 0, £7 10 0 respectively

No. 608.

Cash Prices of No. 607.

12 ft. by 8 ft., 6 ft. 6 in. high at eaves, 10 ft. 6 in. to ridge, £16; floor 34/-
15 ft. ,, 10 ft., 7 ft. 0 in.   ,,   12 ft. 0 in.   ,, 23; ,, 55/-
18 ft. ,, 12 ft., 7 ft. 0 in.   ,,   13 ft. 0 in.   ,, 29; ,, 80/-

Varnished, matchboard, and felt lining to walls extra, £4 10 0,
£6 0 0, £7 15 0 respectively.

School Desks and Fittings, see page 101

Carriage Paid to most Railway Stations in England

ROSE LANE WORKS, NORWICH

No. 466. PARISH ROOM.

ELEVATION

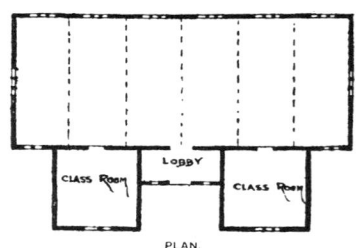

PLAN.

TESTIMONIAL.

From the REV. HENRY EDWARD HODSON.

I hereby certify that your men have erected the Parish Room, Verandah, and Green Room at the Lea Ross to my satisfaction. The building has been most cleverly contrived and thoroughly well carried out. We cannot speak too highly of the men you sent, such men and such work should be a good advertisement for your firm.

No. 467. VILLAGE SCHOOL,

with Master's House adjoining.

FRONT ELEVATION.

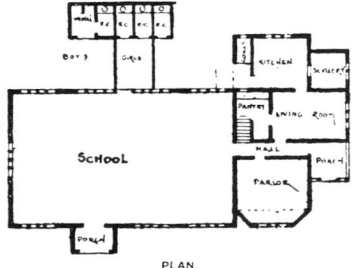

PLAN.

ELEVATION.

This Building could be converted into a **COTTAGE HOSPITAL** with some slight alterations

Estimates and Catalogues free on application.

No. 423. GALVANIZED IRON GYMNASIUM,

with Porch, Lavatory, and Visitors' Gallery.

REGISTERED DESIGN, No. 5989.

Constructed according to the Specification on page 92.

TESTIMONIALS.

From THE HON.
HARBORD HARBORD,
Gunton Park.

I beg to inform you that the Iron House built for me suits admirably for what it is adapted, and very easily moved without detriment to the structure. It is perfect and durable, with the additional advantage of its cost being about half of those usually built as a fixture with bricks and mortar, &c.

From
C. J. PRESCOTT, Esq.,
Wirksworth.

I beg to hand you my cheque for Iron Building which is now erected, and looks very well. If you have any enquiries from this part you can refer them to me, and I will show them your work; they could not have better.

REGISTERED DESIGN, No. 5989.

Interior of No. 423.

TESTIMONIALS.

From
SWAINTON ADAMSON,
Esq., Rugeley.

GENTLEMEN,—I am very pleased with the House; being first fixed on a good foundation, it makes an excellent house.

From
GILBERT W. STRACEY,
Esq., Rackheath Park.

SIRS,—I enclose cheque due to you. Sir Henry Stracey is much pleased with the Riding School.

From
MESSRS. JONES & SON,
Denbigh.

We are very much pleased with the structure, and shall have very great pleasure in recommending any of our friends to your Firm.

SCHOOL DESKS AND FITTINGS.

No. 503. COLLEGE DESKS.

No. 504. BOARD SCHOOL DESKS.

TESTIMONIAL.

From FARRAR RANSON, Esq., Chairman of Finance Committee, Norwich School Board.

June 1st, 1893.

I have much pleasure in stating that the Desks and Fittings supplied by you to our Board are of excellent Manufacture, and have given entire satisfaction, and it is our desire to place further orders with you.

No. 505. NEW BOARD SCHOOL DESK.
Latest Design.

No. 506. INFANTS' SCHOOL DESK.
Latest Design.

Estimates free on application.

SCHOOL AND MISSION ROOM FITTINGS.

No. 500. TEACHER'S SINGLE DESK.

No. 501. HIGH SCHOOL SINGLE DESK.

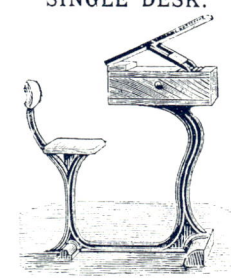

No. 502. As a Seat with Back for Lectures, &c.

No. 502. CONVERTIBLE DESK.

No. 502. As a Table.

No. 502. As a School Desk.

ESTIMATES FREE ON APPLICATION.

No. 509. READING DESK.

Constructed of selected Pitch Pine, varnished.

No. 11. PULPIT OR READING DESK.

No. 12. PULPIT OR READING DESK.

BOULTON & PAUL, MANUFACTURERS,

THE MODEL HYGIENIC COMBINATION CLASS DESKS FOR INFANTS.

Patent applied for.   Provisional protection granted.

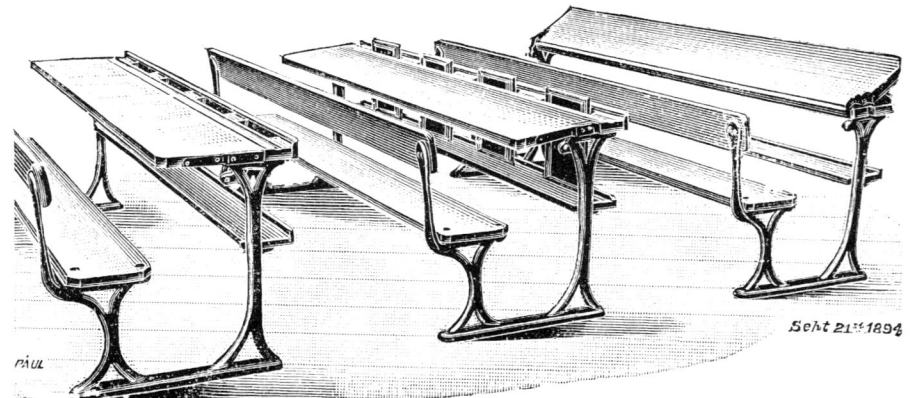

THIS Desk is the latest and most improved for Infant Classes before the public. It is the result of a most careful study of a long-felt want, in all its hygienic requirements. After a severe competition by many manufacturers, the Norwich School Board decided that this was the best Desk in the market, and we are now supplying to their Schools many of our Desks.

The Desk can be used for Kindergarten work, as a Writing Desk, and also as a Reading Desk, as shown in illustration ; it also contains a Slate Rack and Book Board. The elevation to the various positions is so simple in construction that a child of four years can alter it without getting from its seat. The whole of the writing flap is in one piece, and is set to an angle of 15° for that purpose, and when raised up to an angle of 40° for reading, ample room is then given for moving in and out of desks, and also for physical exercises.

We claim for this Desk the following hygienic points as recommended by J. H. Cowham, Esq., Lecturer on School Management, Westminster Training College, viz. :—

1st.—The distance from floor to seat, an easy and solid footing on the ground.

2nd.—The distance from seat to top of writing flap, this point ensures a natural and healthy position of the pupil.

3rd.—The front of seat is in a perpendicular line with the inner edge of writing flap, which causes the pupil to sit in an easy and upright position.

*Special Catalogue of School **Furniture** and **Fittings** free on application.*

ROSE LANE WORKS, NORWICH.

NEW PHOTOGRAPHIC STUDIO.

18 ft. by 12 ft. Containing studio 12 ft. by 12 ft., porch, lobby, and dark room.
Cash Price £53 15 0
In sections, for purchaser to erect on his foundation.

NEW DARK ROOM.

DEVELOPING SINK.

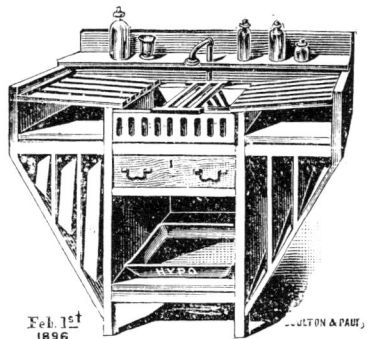

4 ft. by 4 ft. inside, 7 ft. 6 in. high, lined, well ventilated, window fitted with enclosed bay for lamp.
Cash Price £4 17 6
Without Bay for Lamp, £4 10 0

Of pitch pine, varnished, glazed sink, with tap, waste pipe, drawers, shelf, and racks, as shown. Size 4 ft. wide.
Cash Price £3 15 0

PHOTOGRAPHIC WORKROOM AND DARK ROOM.

8 ft. by 6 ft.
Cash Price £12 0 0
Plain building, unlined, £8 10 0
Carriage Paid to most Stations in England.

BOULTON & PAUL, MANUFACTURERS.

No. 648.
BATHING CHALET.

13 ft. by 9 ft. with Entrance Bay in front.

VENETIAN Shutters to windows. Wall framing covered outside with grooved and tongued stained boarding. Painted iron roof, with felt and matchboarding for lining same, wood floor, door and windows glazed with 21-oz. sheet glass. In sections, ready for purchaser to erect.

Cash Price, Carriage Paid, £35 0 0

No. 450.
COVERED SWIMMING BATH.

Estimates for any size free on application.

REGISTERED COPYRIGHT.

REGISTERED COPYRIGHT.

No. 337. BATHING HUT.

DEAL framework, covered with matchboarding, painted three coats; roof corrugated iron lined inside, with boarded floor and side benches; on wheels as shown. In sections, ready for easy erection by purchaser.

Cash Price, Carriage Paid.

7 ft. long, 5 ft. wide £10 0 0

No. 451.
ENCLOSURE FOR SWIMMING BATH.

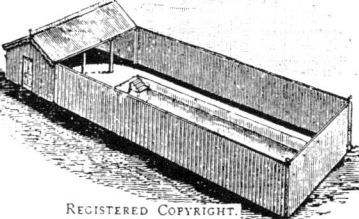

REGISTERED COPYRIGHT.

THIS arrangement is designed to meet the requirements of schools, and can be modified to suit any situation. The Shed can be fitted with lockers or seats if required.

Estimates on application.

REGISTERED COPYRIGHT.

No. 329. PORTABLE IRON BOAT-HOUSE.

THE framework is of wood, covered with best galvanized corrugated iron. Outside woodwork painted three coats. Window glazed with 21-oz. sheet glass. Constructed in sections, ready for easy erection by purchaser.

Cash Prices, Carriage Paid.

12 ft. long; 6 ft. wide . £10 15 0

No. 622. WOOD RESIDENCE.

REGISTERED COPYRIGHT

As erected for LADY SHELLEY on the Cliff, Boscombe, 1891.

Constructed according to the Specification on page 92.

Approximate Price £760.

TESTIMONIAL.

LADY SHELLEY is extremely pleased with the House, and wishes to take this opportunity of saying that the workmen have behaved well throughout.

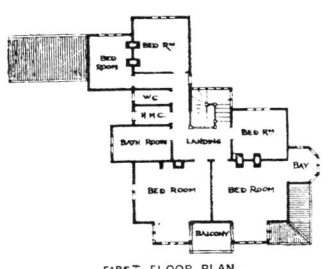

FIRST FLOOR PLAN.

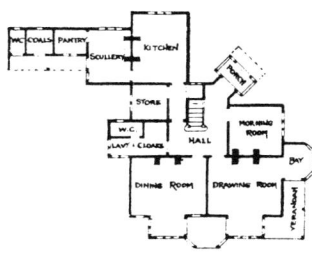

GROUND PLAN.

SIZES OF ROOMS.

Dining and drawing rooms, and bedrooms over, 16 ft. by 12 ft., bay windows in addition.
Morning room and room over 12 ft. by 12 ft.
Hall ... ... 16 ft. by 12 ft.
Kitchen ... ... 15 ft. by 12 ft.

Other rooms in proportion.

No. 407. COMFORTABLE WOOD TWO-STOREY COTTAGE.

REGISTERED, No. 5959.
BEDROOM PLAN.

REGISTERED, No. 5959.
GROUND PLAN.

REGISTERED DESIGN, No. 5959.

Constructed according to the Specification on page 92.

Approximate Price £250.

REGISTERED DESIGN, No. 5963.
View of the interior of Room lined with matchboarding, as usually supplied.

REGISTERED DESIGN, No. 5961.
Interior View of Hall, showing how our Houses can be furnished and decorated.

No. 409a. LODGE

OR GAMEKEEPER'S LODGE.

Special Plans prepared to suit any situation

TESTIMONIAL

From HERBERT KIRKLEY, Esq., Lythe Hall.
April 24th, 1893.
I have pleasure in stating that the work at the new Keeper's Cottage and at the School is satisfactory.

REGISTERED COPYRIGHT.

No. 409. LODGE OR GAMEKEEPER'S COTTAGE.

Consisting of Living Room, Kitchen, Pantry, and Three Bedrooms.

REGISTERED DESIGN, No. 5964.

PLAN.

REGISTERED DESIGN, No. 5964.

Approximate Price £235.

Constructed according to the Specification on page 92, and finished, as shown, with Canadian pattern iron roof and panelled walls.

BOULTON & PAUL, MANUFACTURERS,

No. 413. PORTABLE WOOD BUNGALOW
For Leasehold Property.

As erected for W. HOLLINGWORTH PALMER, ESQ., Byfield.

TESTIMONIAL.

From THE HON. GEOFFREY R. C. HILL, Erwood.

SIRS,—The House is now I believe finished, and I write a line to tell you that the workmen have been very active and have done their work well.

This House can also be constructed in the rustic style with thatch roof

Constructed according to the Specification on page 92 with molded red deal posts and arch pieces to verandah

Approximate Price £600.

TESTIMONIAL.

From RICHARD TERROT, ESQ., Spa

I am very much pleased with the Buildings you have erected for me The Cottage gives great satisfaction, and is remarkably cool this hot weather. I was much pleased with the men you sent to erect the Buildings.

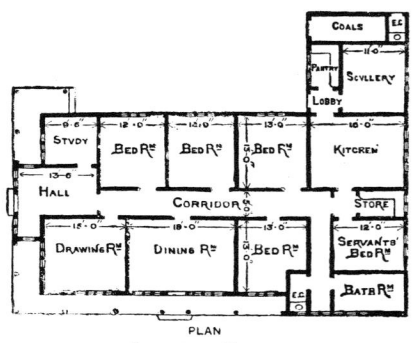

PLAN

REGISTERED, NO. 5975.

Can be modified to suit any situation.

ROSE LANE WORKS, NORWICH. 113A

No. 749. WOOD BUNGALOW.

Erected for Dr. Bond, Dunster.

TESTIMONIAL.

From Dr. BOND, Dunster.

The Bungalow is most delightful inside, and the effect of the light and dark is very good indeed.

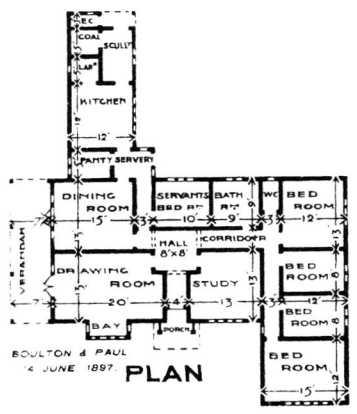

Constructed according to Specification on page 92.

Estimates and Designs to suit any situation free on receipt of particulars of requirements.

BOULTON & PAUL, MANUFACTURERS,

No. 310. BUNGALOW SHOOTING LODGE, &c.

Constructed according to specification on page 92.

Estimates and Special Designs to suit any situation free on receipt of particulars of requirements.

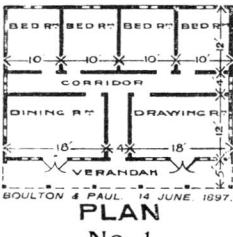

PLAN No. 1.

TESTIMONIAL.

From CAPT. BROADLEY,
Ballyclough.

The House has arrived and we have just finished putting it together. I am most pleased with it. It was most carefully packed, and arrived without any scratches and nothing broken belonging to it.

PLAN No. 2.

ROSE LANE WORKS, NORWICH.

No. 410. SHOOTING BOX.

REGISTERED DESIGN, NO. 5965.

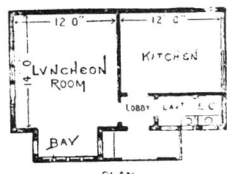

PLAN

REGISTERED DESIGN, NO. 5965.

Constructed according to specification on page 92.

In sections for purchaser to erect.

Cash Price £79 10 0

No. 414. IRON SHOOTING BOXES AND HUNTING LODGE.

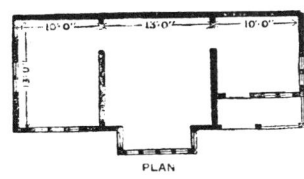

PLAN

REGISTERED NO. 5976.

Constructed as Specification on page 120, but without felt lining.

In sections for purchaser to erect

Cash Price, £90 0 0

REGISTERED DESIGN, NO. 5976.

Carriage Paid to most Railway Stations in England.

No. 411. PORTABLE IRON BUNGALOWS
FOR LEASEHOLD PROPERTY.

REGISTERED DESIGN, NO. 5966.

As erected on Lundy Island.

Constructed according to specification on page 92, but with iron outside walls instead of wood.

Plan No. 5966.   Approximate Cash Price £250.

Carriage Paid, and erected by our men, with Range and Stove.

TESTIMONIALS.

From BUCKLEY HOLMES, Esq., Glanconway.

I am very pleased indeed with the House which answers my purpose admirably. It is evidently very carefully and strongly made.

From H. HARDY, Esq., Uckfield.

The Iron Building, which you have just completed, is perfectly satisfactory. All the work has been well done, and the material is good.

REGISTERED, No. **5966**.

Special Estimates for this Plan.

ROSE LANE WORKS, NORWICH.

No. 752. WOOD BUNGALOW FOR THE TROPICS.

Constructed according to specification on page 92.

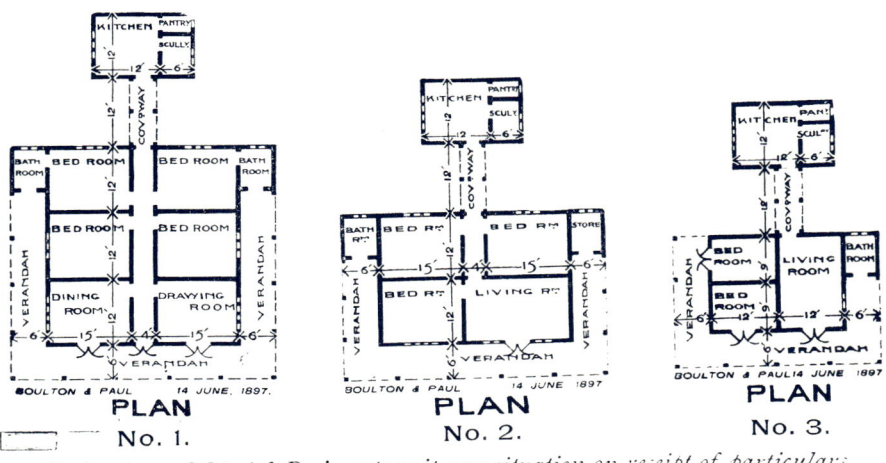

PLAN No. 1.   **PLAN No. 2.**   **PLAN No. 3.**

Estimates and Special Designs to suit any situation on receipt of particulars of requirements.

BOULTON & PAUL. MANUFACTURERS.

No. 635. LODGE.

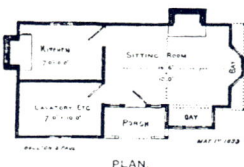
PLAN.

Cash Price, £120 0 0

Constructed according to the Specification on page 92, but with Canadian pattern iron roof; windows glazed with 21-oz. sheet glass in lead diagonal squares.

No. 417. FIVE-ROOMED BUNGALOW COTTAGE.

REGISTERED DESIGN, No. 5983.

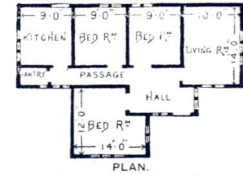
PLAN.
REGISTERED, No. 5983.

Constructed according to Specification on page 92, but with iron outside walls in lieu of wood.

Cash Price, £181 0 0

No. 312. A CONVENIENT COTTAGE.

Suitable for Hunting or Fishing Quarters.

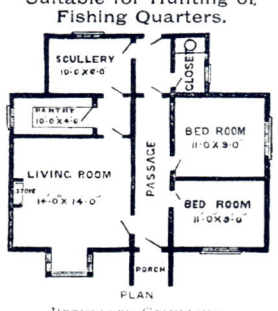
PLAN.
REGISTERED COPYRIGHT.
Constructed as above

REGISTERED COPYRIGHT.

Cash Price, £172 0 0

Estimates on application.

ROSE LANE WORKS, NORWICH.

No. 313. COTTAGE.

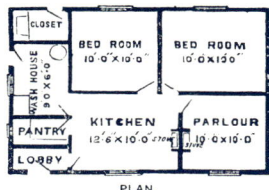

PLAN.

REGISTERED COPYRIGHT.

Cash Price,   £150 0 0

REGISTERED COPYRIGHT.

Constructed according to the Specification on page 120, including range and open fire stove, sink, shelves, and earth apparatus.

No. 318. COTTAGES.

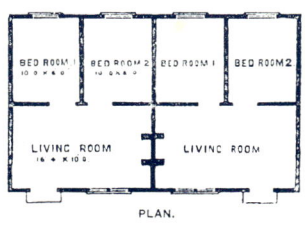

PLAN.

REGISTERED COPYRIGHT.

Constructed like No. 313.

REGISTERED COPYRIGHT.

Cash Price.

Single Cottage	... ...	£75 0 0
Double Cottage	... ...	140 0 0

No. 318a. WOOD COTTAGES.

As erected for the MARQUIS OF NORMANBY.

Constructed according to the Specification on page 92.

Cash Price.

Single Cottage ...		£82 10 0
Double Cottage		154 0 0

No. 318b. IRON COTTAGE.

PLAN.

Cash Price, £96 0 0
Including range and shelves in pantry
Painting roof, cresting, and trellis £6 0 0 extra
Tarred sleepers for foundation, £2 10 0

REGISTERED COPYRIGHT.

SPECIFICATION.

PLAIN Cottages constructed of strong deal framing, covered on the outside with galvanized corrugated iron, lined inside with matchboarding, sheet felted between the iron and the wood. Strong wood floor. Eaves-gutters, down-pipes, locks, and window fasteners included. Erected by our men on purchaser's brickwork foundation, he providing assistant labour. Exclusive of stoves, chimneys, and all inside fittings.

PLANS OF IRON COTTAGES.

No. 314.

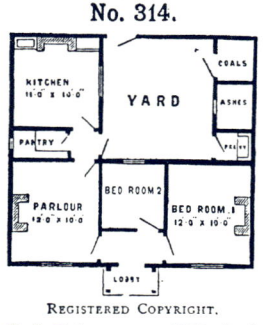

REGISTERED COPYRIGHT.
Cash Price, £130 0 0

No. 411a.

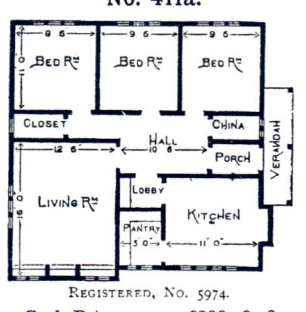

REGISTERED, No. 5974.
Cash Price, £220 0 0

No. 311.

Cash Price, £145 0 0

TESTIMONIAL.

From W. GRAY, ESQ., Maidenhead.

I am glad to be able to inform you that the Cottage (No. 314) your men have just finished for me gives every satisfaction in all parts. The stoves are fixed and work well.

Carriage Paid to most Railway Stations in England.

ROSE LANE WORKS, NORWICH. 121

DWELLING-HOUSES.

No. 482. ROOM.
10 ft. by 12 ft.

Nº 482.
ELEVATION.

Cash Price,   £16 0 0

No. 484. WOOD HOUSE.
WITH BAY.

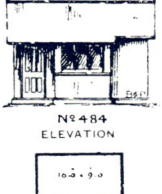

Cash Price, about £22 0 0

Wood walls stained outside, galvanized corrugated iron roof with matchboard lining, wood floor, made in sections for purchaser to erect.

No. 483. ROOM.
15 ft. by 10 ft.

Nº 483.
ELEVATION.

Cash Price,   £19 10 0

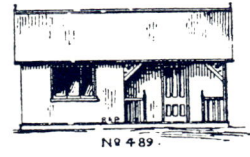

No. 489. WOOD HOUSE.
24 ft. by 12 ft.

Constructed as above.

Cash Price,   £40 0 0

No. 491. WOOD SHOOTING BOX.
28 ft. by 9 ft.

With Bay and Porch constructed as above.

Cash Price,   £43 0 0

Nº 491.
ELEVATION.

No. 495. WOOD COTTAGE.
30 ft by 17 ft.

Constructed as above.

Cash Price,   £70 0 0

Carriage paid to most Railway Stations in England.

DWELLING-HOUSES.

Constructed according to the Specification on page 120.

 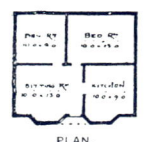

No. 490. WOOD COTTAGE.

Cash Price, £105 0 0

No. 493. WOOD COTTAGE.

Cash Price, £145 0 0

Constructed according to the Specification on page 92.

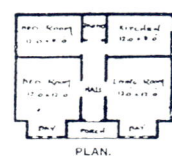 

No. 499. WOOD COTTAGE.

Cash Price £170 0 0

Constructed according to the Specification on page 92.

No. 496. WOOD COTTAGE.

Cash Price, £140 0 0

Constructed according to the Specification on page 92.

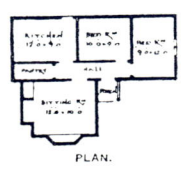 

Carriage Paid to most Railway Stations in England.

STOVES ADAPTED FOR IRON BUILDINGS.

No. 4.

Close Fire Pattern.

Burns 8 to 20 hours, according to size.

CORRUGATED DESIGN, REGISTERED.

THE "GLOW-WORM."
Cash Prices, including Ashes Tray.

No.	Size. High.	Diam.	Heating Power, cubic feet.	Prices. £ s. d.
0	25 in. by	9 in.	4,000	1 3 6
1	26 in. by	10 in.	6,000	1 13 0
2	30 in. by	12 in.	10,000	2 5 0
3	36 in. by	14 in.	17,000	2 15 0
4	25 in. by	9 in.	7,500	1 17 0

THE "SNAIL" SLOW COMBUSTION STOVES.
With Fire-brick Linings.
Specially suited for Mission or School Rooms.

No. 1315 to 1317.

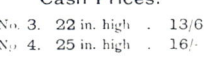

Cash Prices.

No.	Size. High. Square. in. in.	Heating Power, cubic feet.	Prices. £ s. d.	Iron Pans extra. s. d.
1317	25 by 10½	6,000	3 2 6	4 6
1316	32 by 15	12,000	4 5 0	5 9
1315	36 by 15	18,000	5 14 0	8 3

THE "DUMPY" STOVE.
For Shepherds' Huts, &c.

Cash Prices.

No. 3. 22 in. high . 13/6
No. 4. 25 in. high . 16/-

No. 1548. INDEPENDENT STOVE

SPECIALLY suited for Iron Cottages, with fire-brick lining, loose trivet, fender and ashes pan. Fine cast and black.

Cash Prices.
Wide.	High.	
12 in.	31 in.	36/-
14 in.	31 in.	40/-
16 in.	31 in.	43/-

Smoke Flue, 2/- ft., extra.

CAST-IRON SINK FOR IRON COTTAGES.

Size—2 ft. by 1 ft. 6 in. by 4 in. deep
Price 16/6 each.

No. 157. OPEN FIRE STOVE
With Fire-Brick Lining.

SIZE—33½ in. high, 18½ in. at front, 12 in. front to back, 12 in. fire. Flue nozzle at back for 6-in. Chimney pipe.

Cash Price.
Fine cast and black . £2 10 0
Chimney Pipe, 2/- per ft., extra.

TESTIMONIAL.

From W. GRAY, Esq., Maidenhead.

I am glad to be able to inform you that the Cottage your men have just finished for me gives every satisfaction in all parts. The Stoves are fixed, and work well.

No. 136. THE "HOLBORN" PORTABLE SELF-FITTING RANGE.

Specially suited for Iron Buildings.

A NEW and improved self-setting Kitchen Range, constructed on the latest and most generally approved principles. Specially adapted for Iron Buildings.

It combines all that is useful in the English Close Fire Range and the American Cooking Stove.

It is highly finished in all its parts. The oven and boiler are easily heated with a small amount of fuel. The boiler can be lifted out after unscrewing the cock. The fire covers are hinged, and stand upright when lifted (this alone is a very great convenience), while the flue-pipe has a cleaning-door, giving easy access to the flues.

Cash Prices.
With Bright Banjo Fittings, Oven only.
26 in.	30 in.	36 in.	42 in.
34/9	43/6	48/-	64/- each.

Oven and Boiler, with Brass Cock.
30 in.	36 in.	42 in.
48/3	57/-	74/9 each.

Smoke Pipe, 2/- per ft., extra.

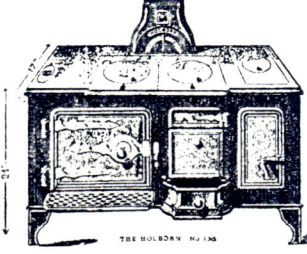

THE QUEEN ANNE MANTEL REGISTERS.

As used in connection with our Iron Buildings when brick chimneys are built.

ROSE LANE WORKS, NORWICH.

ORNAMENTAL WOOD PORCHES.

TESTIMONIAL.

From JAMES SIMMS, ESQ.,
Blackheath, S E.
The Porch has arrived, and I am much pleased with its appearance.

REGISTERED DESIGN, NO. 5984.

No. 426. SPAN-ROOF WOOD PORCH.

6 ft. 6 in. by 4 ft., 7 ft. high at eaves, 10 ft. 6 in. to ridge.
Stained and varnished, painted weatherboard roof.

Cash Price, £10 0 0
Seats, 12/- extra.
In sections ready for purchaser to erect.

TESTIMONIAL.

GLYN MALDEN,
Nov. 1st, 1892.
The Porch has been put up to-day, and MRS. GRIFFITHS is much pleased with it.

REGISTERED DESIGN, NO. 5987.

No. 429. WOOD PORCH.

8 ft. by 4 ft. 6 in., 7 ft. high at eaves.

Stained and varnished, painted iron roof.

Cash Price, £10 10 0

In sections ready for purchaser to erect.

REGISTERED DESIGN, NO. 5985.

No. 427. LEAN-TO WOOD PORCH.

7 ft. by 3 ft., 7 ft. high at eaves.
Stained and varnished, painted weatherboard roof.

Cash Price, £7 0 0
In sections ready for erection by purchaser.

REGISTERED DESIGN.

No. 428a. LEAN-TO WOOD PORCH.

7 ft. 6 in. by 3 ft. 6 in.

Stained and varnished, painted weatherboard roof.

Cash Price, £10 10 0

In sections ready for purchaser to erect.

Carriage Paid to most Railway Stations in England.

BOULTON & PAUL, MANUFACTURERS.

Registered Design, No. 5986.

No. 428.
WOOD VERANDAH OR PORCH.

12 ft. by 4 ft., 7 ft. high at eaves, stained and varnished red deal, painted zinc roof.

In sections, ready for purchaser to erect.

Cash Price, £16 10 0

No. 649. WOOD PORCH.

8 ft. by 6 ft., 9 ft. 6 in. high.

Consisting of outer Porch and closed Lobby.

Wood walls stained and varnished, painted Canadian iron roof.

Cash Price £21 0 0

In sections, ready for purchaser to erect on his floor.

No. 637.
ENCLOSED WOOD PORCH.

8 ft. by 6 ft., roof over door projecting 2 ft. 6 in., 7 ft. high at eaves, 12 ft. to ridge.

Wood walls stained and varnished, Canadian painted iron roof.

Cash Price, £17 10 0

In sections, ready for purchaser to erect on his floor.

Carriage Paid to most Railway Stations in England.

ORNAMENTAL VERANDAHS.

DESIGNS REGISTERED.

CONSTRUCTED in wood and iron in the best manner and style to suit any situation or house. In sections ready for erection by purchaser. The prices include iron bases for bedding in ground to both iron and wood columns. Cast-iron gutters and wood ridge capping in all cases. Cast-iron cresting where shown. All woodwork stained and varnished or painted three coats.

No. 526.
Arcaded, with galvanized iron roof.
Cash Price, 5/9 per foot run, 5 ft. wide.
Turned wood columns 30/- each extra.
Moulded wood arches, 6 ft. wide, with iron cresting and spandrels, 25/- each extra.

No. 521. Wood Verandah with Balustrading.
Galvanized iron roof.
Cash Price, 4/6 per foot run, 5 ft. wide.
Moulded wood columns 15/- each extra.
Shaped heads 10/- each extra.
Moulded balustrading per foot run 2/6 extra.

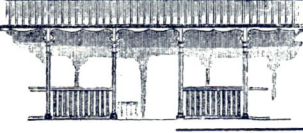

No. 523.
Wood Arcaded and partly enclosed Verandah.
Galvanized iron roof.
Cash Price, 5/3 per foot run, 5 ft. wide.
Moulded wood columns 20/- each extra.
Moulded wood arches 20/- each extra.
Moulded balustrading to columns 10/- each.

No. 527.
Galvanized iron roof.
Cash Price, 6/3 per foot run, 5 ft. wide.
Moulded and turned wood columns 30/- each extra.
Moulded arches 10/- each extra.

No. 524. Double Column Iron Verandah.
Galvanized iron roof.
Cash Price, 4/9 per foot run, 5 ft. wide.
One pair double cast-iron columns and brackets 50/- extra.

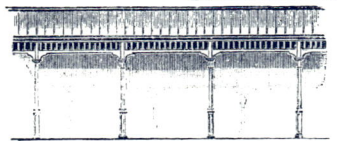

No. 522. Wood Verandah with Glazed Frieze.
Galvanized iron roof.
Cash Price, 5/3 per foot run, 5 ft. wide.
Moulded wood columns 30/- each extra.
Frieze, panelled, and glazed with cathedral glass in tints, 3/6 per foot run extra.

No. 528.
Galvanized iron roof.
Cash Price, 4/3 per foot run, 5 ft. wide.
Iron columns 20/- each extra.
Brackets 5/- each extra.

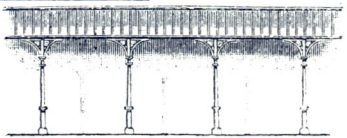

Carriage Paid to most Railway Stations in England.

No. 418. PORTABLE WOOD BILLIARD ROOM.

REGISTERED DESIGN, No. 5991.

Constructed according to the Specification on page 92, but with roof lined with felt, and stained matchboarding in lieu of ceiling, iron painted outside.

Cash Price, £125 0 0

Erected by our men on purchaser's foundation.

Open fire stove, see page 126, with cast-iron smoke pipe, stays, and shield, £3 15 0 extra.

TESTIMONIAL

ST. STEPHEN'S.

Mr. BUNTING is much pleased with the Billiard Room, which he considers has been erected in a most satisfactory manner.

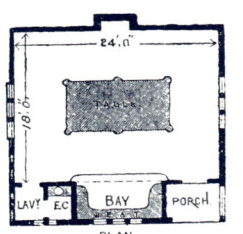

PLAN.

REGISTERED DESIGN, No. 5991.

Billiard Rooms can be specially designed to correspond with any style of house.

Gentlemen waited upon, free of charge, in any part of the country.

Carriage Paid to most Railway Stations in England.

No. 419. PORTABLE IRON BILLIARD ROOM.

REGISTERED, No. 5992.

Constructed as No. 418.
Cash Price £126 0 0
Covered way 4 ft. 6 in. wide, 25/- per foot run extra.

TESTIMONIAL.

From J. W NUTTALL, Esq., Holme Lea.

The Billiard Room you supplied me with four years ago has answered admirably.

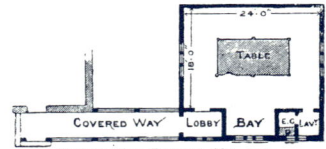

REGISTERED DESIGN, No 5992.

No. 421. BILLIARD AND SMOKE ROOM, OR STUDIO.

REGISTERED DESIGN, No. 5994.

Size—18 ft. by 24 ft., 8 ft. high to eaves, 14 ft. to ridge.

CONSTRUCTED of strong wood framework. Covered on the outside with galvanized corrugated iron. Lined inside with matchboarding. Sheet-felted between the wood and the iron. Strong wood flooring. Eaves-gutters, down-pipes, and locks included. Outside woodwork painted three coats. Windows glazed with 21-oz. sheet glass.

In sections, ready for easy erection by purchaser on his foundation.

Cash Price from £65 0 0, according to finish.

Carriage Paid to most Railway Stations in England.

BOULTON & PAUL, MANUFACTURERS.

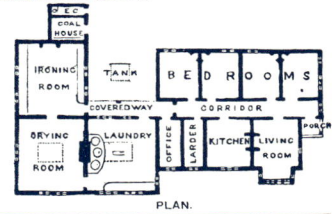

No. 452:
IRON LAUNDRY AND COTTAGE.

REGISTERED DESIGN

Special Estimates given for size of building required.

No. 644. WOOD LAUNDRY AND IRONING ROOM.

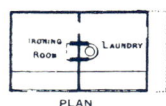

WOOD walls stained outside, galvanized corrugated iron roof with matchboard lining, well ventilated, no lining to verandah roof, without floor. In sections, ready for purchaser to erect.

24 ft. long, roof projecting 3 ft. at end,
15 ft. wide.

Cash Price, £52 0 0

LAUNDRY STOVES.

Flue at Top or Behind.

No. 338.
PORTABLE LAUNDRY.

CONSTRUCTED of deal framework, painted and covered with galvanized corrugated iron, half glass front, louvre ventilator full length on roof, without floor. In sections, ready for easy erection by purchaser. 15 ft. long, 10 ft. wide.

Cash Price £15 10s.
Matchboard lining £5
Floor, £2

REGISTERED COPYRIGHT

Carriage Paid to most Railway Stations in England.

LAUNDRY AND COTTAGE.

SIZE, **40** ft. by **15** ft., containing Laundry, Drying and Ironing Room, Bedroom, Porch, Pantry, Coalhouse, and E.C. with apparatus.

Constructed according to specification on page 92, without flat ceiling, and with floor only to cottage part.

In sections, for purchaser to erect on his foundation.

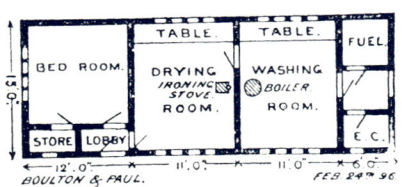

TESTIMONIAL.

From J. H. EASTWOOD, Esq., Todmorden.

Having now begun to use the Laundry you put up for me, I am very much pleased with it.

Cash Price £112 0 0

If walls covered with galvanized corrugated iron, £108 0 0

Laundry Stoves and Fittings extra.

Carriage Paid to most Stations in England.

ROSE LANE WORKS, NORWICH.

PORTABLE WOOD LAUNDRIES FOR LEASEHOLD PROPERTY.

No. 615. PORTABLE WOOD LAUNDRIES.

REGISTERED COPYRIGHT.

20 ft. long by 12 ft. wide, 8 ft. high to eaves, 14 ft. to ridge, with laundry, ironing and drying room.

Cash Price, Carriage Paid, from £49 0 0

For purchaser to erect.

PLAN.

Wood walls stained outside, galvanized corrugated iron roof with matchboard lining, wood floor. In sections, ready for purchaser to erect on his foundation.

BOULTON & PAUL, MANUFACTURERS,

No. 616. PORTABLE WOOD LAUNDRY.

12 ft. by 8 ft., with pent roof over door. 7 ft. 6 in. to eaves, 10 ft. to ridge.

REGISTERED COPYRIGHT.

Cash Price, Carriage Paid, from £14 10 0
For purchaser to erect.

Wood walls stained outside, galvanized corrugated iron roof with matchboard lining, wood floor, made in sections for purchaser to erect.

No. 617. PORTABLE WOOD LAUNDRY.

12 ft. by 8 ft. lean-to against existing wall. 6 ft. 6 in. to eaves, 9 ft. at back.

REGISTERED COPYRIGHT.

Cash Price, Carriage Paid, from £10 10 0
For purchaser to erect.

No. 302. PORTABLE IRON BUILDINGS.

Approximate Prices.

Length. ft.	Width. ft.	Height to Ridge. ft. in.	Unlined £ s.	Lined with Matchboarding £ s.	Planed Flooring extra £ s.
12	10	9 9	11 8	16 10	2 2
20	10	9 9	16 7	24 0	3 11
30	10	9 9	22 10	32 8	5 7
15	12	11 0	16 2	23 7	3 4
20	12	11 0	20 6	28 18	4 16
25	12	11 0	23 0	33 4	6 0
30	12	11 0	26 3	37 14	7 5
25	15	12 3	27 7	39 10	7 11
35	15	12 3	35 9	50 19	10 11
45	15	12 3	43 10	62 8	13 6
25	18	13 9	32 4	46 0	9 1
35	18	13 9	41 12	59 15	12 13
45	18	13 9	50 16	72 15	16 6
35	20	14 9	45 6	64 0	14 2
40	20	14 9	50 4	71 18	16 2
45	20	14 9	56 0	79 14	18 2

FRAMEWORK is of wood covered on the outside with galvanized corrugated iron, outside woodwork painted three coats. Windows glazed with 21-oz. glass. Stock sizes, in sections ready for purchaser to erect on his light brickwork foundations.

No. 303. PORTABLE WOODEN BUILDINGS.

RED deal framing, covered outside with planed 1-in. grooved and tongued matchboarding, painted one coat outside. Galvanized corrugated iron roof. No Gutters. Whitened inside. In sections, ready for easy erection by purchaser. Windows glazed with 21-oz. glass.

Approximate Specimen Prices.

Length. ft.	Width. ft.	Height to Eaves ft. in.	Height to Ridge (about) ft. in.	Price £	Flooring (extra) £ s.	Divisions with doors £ s.
20	10	6 6	9 9	16	3 12	3 0
25	12	7 0	11 0	23	6 0	3 10
30	15	7 6	12 3	31	8 18	5 0

Carriage Paid to most Railway Stations in England.

If roof is lined inside with Matchboarding, Prices £18 15 0, £26 10 0, £36 5 0 respectively.

No. 304. PORTABLE IRON BUILDING.

Approximate Prices.

Constructed as No. 302, with circular roof.

Length. ft.	Width. ft.	Height to Eaves ft. in.	Height to Ridge (about) ft. in.	Price unlined £	Flooring extra £ s.
20	10	6 6	9 9	16	3 12
25	12	7 0	11 0	23	6 0
30	15	7 6	12 3	31	8 18

Carriage Paid to most Railway Stations in England.

ROSE LANE WORKS, NORWICH 135

No. 434. COMBINATION HUT.

CONSTRUCTED of well-seasoned wood, with **double boarded** walls and roof, thus insuring a fresh supply of air continually passing between outer and inner boards. Painted outside three coats. The upper and lower part of sides fitted with perforated zinc ventilators and shutters, improved notched flap shelves, the ridge supplied with hooks for venison or joints of meat, door fitted with good lock, planed wood floor, mounted on wheels. Including stove and chimney pipe. In sections, ready for purchaser to erect. Size—8 ft. by 6 ft.

Cash Price, Carriage Paid, £17 15 0
If without Game Larder arrangement, £15 10 0   Portable Boiler extra.

FOR SHEPHERD'S USE,
In February and March.

FOR KEEPER'S USE,
In April, May, and June.

GAME LARDER.
In August, September, & October

REGISTERED COPYRIGHT.   REGISTERED COPYRIGHT.   REGISTERED COPYRIGHT.

No. 336.
PORTABLE IRON HOUSE.
For Cricket-Grounds, Potting House, Workshop, Out-House, &c.

No. 432.
SHEPHERD'S LAMBING HUT.

No. 325.
PORTABLE WOOD HOUSE.
For Shooting Boxes, Gamekeepers' Huts, Out-houses, Children's Play-houses.

REGISTERED COPYRIGHT.   REGISTERED COPYRIGHT.   REGISTERED COPYRIGHT.

THE framework of wood, painted and covered with galvanized corrugated iron. Without floor. In sections, ready for easy erection by purchaser.

Cash Price.

12 ft. long, 8 ft. wide, 6 ft. 6 in. high at eaves, 9 ft. high to ridge, **£10 10 0**
If lined with matchboarding inside, **£3 10 0** extra.
Wood floor, if required, **£1 2 0** extra

WOOD framework, covered with galvanized corrugated iron, lined inside with matchboarding, mounted on wheels, with boarded floor. In sections, ready for purchaser to erect. Size—10 ft. long, 7 ft. wide.

Cash Price, £16 10 0

If with Ash Shafts and Fore Carriage, **£5 10 0** extra.

Stove, **27/6** extra.

WOOD framework covered with well-seasoned grooved and tongued matchboarding, painted three coats outside. Corrugated iron roof. Doors and windows can be made to suit the situation. In sections, ready for easy erection by purchaser.

Cash Price.

8 ft. long, 6 ft. wide, 8 ft. high to ridge, **£7 0 0**
10 ft. long, 8 ft. wide, 8 ft. high to ridge, **£8 17 6**
Wood Floors extra, **12/6** and **18/6** respectively.

Carriage Paid to most Railway Stations in England.

ROSE LANE WORKS, NORWICH.   135A

No. 336. PORTABLE IRON HOUSE.

REGISTERED COPYRIGHT.

This House is suitable for Potting House, Tool House, Coal House, Out-House, Bicycle House, and Children's Play House, etc. The framework is of wood, painted and covered with galvanized corrugated iron. In sections, ready for easy erection, with bolts and nuts, not nailed.

Cash Price, Carriage Paid.

12 ft. long, 8 ft. wide, 6 ft. 6 in. high at eaves, 9 ft. high to ridge, £10 10 0
If lined with matchboarding inside, £3 10 0 extra.
Wood floor, if required ... ... 1 2 0 ,,

A CHEAPER PATTERN OF THE ABOVE HOUSE

with one window and one door in end, woodwork painted or stained one coat. Made in sections, ready for easy erection by purchaser.

Cash Prices, Carriage Paid.

	£	s.	d.
6 ft. by 4 ft., 6 ft. high	2	14	6
7 ft. ,, 6 ft., 7 ft. ,, ...	3	0	0
8 ft. ,, 6 ft., 7 ft. ,, ...	3	15	6
8 ft. ,, 8 ft., 9 ft. ,, ...	4	18	0
10 ft. ,, 8 ft., 9 ft. ,, ...	5	5	6
12 ft. ,, 8 ft., 9 ft. ,, ...	6	10	0

Floor and Lining extra.

No. 332. KEEPER'S WATCH HUT OR CONTRACTOR'S HUT.

Wood framework, covered with matchboarding, painted three coats; with roof of corrugated iron, lined inside, boarded floor, and mounted on wheels. In sections, ready for easy erection by purchaser.

Price, Carriage Paid.

8 ft. long, 6 ft wide £11 0 0

Portable Cooking Stoves, see page 126.

TESTIMONIAL.

CROUGHTON COURT.

The Portable Wooden House you made for me gives me the greatest satisfaction, and meets my expectations in every way. I do not know what I should do without it when rearing my pheasants.

REGISTERED COPYRIGHT.

No. 433. LEAN-TO POTTING HOUSE FOR GARDENS.
For placing against a wall.

Wood framework, painted, and covered with best galvanized corrugated iron, unlined, glazed sashes in front, without floor. In sections, ready for purchaser to erect.

12 ft. long, 8 ft. wide, 6 ft. high at eaves, 8 ft. at top. £9 15 0
Also made with Span-roof and back wall to stand independent. £11 10 0
Flooring and matchboard Lining, £4 and £3 extra respectively.

TESTIMONIAL.

From WILLIAM WILSON, ESQ., Parkholme.

The Potting Shed which I ordered from you some time ago is now giving entire satisfaction.

REGISTERED DESIGN, No. 6009.

No. 323. INDEPENDENT PORTABLE OUT-HOUSE.
COMPLETE IN ITSELF.
For Tricycles, Bicycles, &c.

REGISTERED COPYRIGHT.

Wood framework, covered with galvanized corrugated iron. Door is made of wood and in two halves. Can be erected in a few minutes.

Cash Price.

6 ft. long, 3 ft. 9 in. wide, 4 ft. high at eaves,
4 ft. 9 in. high to ridge,

Wood floor, extra.

No. 333. IRON GARDEN HOUSE, TOOL HOUSE, WATCHMAN'S HUT, &c.

REGISTERED COPYRIGHT

Size, 5 ft. 6 in. by 3 ft. 6 in. Constructed in sections, ready for easy erection by purchaser.

Cash Price £2 10 0

Floor **7/6** extra
Matchboard lining **20/** extra
Earth Closet Apparatus "Pull out," with iron rim and deal seat, with pail to hold 20 charges, £2 10 0 extra.

No. 322. LEAN-TO PORTABLE OUT-HOUSE.
FOR PLACING AGAINST A WALL.
For Tricycles, Bicycles, &c.

REGISTERED COPYRIGHT.

Wood framework, covered with galvanized corrugated iron. Can be erected or removed in a few minutes.

Cash Price.

6 ft. long, 3 ft. 9 in. wide, 4 ft. high at eaves, 4 ft. 9 in. high to ridge,

Wood floor,

Carriage Paid to most Railway Stations in England.

ROSE LANE WORKS, NORWICH.

NEW SPAN-ROOF BICYCLE HOUSE.

CONSTRUCTED of strong red deal framing, covered with rustic joint weather boarding, stained two coats with our preparation. Framed door, strong lock, and hinges.

Cash Prices.

6 ft. 6 in. by 3 ft. 6 in., 5 ft. 6 in. high   £3 0 0
Floor, **6/-**
6 ft. 6 in. by 4 ft. 6 in., 5 ft. 6 in. high   3 10 0
Floor, **7/6**
6 ft. 6 in. by 6 ft., 5 ft. 6 in. high .. 4 15 0
Floor, **12/-**

If with ledged door and ledged match-boarded sides, stained one coat with our preparation,

Cash Prices respectively,
£2 10 0   £2 17 6   £4 0 0
Floor extra as above.

REGISTERED COPYRIGHT.

NEW LEAN-TO BICYCLE HOUSE.

CONSTRUCTED of strong red deal framing, covered with weather boarding, stained two coats with our preparation. Framed door, strong lock, and hinges.

Cash Prices.

6 ft. 6 in. by 3 ft. 6 in., 4 ft. high in front £2 5 0
Floor, **6/-**
6 ft. 6 in. by 4 ft. 6 in., 4 ft. high in front   2 14 0
Floor, **7/6**
6 ft. 6 in. by 6 ft., 4 ft. high in front ..   3 15 0
Floor, **12/-**

If with ledged door and ledged match-boarded sides, stained one coat with our preparation,

Cash Prices respectively,
£1 17 6   £2 5 0   £3 0 0
Wood Back **10/6** extra in each case.
Floor extra as above.

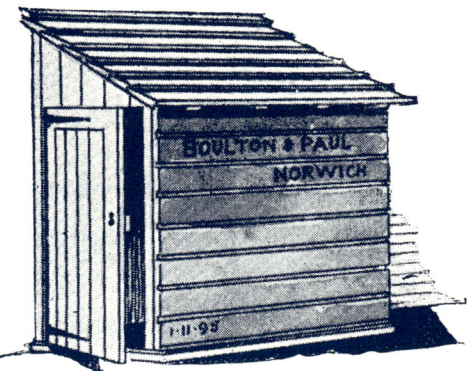

REGISTERED COPYRIGHT.

WOOD BICYCLE STAND.

MADE of strong well seasoned red deal, and arranged to accommodate six machines.

Size—4 ft. by 2 ft. 7 in. at base.

Cash Price 30/-

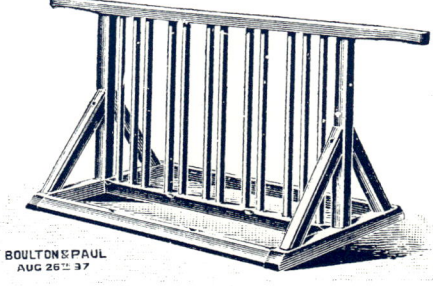

Carriage Paid to most Stations in England.

REGISTERED COPYRIGHT.

No. 650. ORNAMENTAL WOOD STABLE.

REGISTERED COPYRIGHT.

TESTIMONIAL.

From OSWALD E. PART, Esq., Browsholme Hall.
My Stable you erected for me is most satisfactory.

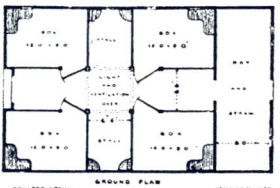

This Design can be modified to suit any situation or requirements.

Special Plans prepared, and Gentlemen waited upon free of charge.

Special Estimates free on application.

No. 651. WOOD STABLE.

GROOVED and tongued weather-board walls, stained outside with our preparation. Galvanized corrugated iron roof, with felt and matchboard lining for same. Ventilating windows protected with wrought-iron bars. Ridge ventilator. In sections, ready for purchaser to erect.

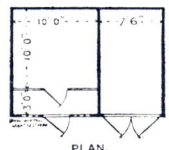

Cash Price, £35 0 0.
Loose Box Division, and Door and Manger, £5 0 0 extra.

No. 652. WOOD STABLE.

Constructed as above. Floor only to Loft.

Cash Price, £55 10

No. 653. ORNAMENTAL WOOD STABLE.

Constructed as above and finished as shown. Loft over Coach-house.

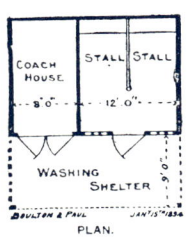

Cash Price, £72.

Stall Divisions and Mangers, £7 10 extra.

Carriage Paid to most Railway Stations in England.

ROSE LANE WORKS, NORWICH

REGISTERED DESIGN, No. 5945.

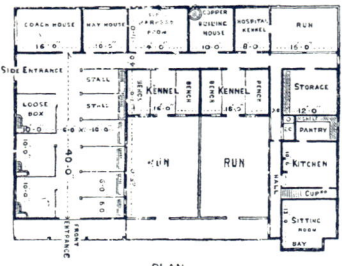

PLAN.
REGISTERED COPYRIGHT.

HUNTING ESTABLISHMENT.

Estimates free.

Special designs prepared and Gentlemen waited upon in any part of the Country.

TESTIMONIAL.

From R. FORREST TOD, ESQ., Rabley.

I hereby certify that your men have erected the Summering Boxes at Rabley to my satisfaction.

REGISTERED COPYRIGHT.
As erected for E. S. CAMERON, ESQ., Orkney Isles.

No. 307. STABLE.

Coachman's Room or Loft over Coach-house.

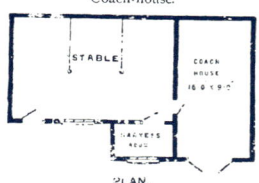

PLAN.

COMPRISING Stable 18 ft. by 14 ft., Coach-house 18 ft. by 9 ft., and Harness Room 7 ft. by 4 ft., Loft or Coachman's Room over Coach-house, constructed of strong deal framing, covered outside with galvanized corrugated iron sheets. Lined inside with felt and matchboarding, gutters to eaves, &c. Outside woodwork painted three coats. Floor only to loft. Wood stall divisions and No. 6 mangers. Walls in sections floor and roof bundled. Ready for purchaser to erect.

Cash Price from £90 0 0

Carriage Paid to most Railway Stations in England.

ROSE LANE WORKS, NORWICH.

141A

No. 750. HUNTING ESTABLISHMENT,

containing loose boxes, barn, and lofts for hay, straw, and corn, grooms' mess rooms, stud-groom's house.

Plans and Estimates to suit any situation or requirements on application.

We erected a Range of Buildings similar to the above for GRANVILLE FARQUHAR, ESQ., Oakham.

BOULTON & PAUL, MANUFACTURERS,

No. 614. RANGE OF WOOD SUMMERING BOXES FOR HUNTERS.

GROOVED and tongued weather-board walls stained outside with our preparation. Galvanized corrugated iron roof with felt and matchboard lining for same. Ventilating windows protected with wrought-iron bars. Ridge ventilator. In sections, ready for purchaser to erect.

Cash Prices, Carriage Paid.

One Box	10 ft. by 10 ft.	£12 10 0	One Box	10 ft. by 12 ft.	£15	One Box	12 ft. by 12 ft.	£16
Two Boxes	20 ft. by 10 ft.	24 0 0	Two Boxes	20 ft. by 12 ft.	27	Two Boxes	24 ft. by 12 ft.	29
Four Boxes,	40 ft. by 10 ft.	47 10 0	Four Boxes,	40 ft. by 12 ft.	49	Four Boxes,	48 ft. by 12 ft.	55

If lined with 1 in. matchboarding, 4 ft. high, £2 10 0 each box extra.

No. 661. WOOD COACH-HOUSE.

REGISTERED COPYRIGHT.

32 ft. by 14 ft., 8 ft. high to eaves, 13 ft. to ridge.
Constructed as above ; but with Canadian pattern galvanized iron roof ; ridge cresting.

Cash Price £49 0 0.

No. 751. RANGE OF LOOSE BOXES.

This arrangement contains six loose boxes, harness room, with loft or mess room over.

Special plans and estimates on application.

No. 83. PIGGERIES.

Range of eight Piggeries with fodder store, and swill house in centre.

No. 308. IRON STABLE AND COACH-HOUSE.

RED deal framework, covered with best galvanized corrugated iron, lined inside with felt and matchboarding. Outside woodwork painted three coats. Loft over Coach-house In sections, ready for purchaser to erect

Cash Price, Carriage Paid.
20 ft. long, 14 ft. wide £40 0 0

No. 435. WOOD STABLE AND COACH-HOUSE.

As supplied to the RIGHT HON. COUNTESS OF SHAFTESBURY, A. NEWBOLD, Esq., Crawley, and others.

Constructed according to specification on page 142 ; but stable walls lined
For Stable Fittings, see pages 181 to 188.

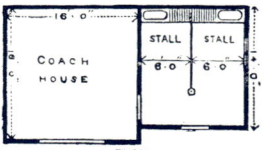

Cash Price. £65 0 0

No. 305. IRON COTTAGE, STABLE, AND COACH-HOUSE.

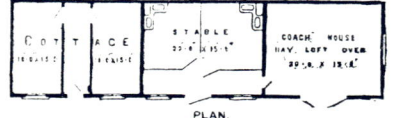

Erected for MRS. GRAHAM SMITH, The Fishery, Maidenhead.
Constructed as No. 308 ; Floor only to Cottage.

Estimates on application.

Carriage Paid to most Railway Stations in England.

No. 654. STABLE, COACH-HOUSE, & HARNESS ROOM.

REGISTERED COPYRIGHT

To face South and West.

Constructed as No. 614, with floor to Loft over Coach-house.

Cash Price,   £80

In sections, ready for your men to erect on your foundation.

TESTIMONIAL.

From A. H. JEFFERIS, ESQ., Manchester.
Sept. 8th, 1893.

I hereby certify that your men have erected the Stables and Coach-house at Withington to my satisfaction.

REGISTERED COPYRIGHT.

No. 309.
PORTABLE IRON COACH-HOUSE OR WAREHOUSE.

Wood framework covered with best galvanized corrugated iron. Outside woodwork painted three coats. In sections, ready for purchaser to erect.

Cash Prices, Carriage Paid.

16 ft. long, 12 ft. wide	.. ..	£17 10 0
If lined with matchboarding	..	25 0 0

This Building can be erected or removed by any ordinary workman.

No. 406. STABLE FOR TWO LOOSE BOXES.

REGISTERED DESIGN, No. 5958.

FRAMEWORK of deal, covered with galvanized corrugated iron, lined with matchboarding. Eaves-gutters, and down-pipes included. Outside woodwork painted three coats. Windows glazed with 21-oz. glass. In sections ready for purchaser to erect,

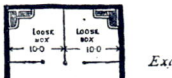

PLAN.

Cash Price, Carriage Paid.

21 ft. by 14 ft.   ...   £39 0 0

Exclusive of all divisions and mangers, floor, and drains.

For Fittings, see pages 181 to 188.

Carriage Paid to most Railway Stations in England.

ROSE LANE WORKS, NORWICH.

No. 334. STABLE AND SHED.
For Cricket-Fields, Hunting Districts, Paddock, &c.

DEAL framing, covered outside with galvanized corrugated iron. Stable 6 ft. 6 in. by 8 ft., lined inside with boards. Without floor. In sections, ready for easy erection by purchaser.

Approximate Cash Prices, Carriage Paid.

15 ft. long, 8 ft. wide, 6 ft. 6 in. high to eaves . £15 10 0
20 ,, 10 ,, 6 ,, 6 ,, ,, 21 0 0
Larger sizes at proportionate prices.

REGISTERED COPYRIGHT.

No. 655. LEAN-TO SHED AND TOOL HOUSE.

CONSTRUCTED of strong red deal framing and galvanized corrugated iron, red deal posts.

Special Estimates for Tool House or Shed on application.

REGISTERED COPYRIGHT.

No. 656. SPAN-ROOF CART SHEDS.

Constructed as above, made to any dimensions.

Special Estimates free on application.

TESTIMONIAL.

From CHAS. DIX, ESQ., Slough.

I have pleasure in stating that your men have erected the Iron Work Room to my satisfaction.

REGISTERED COPYRIGHT.

Carriage Paid to most Railway Stations in England.

WOOD HUNTING ESTABLISHMENT.

FOR KENNELS.

Houses constructed of red deal framing, covered outside with stained weather boarding, lined corrugated iron roof, whitened inside, wrought-iron kennel railing.

Estimates and Specifications free on application.

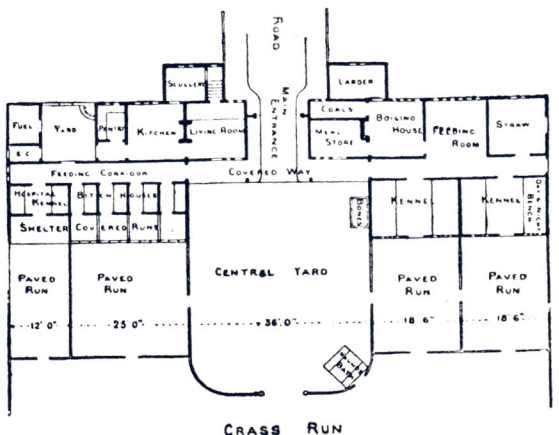

This plan can be modified to suit any situation, and special designs will be prepared free of charge.

Sites surveyed and Gentlemen waited upon in any part of the country.

ROSE LANE WORKS, NORWICH.

No. 470. SINGLE LOOSE BOX.
10 ft by 10 ft
Cash Price, £12 10 0

No. 506. UNLINED IRON SHED.
15 ft. by 8 ft.
Cash Price, £13 0 0

No. 471. TWO STALL STABLE.
15 ft by 13 ft.
Cash Price, £20 0 0

SPECIFICATION.

Walls. — Constructed of strong planed deal framing, morticed and tenoned, and covered outside with rustic joint weather-boarding, stained outside.

Doors. — Half-hatch framed doors, painted, fitted with strong hinges, good locks, and latches.

Windows. — Made in casements, glazed with 21-oz. sheet glass. Top casements to open for ventilation, bottom protected by iron bars.

Roof. — Consisting of strong deal principals, purlins, and best galvanized corrugated iron, 24 gauge. Matchboarding and felt lining. Capping to ridge. Gutters to eaves, and down-pipes to the ground.

Carriage. — Carriage Paid to any principal goods station in England, or to Docks for Shipment. In sections numbered, ready for easy erection by purchaser on his foundation and floor. Packing will be charged for, but the amount will be allowed in full if cases are returned here carriage paid.

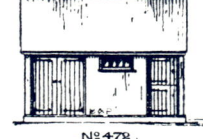

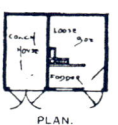

No. 472. WOOD STABLE.
20 ft. by 14 ft.
Cash Price, £35 10 0
Constructed according to the Specification above.
For Fittings, see pages 151 and 181.

No. 477. WOOD STABLE.
30 ft. by 15 ft.
Constructed according to the Specification above.
Cash Price, £52 10 0
Lining walls with ⅜-in. matchboarding, £10 extra.
For Fittings, see page 181.

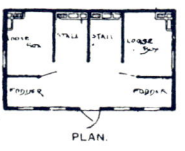

No. 476. WOOD STABLE.
31 ft. by 15 ft.
Constructed according to the Specification above.
Cash Price, £53 10 0
Lining walls with ⅜-in. matchboarding, £10 10 0 extra.
For Fittings, see page 181.

STABLES AND COACH-HOUSES.

No. 473. WOOD STABLE.

20 ft. by 14 ft.

Constructed according to the Specification on page 147.

Cash Price, £35 0 0

Lining outside walls of Stable with ⅜-in. matchboarding, £4 0 0 extra.

For Fittings, see pages 152 and 181.

No. 474. WOOD STABLE.

25 ft. by 15 ft.

Constructed according to the Specification on page 147.

Cash Price, £39 0 0

⅜ in. matchboard lining to walls, £9 0 0 extra.

For Fittings, see page 181.

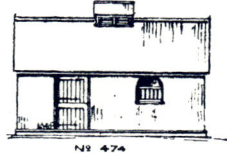

No. 475. WOOD STABLE.

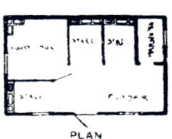

30 ft. by 18 ft.

Constructed according to the Specification on page 147, wood floor to Harness Room.

Cash Price, £62 10 0

⅜ in. matchboard lining to walls, £11 extra.

For Fittings, see page 181.

No. 479. WOOD STABLE.

36 ft. by 17 ft.

Constructed according to the Specification on page 147, wood floor to loft over Stable.

For Fittings, see page 181.

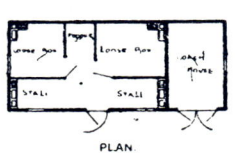

Estimates and Catalogues free on application.

ROSE LANE WORKS, NORWICH.

STABLES AND COACH-HOUSES.

No. 478. WOOD STABLE.

For Fittings, see page 181.

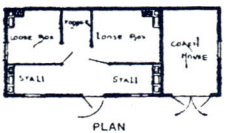
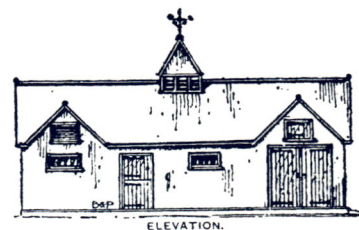

Constructed according to the Specification on page 147; with floor only to Loft over Coach-house. Louvre Ventilators in stable gables and in roof turret.

No. 481. WOOD HUNTING STABLE.

For Fittings, see page 181.

Constructed according to the Specification on page 147; wood floor only to Harness Room.

Special arrangements for ventilation.

No. 480. WOOD STABLE.

For Fittings, see page 181.

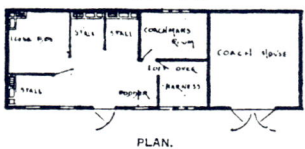
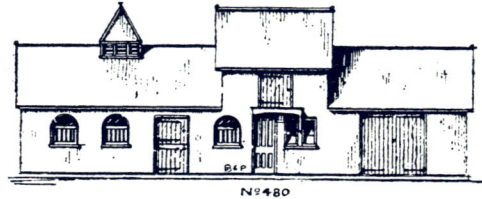

Constructed according to the Specification on page 147; with floor only to Harness and Coachman's Room, and to Loft over same. Turret, Window, and Louvre Ventilators.

Estimates and Catalogues free on application.
Carriage Paid to most Railway Stations in England.

BOULTON & PAUL. MANUFACTURERS.

No. 404. CARRIAGE WASHING SHED.

REGISTERED DESIGN, No. 5956.

CONSTRUCTED of red deal framework with galvanized corrugated iron roofing sheets, 24 ft. by 8 ft.

Ornamental cresting and barge boards, moulded wood posts with projecting prepared deal bases.

Eaves, gutters, and down-pipes, ready for purchaser to erect. Woodwork painted three coats.

Cash Price, Carriage Paid.

Size, 16 ft. by 12 ft. £14 0 0

Roofing for Covering Stable Yards, see page 177.

PORTABLE WOOD HARNESS ROOMS.

Constructed according to the Specification on page 147; but with panelled doors and wood floors, finished as shown.

No. 615.

20 ft. by 12 ft., 8 ft. high to eaves, 14 ft. to ridge, with Store and Coachman's Room.

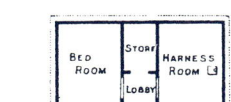

Cash Price, £49 0 0

No. 616.

12 ft. by 8 ft., with pent roof over door.

Cash Price, £14 10 0

HARNESS ROOM STOVE WITH BOILER.

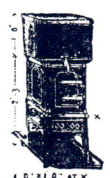

Cash Price 45/-

Smoke Pipe, 2/- per foot run.

No. 617.

12 ft. by 8 ft., lean-to against existing wall.

Cash Price, £10 10 0

TESTIMONIAL.

From H. STUDDY, ESQ., Brixham.   24th February, 1893.

I have tried your Harness Room Stove and find it most useful; I am very much pleased with it.

IMPROVED STABLE FITTINGS.

First Class Prizes have been awarded to these Stable Fittings at the principal Agricultural Shows in England, Scotland, and Ireland.

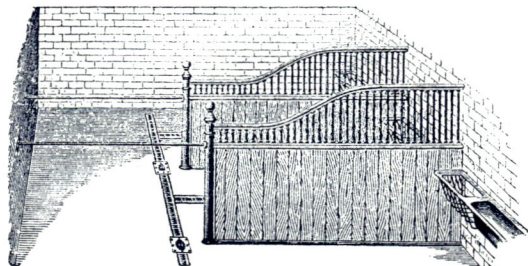

TESTIMONIALS.

Shuckburgh.
LADY SHUCKBURGH is happy to be able to say that the Stable Fittings are entirely satisfactory.

From ALEX. LYON, ESQ., Ellenhall.
The Stables meet my expectations in every way, and I am very well pleased with them. They are just the thing.

Sketch, showing mode of fitting Stall Divisions with Sliding Barrier.

TESTIMONIAL.

From
REV. F. KEPPEL, The Rectory, Winfarthing.

Your men have just finished erecting the stables, and so far as I have seen, they appear to have executed the work in every way satisfactorily. I have been pleased with the men you sent.

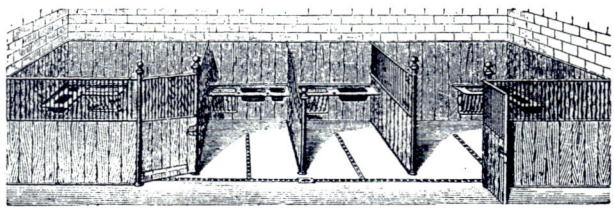

Sketch, showing arrangement for Two Stalls and Two Loose Boxes, including Four Manger Sets, and Floor Gutters.

IMPROVED CATTLE FITTINGS.

TESTIMONIAL.

From
J. HARTOP, ESQ., Barnburgh Hall.

Your man has erected the Stables, and it gives me pleasure to inform you that I am much pleased with the manner in which the work has been completed.

In ordering Stable or Cattle Fittings to suit any particular building, it is desirable to give a Sketch in Plan, with dimension and relative description.

SEPARATE CATALOGUE, with upwards of 350 Illustrations and Plans of Stable and Cow House Fittings sent free on application.

Gentlemen about to make alterations or extensions should send for our Catalogues.

No. 453. IRON GAME LARDER AND GUN ROOM.

Size—20 ft. by 14 ft., 7 ft. high to eaves, 12 ft. to ridge.

Gun room 8 ft. wide, larder 12 ft. wide

Constructed as below, with floor for Gun Room

Cash Price.   £48 0 0

For Fittings, see following page.

TESTIMONIAL.

From SAMUEL J BROWN, Esq., Knaresboro'

I beg to state that the Iron House and Game Larder you last year sent for me to Darness give entire satisfaction.

REGISTERED DESIGN.

As erected for the Hon. AILWYN FELLOWES, Honingham Hall.

CONSTRUCTED of strong deal framing, covered outside with galvanized corrugated iron sheets, lined inside with matchboarding, having part of framework covered outside with fly-proof perforated zinc, and also ventilated by louvre and perforated zinc ventilator on roof; galvanized corrugated iron roof with felt and matchboarding for lining same. Capping to ridge, gutters to eaves. In sections, ready for purchaser to erect.

No. 454. IRON DAIRY AND GAME LARDER.

REGISTERED DESIGN.

Size—25 ft. by 15 ft., 7 ft. 6 in. high at eaves, 12 ft. 6 in. to ridge.

Constructed as above, with Louvres to walls in addition to the perforated zinc.

Cash Price.   £60 0 0

For Fittings, see following page.

Estimates on application.

TESTIMONIALS.

From L. UNWIN, Esq., Strathdon.

I think the House very satisfactory, and just what I required.

From Mrs. GARFIT, Kenwick Hall.

I beg to enclose cheque for Iron Building, which has given great satisfaction.

Carriage Paid to most Railway Stations in England.

BOULTON & PAUL, MANUFACTURERS,

NEW PORTABLE LARDER.

CONSTRUCTED of strong red deal framing covered with perforated zinc, and having weather-board base, and weather-board roof, painted three coats. Strong lock and hinges to door, rebated wood floor.

Cash Prices, Carriage Paid.

4 ft. by 6 ft.	...	£8 0 0
5 ft. ,, 7 ft.	...	10 0 0
6 ft. ,, 8 ft.	...	13 0 0

If with Zincwork carried down to floor, and without the projections at eaves.

4 ft. by 6 ft.	...	£5 10 0
5 ft. ,, 7 ft.	...	7 10 0
6 ft. ,, 8 ft.	...	9 0 0

REGISTERED COPYRIGHT.
Illustration of No. 3 size.

OUTDOOR PORTABLE LARDER OR MEAT SAFE.

EXCELLENT for preserving poultry and game. Wooden Frame covered with perforated zinc. Neatly painted and varnished.

Cash Prices.

No.	Width.	Height.	Back to Front.	
1	1 ft. 9 in.	5 ft. 3 in.	1 ft. 9 in.	£1 11 0
2	2 ft. 0 in.	5 ft. 3 in.	2 ft. 0 in.	1 16 0
3	3 ft. 0 in.	5 ft. 3 in.	1 ft. 9 in.	2 10 0

Larger sizes made to order.

INDOOR PORTABLE LARDERS OR MEAT SAFES.

WOOD frame with perforated zinc panels, with lock and key to No. 6 size. Neatly painted.

Cash Prices.

No.	Width.	Height.	Back to Front.	
4	1 ft. 6 in.	1 ft. 8 in.	1 ft. 3 in.	£0 15 0
5	2 ft. 0 in.	2 ft. 0 in.	1 ft. 5 in.	0 17 6
6	3 ft. 0 in.	3 ft. 0 in.	1 ft. 6 in.	1 10 0

No. 4 size.

No. 6 size.

Carriage Paid on all Orders above **40/-** *value to the principal Railway Stations in England and Wales.*

No. 636. WOOD AND IRON DAIRY AND SCULLERY.

TESTIMONIAL.
From B. HODFELLS, ESQ.,
Fernhurst, Sussex.
The Dairy has arrived, is in good condition, and very satisfactory.

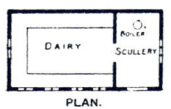

PLAN.

Constructed to any size.

TESTIMONIAL.
From GEO. J. COOKSON, ESQ., East Harling.

Your man finished erecting the Game Larder this afternoon. I think the design very good and the best one I have seen. The material seems good and the work well done.

GUN ROOM FITTINGS.
REGISTERED DESIGNS.

No. 10. RACK.

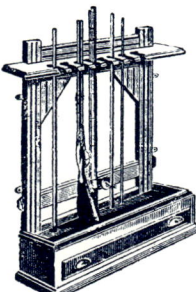

No. 10. RACK.

2 ft. 6 in. by 3 ft. 6 in. high.

For eight guns or rods, with drawer beneath tray. Pitch pine varnished.

Cash Price, £3 0 0

For Stoves, see page 126.

TESTIMONIAL.
From
R. TOWSE, ESQ., Savernake Forest.
We have the Gun Room fitted up and it is giving great satisfaction.

No. 11. TABLE.

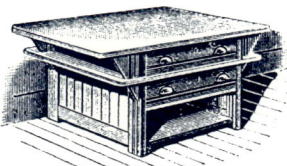

3 ft. 6 in. by 2 ft. 6 in. on top.

With two drawers fitted with locks and brass handles; shelf at side and bottom. Pitch pine varnished.

Cash Price, £3 10 0

GAME LARDER FITTINGS.
Cash Prices.

Ground Game Trestles, pitch pine varnished, 6 ft. long and 4 ft. high, with three rails for about fifty brace, 20/- each.

Game Bars, pitch pine, up to 10 ft. long, fitted with tinned hooks, six inches apart on both sides, to mock, 1/- per running foot.

Carriage Paid to most Railway Stations in England on Orders over 40/- *in value.*

BOULTON & PAUL, MANUFACTURERS,

PORTABLE WOOD DAIRIES.

Constructed on the most approved principles.

We were awarded the only Prize for Model Wood Dairy at the Dairy Show, Islington, 1895.

TESTIMONIAL

From
E. S. WOODIWISS, Esq.,
Upminster.

January, 1896.
I am extremely pleased with the Dairy, and am sure shall find it well suited to my requirements, as well as being most ornamental.

This Dairy has been erected at Broome Park and at Cranham.

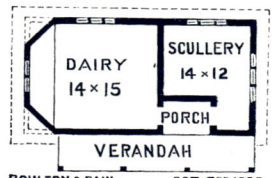

Cash Prices.

In sections, ready for purchaser to erect £110 0 0
Without Pigeon Loft 104 0 0

---

18 ft. by 12 ft., containing Porch, Scullery, and Dairy 12 ft. by 12 ft., constructed as above, and well ventilated.

In sections, ready for purchaser to erect on his foundation floor.

Cash Price £45 0 0

Estimates and Designs given for Dairies of any size.

Carriage Paid to most Stations in England.

BOULTON & PAUL, MANUFACTURERS.

PORTABLE WOOD GARDEN HOUSES.

No. 435.

REGISTERED DESIGN.

No. 436.

REGISTERED DESIGN.

WALLS stained outside with our preparation, weather-board roof, painted; wood floor. Made in sections, for purchaser to erect.

Cash Prices.

No. 435 ... £19 5 0
No. 436 ... 38 10 0

Upper Room and Balcony, 15 ft. by 15 ft.

With Tool House under, 12 ft. by 12 ft.

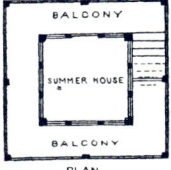

PLAN.

REGISTERED DESIGN, No. 5988.

REGISTERED DESIGN, No. 5988.

No. 430.

Constructed as above, in sections, ready for purchaser to erect.

Cash Price from £90 0 0

TESTIMONIAL.

From CHAS. M. BARTLETT, ESQ., Broadwater.

I have the pleasure of enclosing cheque for Summer House, which I like very much.

No. 437.

REGISTERED DESIGN.

No. 438.

REGISTERED DESIGN.

Constructed as above.

Cash Prices.

No. 437 ... £30 5 0
No. 438 ... 43 0 0

TESTIMONIAL.

From S. MORLEY, ESQ., Tunbridge.

The Garden House you sent me has quite met my requirements, and is, I think, extremely well adapted to the purpose.

Carriage Paid to most Railway Stations in England.

ROSE LANE WORKS, NORWICH. 155

No. 441.
SHELTER FOR BAND, OR VILLAGE PUMP.

9 ft. diameter, 7 ft. high at eaves.

WALLS stained with our preparation, weather-board roof painted. Made in sections, for purchaser to erect.

Cash Price, £21 0 0
Exclusive of pump.

REGISTERED DESIGN.

REGISTERED DESIGN.

No. 439.
PORTABLE BAY WINDOW WITH BALCONY.

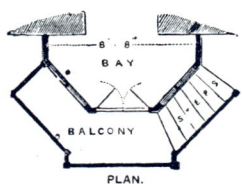

PLAN.

TESTIMONIAL.

From H. BECK, ESQ., Harpley, Swaffham.
October 19th, 1890.

I am well pleased with the Bay Smoking Room you have put over the portico at my house, also with the workmanship of those employed in the erection.

WALLS stained with our preparation, varnished inside, weather-board roof painted, wood floor. Made in sections, for purchaser to erect.

Cash Price, £33 0 0

No. 440.
CABMEN'S SHELTER AND URINAL.

WALLS stained with our preparation, weather-board roof painted, floor for Shelter. Made in sections, for purchaser to erect

Cash Price, £32 10 0

11 ft. 6 in. by 10 ft. 6 in., 7 ft. high at eaves.

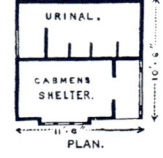

PLAN.

REGISTERED DESIGN.

Carriage Paid to most Railway Stations in England.

BOULTON & PAUL, MANUFACTURERS.

PORTABLE RUSTIC GARDEN HOUSES.

REGISTERED DESIGNS.

No. 443.

13 ft. by 9 ft., 7 ft. high at eaves.
Cash Price, £30 5 0

No. 444.

12 ft. by 12 ft., with open porch 6 ft. 6 in. by 3 ft., 7 ft. high at eaves.
Cash Price, £43 0 0

No. 445.

9 ft. by 9 ft., 7 ft. high at eaves.
Cash Price, £34 5 0

GENERAL SPECIFICATION.

CONSTRUCTED of red deal framing, covered outside with match or weatherboarding; stained with our preparation.

Wood floor; galvanized iron roof painted; no lining.

Windows glazed with 21-oz. sheet glass.

In sections, ready for erection by purchaser.

TESTIMONIALS.

From MISS PANTON, Bournemouth.

The Garden House arrived quite safely. It is now put up, and is very complete in every way, and Miss Panton is much pleased with it.

From GEO. ALLAN, ESQ.

I have pleasure in stating that your man has erected the Summer House at Strangeways to my satisfaction. Your man is an intelligent workman, and has done his work expeditiously and satisfactorily.

No. 446.

13 ft. by 9 ft., with porch 8 ft. by 3 ft., 7 ft. high at eaves.
Cash Price, £44 0 0

No. 447.

9 ft. by 9 ft., with semi-octagonal porch 4 ft. 6 in. by 9 ft., 7 ft. high at eaves.
Cash Price, £33 0 0

Carriage Paid to most Railway Stations in England.

ROSE LANE WORKS, NORWICH. 157

IMPROVED PORTABLE RUSTIC GARDEN HOUSES.

No. 529.

8 ft. by 6 ft., with projecting iron roof.
Cash Price £12 0 0
No. 24 Pattern Seat 20/- extra.

CONSTRUCTED of red deal, in sections ready for easy erection by purchaser. Walls sized, stained, and varnished. Weather-board roofs painted. **Wood floor**, if required, 20/- extra in each case.

No. 530.

Cash Price, 9 ft. by 6 ft., with iron roof, £19 10 0

No. 657.

No. 531.

8 ft. by 6 ft., with weather-boarded roof.
Cash Price £14 10 0

No. 532.

8 ft. by 6 ft., with lean-to iron roof.
Cash Price, £8 0 0

KIOSKS OR BAND STANDS.

REGISTERED
No. 519.

Cash Price, 20 ft. diameter, £105

No. 518. Cash Price, 20 ft. by 13 ft. 4 in., **£85**
CONSTRUCTED of wood and iron with galvanized iron roofs, well finished with mouldings, finials, and cresting. Woodwork stained in tints and varnished, or painted; ironwork painted approved tints. Made in sections, ready for erection by purchaser. Without floor.

COPYRIGHT.
No. 520.

Cash Price, 20 ft. diameter, £80

Carriage Paid to most Railway Stations in England.

BOULTON & PAUL, MANUFACTURERS.

ORNAMENTAL SHELTERS FOR PUBLIC PARKS AND GARDENS.

No. 645.

As supplied to the London County Council.
20 ft. by 20 ft., 8 ft. high to eaves,
15 ft. to ridge.

Constructed like No. 646.

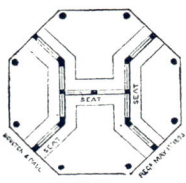

Prices on application

No. 646.

CONSTRUCTED of selected red deal, painted, and the upper portion glazed with 21-oz. sheet glass, galvanized iron roof, matchboard ceiling. In sections, ready for easy erection by purchaser on his floor-foundation. Without floor. Roof covers space 22 ft. by 12 ft.

20 ft. by 20 ft., 8 ft. high to eaves.
15 ft. to ridge.

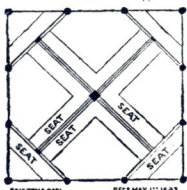

Estimates on application.

No. 647.

20 ft. by 15 ft., 8 ft. high to eaves
15 ft. 6 in. to ridge.

Constructed like No. 646.

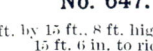

Estimates on application.

Carriage Paid to most Railway Stations in England.

ROSE LANE WORKS, NORWICH. 159

No. 339. ARBORETUM SHELTER

For Watering Places, Public Grounds, Promenades, &c.

CONSTRUCTED of selected red deal, painted, and the upper portion glazed with 21-oz. sheet glass. In sections, ready for easy erection by purchaser on his foundation. Without floor. Canadian pattern iron roof covers space 22 ft. by 12 ft.

REGISTERED COPYRIGHT.
Jubilee Memorial as erected by us at Herne Bay.

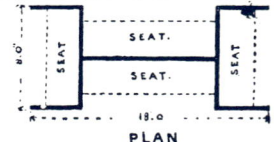

PLAN
REGISTERED COPYRIGHT.
Cash Price, £42 0 0
Seats, £7 0 0 extra.

No. 431. NEW SHELTER.

For Watering Places, Parks, Schools, &c.

Cash Prices.
20 ft. by 8 ft. .. **£24 0 0**
30 ft. by 8 ft. .. **38 0 0**
Including Seats.

ROOF projects 2 ft. in front; 8 ft. high at eaves. Deal framework, covered with matchboarding, unlined, painted three coats, roof corrugated iron. In sections, ready for easy erection by purchaser.

If plain roof, £21 15 and £33.

REGISTERED COPYRIGHT.

REGISTERED COPYRIGHT.

No. 324. MOVABLE CABMEN'S SHELTER, WAITING ROOM, WEIGHING-MACHINE ROOM, &c.

CONSTRUCTED of strong deal framework, covered with galvanized corrugated iron, half-glass front, strong wood floor. In sections, ready for easy erection by purchaser.

Cash Prices.

15 ft. by 10 ft.	20 ft. by 12 ft.	25 ft. by 15 ft.	25 ft. by 18 ft.
£17 0 0	£27 0 0	£37 0 0	£43 0 0

Estimates on application

Carriage Paid to most Railway Stations in England.

BOULTON & PAUL, MANUFACTURERS.

No. 662. GRAND STAND.
FOR RACE-COURSES AND RECREATION GROUNDS.

REGISTERED COPYRIGHT.

Estimates on application.

No. 663. ESPLANADE SHELTER.

Estimates or Special Designs free on application.

REGISTERED COPYRIGHT.

ROSE LANE WORKS, NORWICH. 161

REGISTERED COPYRIGHT.

No. 658.
WOOD CLOAK ROOM AND LAVATORY.

Made specially to suit any situation.

Estimates free on application.

No. 659.
ESPLANADE SHELTER.

Made to any dimensions.

Estimates free on application.

REGISTERED COPYRIGHT.

REGISTERED COPYRIGHT.

No. 660.
ARBORETUM OR ESPLANADE SHELTER.

Estimates free on application.

BOULTON & PAUL, MANUFACTURERS.

Registered Design.

No. 23.
ORNAMENTAL BRIDGE.

For connecting Park with Garden. Supported on arched lattice girder. Kiosk in centre, with curved roof projecting on brackets over the sides of bridge, to shelter the rounded bays enclosing the seats. Finished in any style of decoration.

Special Designs prepared for Rustic Work or other Bridges

Gentlemen waited upon in any part of the country.

No. 20.
WOOD FOOT BRIDGE.

Bridge constructed of wood, with supporting posts in centre. This style of Bridge is very suitable for crossing streams having marshy borders, as it can be made to any length, being supported at intervals on posts.

Registered Design.

No. 21.

Registered Design.
This design can be made up to 20 ft. span.

No. 22.

Registered Design.
Garden Bridge can be made up to 30 ft. span.

Estimates on application.

BOULTON & PAUL, MANUFACTURERS.

No. 753. CONCERT PAVILION & REFRESHMENT ROOMS.

As erected for the MAATSCHAPPIJ ZEEBAD at Scheveningen, Holland.

Estimates and Special Designs to suit any situation free on receipt of particulars of requirements.

TESTIMONIAL.

This is to certify that the Bar and Bathing Pavilions and the two Shelters, erected for the Maatschappij Zeebad at Scheveningen by Messrs. Boulton and Paul of Norwich, meet the requirements in every respect. The materials employed are of excellent quality, and the architecture and construction are well and tastefully devised.

(*Signed*) BD. GOLDBECK, Director General.

ROSE LANE WORKS, NORWICH. 163B

No. 743.
GOLF PAVILION.

13 ft. to ridge.

Price £160 0 0

Cheaper construction, £130 0 0

In sections, ready for easy erection by purchaser on his foundation.

Carriage Paid to nearest Station.

No. 728. CRICKET PAVILION.

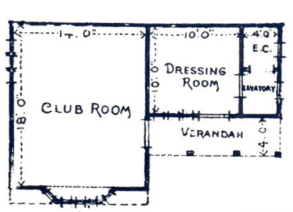

Price £99 0 0

Cheaper Construction, £78 0 0

In sections, ready for purchaser to erect on his foundation.

Carriage Paid to nearest Station.

BOULTON & PAUL. MANUFACTURERS.

No. 639. WOOD AND THATCH PAVILION.
For Golf, Cricket, or Tennis.

Constructed according to the Specification on page 92, but without iron roof.

No. 422 GALVANIZED IRON CLUB HOUSE.

TESTIMONIALS.

From JAMES D. FAWCETT, Esq., Sheffield.

The Club House was duly delivered last week, and I have to compliment you on the perfect way it has been sent out. It was beautifully packed, and not a thing missing.

From G. M. HARDY, Esq., Danehurst, Uckfield.

The Iron Building which you have just completed, is perfectly satisfactory, and the work has been well done, and the material is good.

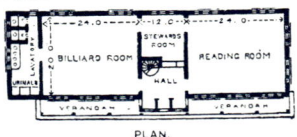

REGISTERED DESIGN, No. 5990.

Constructed according to the Specification on page 92, but with walls covered outside with galvanized corrugated iron instead of wood.

Estimates free on application.

No. 641. VILLAGE CLUB HOUSE.

As erected at Tyn-y-Graig for Miss Thomas.

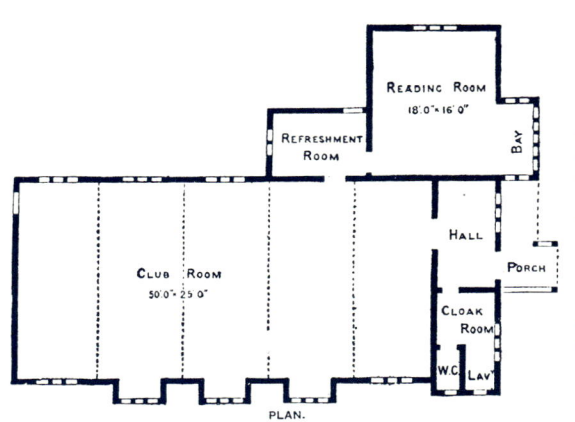

TESTIMONIALS.

From WETHERED & DAVIES, Stroud.

We are glad to say that we are pleased with the Building, and that it meets our requirements in every way. The materials are sound and strong, and the construction so well devised that it promises to need no repair for years to come.

From GEO. P. FISHER, Esq., Slough.

I should like to take this opportunity of expressing my great satisfaction with the Schoolroom you have erected for me. Should I ever want anything of the kind done again, I should always desire you to do it

Constructed according to the Specification on page 92, but having walls covered outside with galvanized corrugated iron in lieu of wood.

Special Estimates on application.

WOOD GOLF HOUSES.

Registered. Copyright.

No. 664.

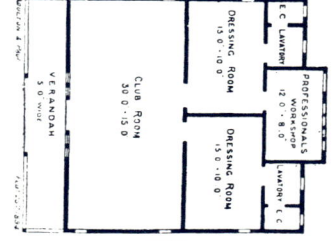

PLAN.

Special Estimates free on application.

Gentlemen waited upon in any part of the country.

No. 665.

As above plan, but 33 ft. by 22 ft., without Lavatories, Closets, and Professionals' Room.

Constructed according to the Specification on page 92, without lining to walls, roof lined with felt, and matchboarding in lieu of ceiling.

Cash Price, £120

Carriage Paid to most Rail. Stations in England.

No. 640. WOOD GOLF PAVILION.

This Pavilion was erected by us at Rhyl, 1893.

Constructed according to the Specification on page 92.

Estimates on application.

GOLF LOCKERS.

Each 12 in. by 9 in., 4 ft. 6 in. high.

Estimates on application.

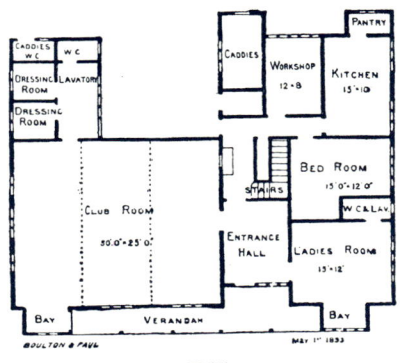

PLAN.

TESTIMONIAL.

From CHARLES F. HUTTON, ESQ.

I hereby certify that your men have erected the Pavilion for the Grammar School at Pocklington to my satisfaction. The work has been much admired.

No. 424. CRICKET OR GOLF PAVILION.

TESTIMONIAL.

From BURCHALL HELME, Esq.,
Warminster.

I have now seen the Pavilion you have put up for me to act as a reading room in the winter and a cricket pavilion in the summer, and I am quite satisfied with it in every respect.

REGISTERED DESIGN, No. 5995.

Constructed according to Specification on page 92, without lining to walls.

REGISTERED DESIGN, No. 5995.

As erected by us at Calshot and other places.

Cash Price,   £198   0   0

No. 425. NEW CRICKET OR GOLF PAVILION.

TESTIMONIAL.

From A. PERCY ECCLES, Esq.,
University "Pitt" Club, Cambridge.

The Iron House arrived on Saturday safely. We are very pleased with it, and think it will suit our purpose very well indeed.

REGISTERED DESIGN, No. 5996.

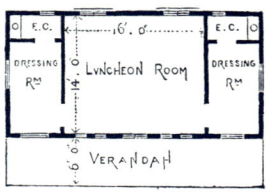

REGISTERED DESIGN, No. 5996.

Constructed as above.

Cash Price,   £117   0   0

TESTIMONIAL.

From EDWARD FORD, Esq., West Malvern.

The Cricket Pavilion which you sent me last summer has answered well, and has met my expectations in every way.

Carriage Paid to most Railway Stations in England.

BOULTON & PAUL, MANUFACTURERS.

No. 328.
PORTABLE CLUB HOUSE OR GARDEN PAVILION.

As erected for the CORINTHIAN YACHT CLUB, Southsea.

Size 18 ft. by 12 ft.

WOOD walls stained outside, galvanized corrugated iron roof with matchboard lining, wood floor. In sections, for purchaser to erect.

Cash Price, £31 0 0

No. 331. PORTABLE PAVILION.

24 ft. by 12 ft.   Each Dressing-room 7 ft. wide.

WOOD walls stained outside, galvanized corrugated iron roof with matchboard lining, wood floor. In sections, for purchaser to erect.

Cash Price, £37 0 0

No. 326.   CRICKET OR TENNIS PAVILION.

18 ft. by 22 ft.

COMPRISING General Room 18 ft. by 12 ft., Dressing-room 15 ft. by 5 ft., Closet 5 ft. by 3 ft., and Verandah 18 ft. by 5 ft., 7 ft. high at eaves, with four verandah columns, one door, and two windows in front.

Constructed as above, no lining for verandah roof.

Cash Price, £52 0 0   Seats extra.

No. 335.   CRICKET OR TENNIS PAVILION.

18 ft. long by 15 ft. wide, elevated on piles.   Verandah in front 3 ft. wide.

Constructed as above.

Cash Price, £48 0 0

TESTIMONIAL.

From H DANDY, ESQ., Malpas.
The Pavilion meets with general approval.

No. 330.   PORTABLE PAVILION.

THE upper portion of the front opens when the Pavilion is in use, and forms a Verandah, and shuts down and locks when not in use.

In sections, ready for purchaser to erect.

Cash Prices, Carriage Paid,

	Including floor.	Corrugated Iron, Unlined.	Lined with Matchboarding.
12 ft. long, 8 ft. wide, 8 ft. 6 in. high to ridge	£13 0 0		£17 10 0
20 ,, 12 ,, 9 ,, 6 ,, ,,		25 0 0	35 0 0
30 ,, 15 ,, 10 ,, 6 ,, ,,		40 0 0	58 0 0

Partition, with door, 5s. per foot run, extra.   Cresting 1s. per foot run, extra.

Carriage Paid to most Railway Stations in England.

Portable Farm Buildings

AND

Covered Yards for Cattle.

The following is an extract from The Agricultural Gazette, *January 31st, 1887:—*

"*Waste* is a sin in any branch of trade, or in any class of business. Whether it is that we are more interested in farming, or that we have more to do with that occupation than any other, however that may be, we are persuaded that *waste* is more prevalent among farmers than any other class of men. We consider that the style of feeding cattle in large yards only furnished with one or, it may be, two side roofs, is wasteful to a degree. Upon English farms, especially those where corn farming is the main characteristic, and where the tenant's capital is only half what it should be, waste by means of large open yards, and resulting in the making down of a large quantity of straw, with a few cattle, cannot be too strongly deprecated. There are those who cry out against covered yards, and in favour of open courts, this we put down as the cry of expediency. If properly-made manure is wanted, straw, even in the case of a covered yard, must be sparingly applied; firefanged manure is but little good, and firefang will result if straw is improperly used. A tenant on a farm furnished with covered yards, and with a thorough knowledge of the system of feeding cattle, will keep *three times as many beasts*, as will a farmer with wet open yards and upon the old-fashioned heavy turniping system; nay, more, he will keep them better and to a larger profit per head. His manure will be veritably manure, and not simply wet straw. It is the quality more than the quantity of manure which makes it valuable; and when we see the straw grown upon, say, 250 acres trampled down by twenty-five or thirty small bullocks in open yards almost devoid of roofing, and without a spout to catch rain water, we cannot refrain from exclaiming, '*Sheer waste: waste of straw, waste of food, waste of money*, and positively *waste of water*.' If the landlord of that farm would cover these yards, and the tenant properly use them, 100 cattle would not be too many, and his farm and his crops would soon tell where lay the difference between well-made manure and wet straw. Of course, this is largely a question of capital upon the part of the tenant. We are persuaded that for a tenant possessed of ample or sufficient capital no landlord would hesitate to cover cattle yards; the benefit to all parties is so apparent that it could hardly be denied. Cheap roofs, such as **Galvanized Corrugated**, answer the purpose quite well, and for £50 it is really surprising to see the extent of roofing to be got. It is with covered yards or covered boxes that justice is done to cattle, to food, to manure, and therefore to farmer and to farm."

BOULTON & PAUL MANUFACTURERS.

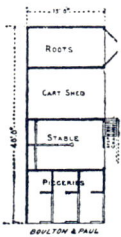

PLAN.

No. 618.
PORTABLE FARM BUILDINGS.

TESTIMONIAL.

From J. J. D. RAWLINS, Esq., Lymington.
All the work you have done for me has given me great satisfaction.

Intending purchasers waited upon in any part of Great Britain.

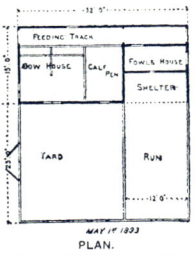

PLAN.

No. 306.
COMBINATION IRON COW HOUSE, PIGGERY, AND FOWLS' HOUSE.

REGISTERED COPYRIGHT.

LINED inside with matchboarding; perches and nest boxes for fowls' house; no floor.

Cash Price,   £60 0 0

In sections, ready for easy erection by purchaser on his foundation.

Mangers, Troughs, and Cow House Division extra.
See page 151.

TESTIMONIAL.

From C. McIVER, Esq., Beechfield.
I erected the Iron House with my own men easily, and it is a thoroughly satisfactory building.

Carriage Paid to most Railway Stations in England.

BOULTON & PAUL, MANUFACTURERS,

COMBINATION PIGGERY AND FOWL HOUSE.

EXTERIORS of galvanized corrugated iron. Houses cladded inside with 1-in. boards, wood floor raised 6-in. above the ground.

Run 6 ft. by 5 ft. House 5 ft. by 5 ft., with Fowls' House over. Fitted with perches and nest boxes.

Price £9 0 0

TESTIMONIALS.

From CHARLES RICE, ESQ.,
Bushey, Hertfordshire.
The Pigsty I had of you in the Spring of last year wears remarkably well.

From
MRS. COLERIDGE KENNARD,
Stockton House.
Mrs. Coleridge Kennard was much pleased with the Double Piggery supplied her by Messrs. Boulton and Paul, and she has found it answer its purpose well.

No. 80. PORTABLE PIGGERIES.

11 ft. long, 5 ft. wide, 5 ft. high

Cash Prices.

Single Sty	... £6 5 0	Double Sty ... £11 0 0
	If Houses are not boarded.	
Single Sty	... £5 5 0	Double Sty .. £9 10 0

No. 81. MOVABLE PIGGERY.

Cash Price.

House 5 ft. by 4 ft., with Run 8 ft. long, £4 10 0

Special Prices quoted for quantities.

Carriage Paid to most Railway Stations in England.

No. 83. PIGGERIES.

Range of eight Piggeries with fodder store, and swill house in centre.

ROSE LANE WORKS, NORWICH.   175A

CORRUGATED IRON FENCING.

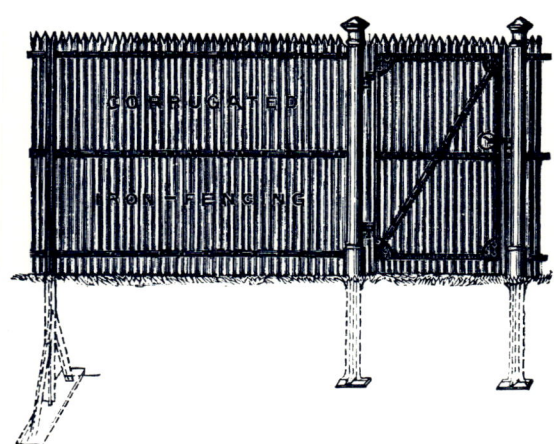

A VERY efficient Fencing for enclosing Parks and Gardens, etc. Is practically unclimbable, very substantial, and can be erected by ordinary labourers. It is made 4 ft., 5 ft., 6 ft., and 7 ft. high, with either wood or iron rails and standards.

Illustration shows Fence with iron rails and posts.

TESTIMONIAL.   MANCHESTER.

GENTLEMEN,—The 457 yards of Corrugated Iron Fencing and Gates you sent to St. Asaph for Cricket and Lawn Tennis Ground have given great satisfaction. It is substantial, and was easily fixed by local men.
(*Signed*) WILLIAM DAWES, Architect.

Enlarged block of Serrated Sheet.

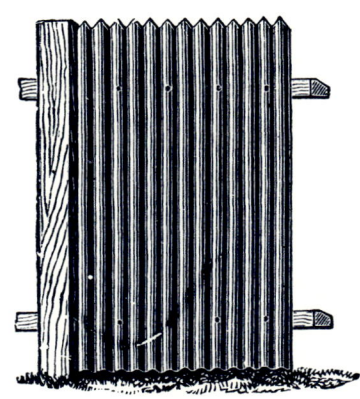

With wood rails and posts.

Special Estimates given upon receipt of particulars of requirements.

BOULTON & PAUL, MANUFACTURERS,

PORTABLE BOILER

For preparing Food for Cattle, Kennels, Poultry Yards, Piggeries, Dairy or Wash House.

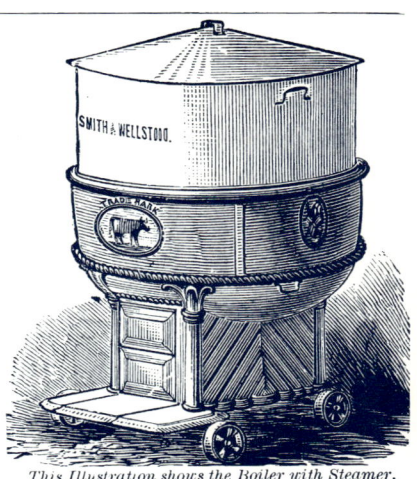

OUR Improved Portable Boiler, with its self-contained furnace, is by far the most useful and convenient in every respect of anything of the kind ever invented; complete within itself, ready for immediate use in any convenient position or place, either inside the house or out in the open air. It is constructed entirely of cast-iron, the fire chamber being lined with fire-brick, and for the purposes of cleaning the boiler can be lifted out of its frame or furnace at any time.

This Illustration shows the Boiler with Steamer, and on wheels.

Cash Prices, Carriage Paid.

Size.	With Plain Boiler. £ s. d.	With Galvanized Boiler. £ s. d.	Extra for Galvanized Iron or Tinned Steamer to fit the Boiler. £ s. d.
15 gallons	2 1 0	2 11 6	15 gallons 1 5 0
30 ,,	3 1 0	3 17 0	30 ,, 1 10 0
45 ,,	3 15 0	4 16 6	45 ,, 1 15 0
60 ,,	4 14 0	6 4 0	60 ,, 2 5 0

Smoke Funnel, for connecting with a chimney flue, or for giving the necessary draught if used in the open air, furnished at **1/-** to **1/6** per foot.
If mounted on wheels, **13/-** extra.   Brass draw-off tap fitted, extra **16/-** each.

THE LUDGATE BOILER.

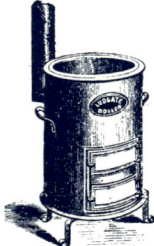

THIS is a cheaper made Boiler than the above; it is well adapted for small Poultry Keepers.

With 4½ ft. of smoke funnel all ready for immediate operation.

Cash Prices.

No. 5.	5 gallons	... ...	£1 4 0
No. 7.	7 ,,	... ...	1 7 6
No. 9.	9 ,,	... ...	1 10 6
No. 12.	12 ,,	... ...	1 15 0

Carriage Paid on Orders of **40/-** *value.*

CATTLE SHED & YARD, WITH STOCK KEEPER'S COTTAGE.

TESTIMONIAL.

From
CAPT. BACON,
Raveningham Hall.

The Span and two Lean-to Cattle Sheds which you erected here last year have given very great satisfaction. The ventilating doors in the end of Span-roof Shed are a great success in hot weather.
(Signed)
John Mendham,
Agent.

REGISTERED COPYRIGHT.

Erected for J. J. COLMAN, ESQ., M.P.

TESTIMONIAL.

From CAPT.
G. HEAVISIDE,
Norwich.

The long Corrugated Iron Roof over our twenty bullock boxes and turnip house at Salhouse, for Sir Edward Stracey, does capitally. The bullocks did well under it, and I don't see the least flaw after the last trying winter, which was so disastrous to many of our tiled roofs.

RANGE OF COVERED YARDS.

TESTIMONIAL.

From the REV.
C. H. LIPSCOMB,
Howe Rectory.

I have much pleasure in informing you that the Covers you put over my Farm Yards have worn remarkably well, and have given me great satisfaction. I can thoroughly recommend them as being a great saving in straw and in preserving the quality of the manure.

REGISTERED COPYRIGHT.

Erected for R STROYAN, ESQ., Bixley.

TESTIMONIAL.

From
H. HEPBURN, ESQ.,
Scarisbrick Hall.

DEAR SIRS,—The Cattle Shed is fixed to my satisfaction. Your man sent has been very obliging. I hope if I should have more work done by you, that I get the same man to fix it.

COVERED CATTLE YARD.

TESTIMONIAL.

From MESSRS.
J. CRISP & SON,
Beccles.

We have pleasure in testifying to the satisfactory manner in which you erected a Covered Yard at Hales for us, the arrangement of the arch being specially convenient for sheltering loads of hay, corn. etc.

REGISTERED COPYRIGHT.

Erected for MESSRS. J. CRISP AND SON, Beccles.

TESTIMONIAL.

From
JOHN COSSEY,
ESQ., Church Grove,
London.

The Roofing erected on the farm at Raveningham answers my purpose admirably. I think other land-owners would do well to give their tenants better accommodation for the protection of their stock during the winter.

Iron Roofing, Shedding, and other Estate Work carried out in the most economical way. Estimates free.

IRON ROOFING FOR HAY AND CORN STACKS, COVERED YARDS, CATTLE SHEDDING, &c.

No. 150.
HAY BARNS, CORN STORES, &c.

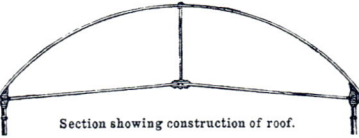

Section showing construction of roof.

THE above illustration represents the simplest form of Corrugated Roofing, as the sheets being curved are self supporting.

Materials consist of necessary galvanized corrugated sheets, No. 22 gauge, wrought-iron tie and king rods, bolts, washers and guttering. Iron columns and sills extra.

No. 151.
LEAN-TO CART OR CORN SHED.

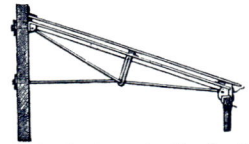

Section showing construction of roof.

Materials consist of necessary galvanized corrugated sheets, No. 22 gauge, wrought-iron trusses about 10 ft. apart, angle-iron and deal purlins, and deal wall plates, guttering, bolts, etc. Iron columns and sills extra.

No. 152. SPAN-ROOF SHED.

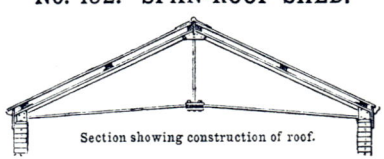

Section showing construction of roof.

TESTIMONIALS.

From
M. PARRY, ESQ.,
Kitemore, Faringdon,
Berks.

The Shed arrived safely. It is beautifully finished and gives great satisfaction.

From
R. A. LONG, ESQ.,
Newton Flotman,
Norfolk.

I am very well pleased with the Shed, 30 ft. by 21 ft., for Implements, erected by you.

TESTIMONIAL.

NORWICH.

GENTLEMEN,—I have much pleasure in informing you that the Covered Cattle and Horse Yards you have erected under my instructions at Bixley, Arminghall, Whinburgh, Buxton, and other places have answered well and given every satisfaction.

(*Signed*)
J. B. PEARCE, F.R.I.B.A.

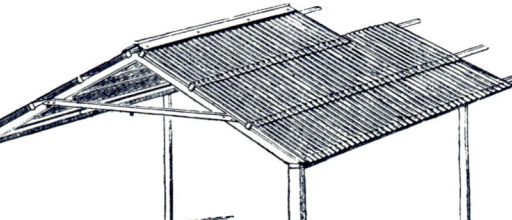

Sketch showing mode of fixing the sheets to wood framing.

Special quotations free on receipt of particulars of requirements.

BOULTON & PAUL, MANUFACTURERS.

WROUGHT-IRON ROOF PRINCIPALS.

For Galvanized Corrugated Iron or other covering   For fixing to Walls, Wood Posts, or Iron Columns.

No. 345.   No. 343.

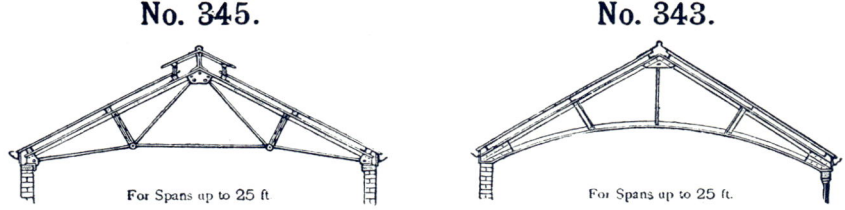

For Spans up to 25 ft.   For Spans up to 25 ft.

TESTIMONIALS.

From PALMER LEEDER, ESQ., Brooke.

I have much pleasure in saying that I am perfectly satisfied with the Curved Roof you put up for me at Haddiscoe last summer. I believe it to be a most economical way of converting open yards into covered ones.

From J. HENRY DUGDALE, ESQ., Rowney Abbey.

As far as I am able to judge the Covered Yard and Corrugated Iron Fencing erected by you here is a very satisfactory structure.

No. 342.

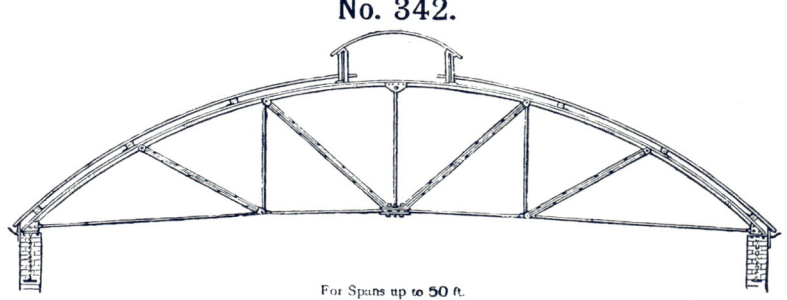

For Spans up to 50 ft.

TESTIMONIALS.

From THOMAS SLIPPER, ESQ., Braydestone Hall.

I am well satisfied with the Corrugated Iron Roofing you supplied me with last year. The objections to it, so commonly talked about, I find are without foundation, and for a light, indeed any other roof, I should think would be hard to beat. I have just issued order for another Lean-to, 110 feet by 30 feet.

From JAMES E. PLATT, ESQ., Brentwood, Cheadle, Cheshire.

I have great pleasure in being able to inform you that the Shed I had from you has given me every satisfaction.

No. 344.   No. 341.

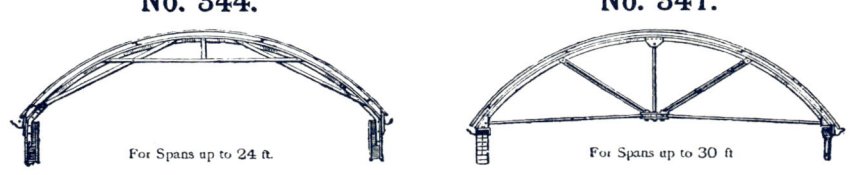

For Spans up to 24 ft.   For Spans up to 30 ft.

Estimates free on application.   *Gentlemen waited upon in any part of the country.*

ROSE LANE WORKS, NORWICH

MATERIALS FOR IRON ROOFING.

No. 6. Galvanized Cone-headed Screw.

No. 5. Galvanized Cone-headed Nail.

CONTRASTED with slates and tiles, the saving in timber to carry our Corrugated Iron Sheets is very considerable, the scantlings throughout being of a much lighter character and less in quantity. The principals may be placed about 10 ft. apart, and the purlins from 4 ft. apart. This class of roofing is well understood by country carpenters, and the erection is a matter of great simplicity, and is about the cheapest roof that can be erected.

The sheets we supply are of the best quality only, and thoroughly well galvanized. They are rolled with eight 3-in. deep corrugations, and are gauged **before** galvanized, so that a 24 gauge sheet, in reality, gauges nearly 22 gauge. These sheets should not be compared with the low-priced sheets usually advertised, which are not efficiently galvanized, and are known to the trade as "wasters." Straight corrugated Sheets, 4 ft. to 8 ft. length, always in stock.

Made in lengths of 5 ft., 6 ft., 7 ft., 8 ft., 9 ft., and 10 ft., 24 and 22 gauge.
EACH SHEET COVERS TWO FEET WHEN FIXED.
For Prices see Special Price Sheet.

No. 3. Curved Corrugated Sheet.

No. 2.
O.G Corrugated Sheet.

Section showing mode of fixing

No. 21.
Cast-iron
Column.

No. 22.
Wrought-iron
Stanchion
and Plate.

CORRUGATED SKYLIGHT

No. 4. Galvanized Ridge Capping.

No. 7.
Galvanized Bolt
and Nut.

No. 9. Galvanized Washer

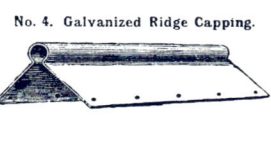

No. 8.
Galvanized Rivet.

For Prices see Special Price List.

BOULTON & PAUL, MANUFACTURERS.

MATERIALS FOR IRON ROOFING.

No. 10. Galvanized Sheet-iron Half-round Gutter. No. 11. Galvanized Sheet-iron O.G. Moulded Gutter.

No. 12. Galvanized Sheet-iron Angle Piece. No. 14. Galvanized Sheet-iron Cistern Head. No. 13. Galvanized Sheet-iron Angle Piece.

  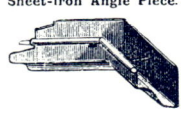

No. 15. Galvanized Sheet-iron Leading-down Pipe and Shoe.

No. 16. Galvanized Gutter Bracket.   No. 19. Galvanized Hollow Washer.   No. 20. Cast-iron Base for Wood Posts.   No. 18. Galvanized Pipe Hook.   No. 17. Galvanized Gutter Bracket.

   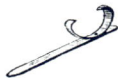

Estimates on application.

GALVANIZED CORRUGATED IRON RICK COVERS.

PITCH SPAN RICK COVER. CIRCULAR SPAN RICK COVER.

These Rick Covers are made of galvanized iron in two designs as shown, viz., the "Pitch Span" and the "Circular Span" Covers.

The "Pitch Span" Cover has a ventilator running along the ridge, to which are attached galvanized corrugated sheets, so arranged that any one can be easily removed leaving the others quite secure.

The "Circular Span" Cover is made with curved galvanized corrugated sheets, fastened to wood scantlings placed on the rick, thus providing ample ventilation.

The sheets used each cover 2 feet in width, and are attached to the rick by improved screw fasteners, which hold them quite securely.

The advantages of these galvanized iron Rick Covers over the ordinary thatched covers will be at once apparent,—they are less expensive, and the hay cuts out in much finer condition, the covers being impervious to rain and damp, and any intelligent labourer can fix them in a very short time.

Prices for any particular sizes free on application.

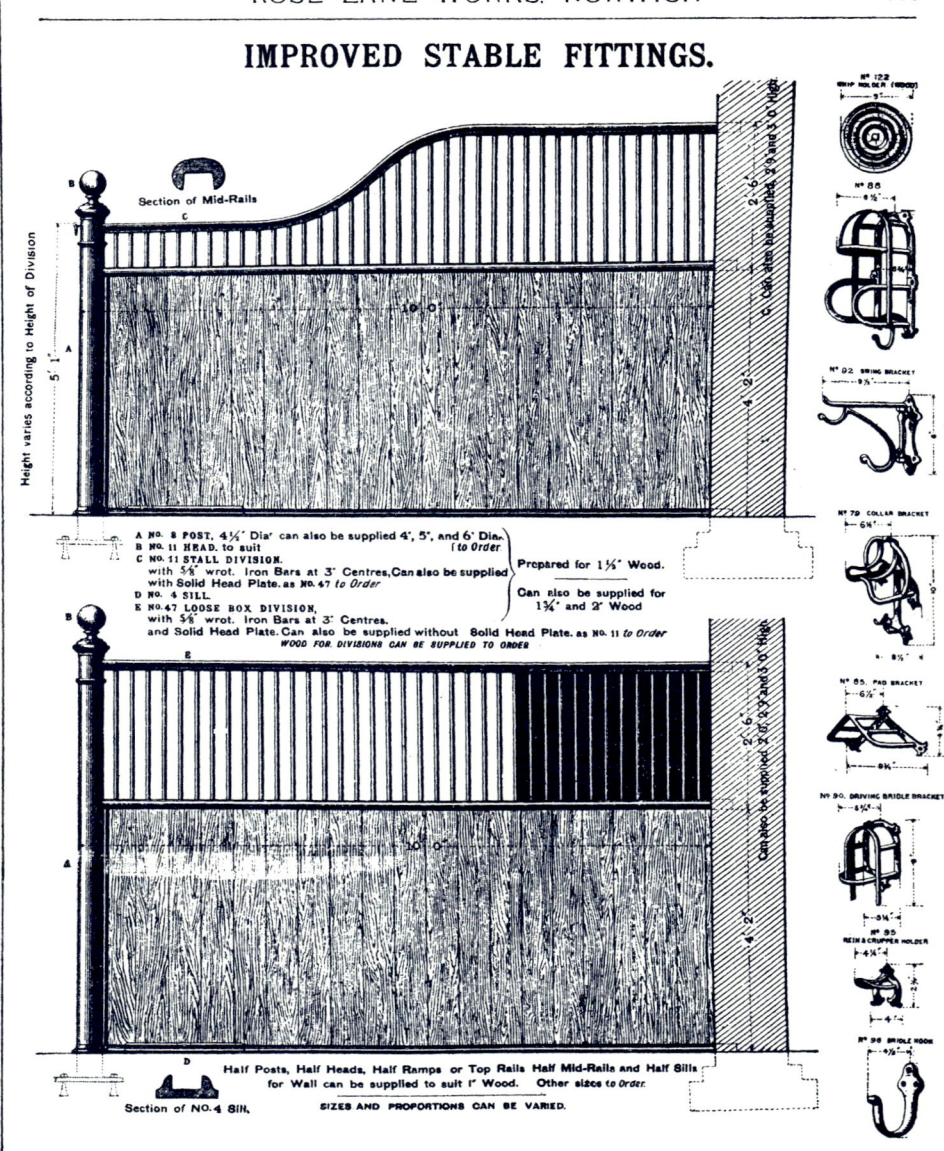

WOOD LOOSE BOX DIVISIONS.

As supplied in our Stables.

Red deal chamfered posts with turned tops and cast-iron bases, **25/-** each.
Red deal grooved and chamfered rails, filled in at bottom with match-boarding, and above with wrought-iron bars, **5/-** per foot run.
Red deal framed doors with strong hinges and bolt, **30/-** each.
Woodwork stained two tints.
Ironwork painted black.

WOOD STALL DIVISIONS.

As supplied in our Stables.

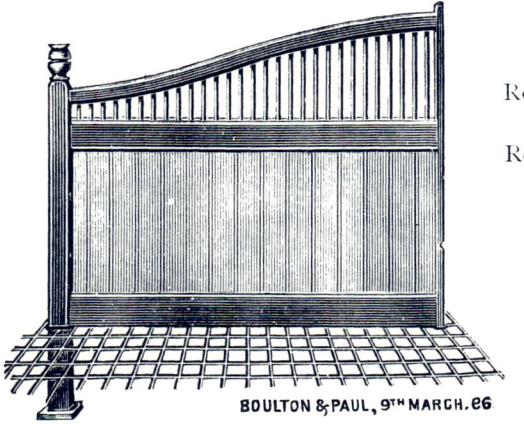

Red deal chamfered post with turned top and cast-iron base.

Red deal grooved and chamfered bottom and middle rails, ramped top rail, bottom filled in with match-boarding, and above with wrought-iron bars. **Price £3.**

Woodwork stained two tints.

Ironwork painted black.

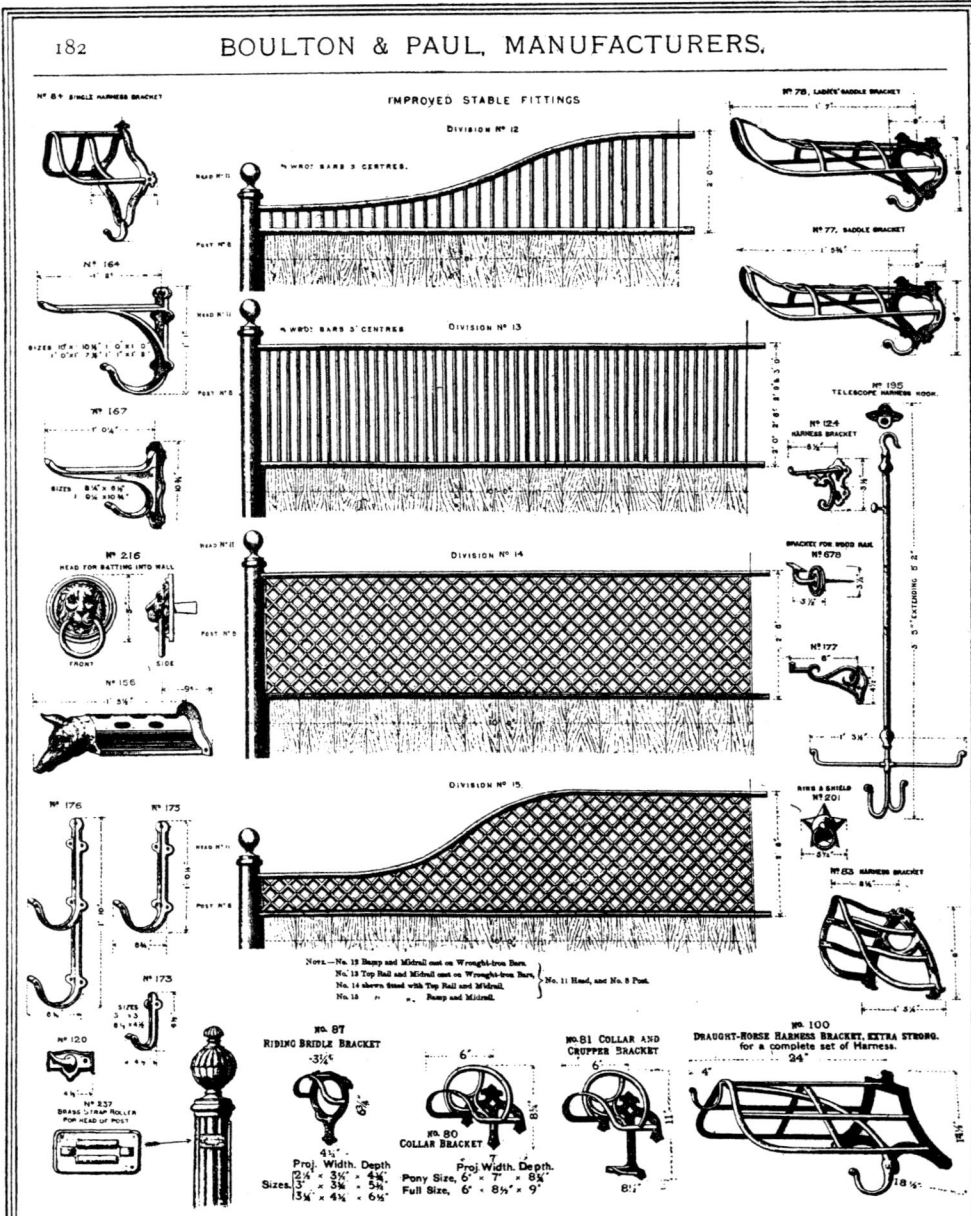

ROSE LANE WORKS, NORWICH

IMPROVED STABLE FITTINGS.
STALL AND LOOSE BOX MANGER SETS to Match.

Consisting of Wrought Iron RAISED HAY RACK,(Right or Left Hand) with Seed Pan and Grating, extra large MANGER TROUGH (opening), 24¼"×11¼"×9" deep, with Food Guards at ends, and large Safety Front, Skirting Prepared for Tiles; can also be had without Skirting to Order.
FASTENING SHOES, Charged Extra

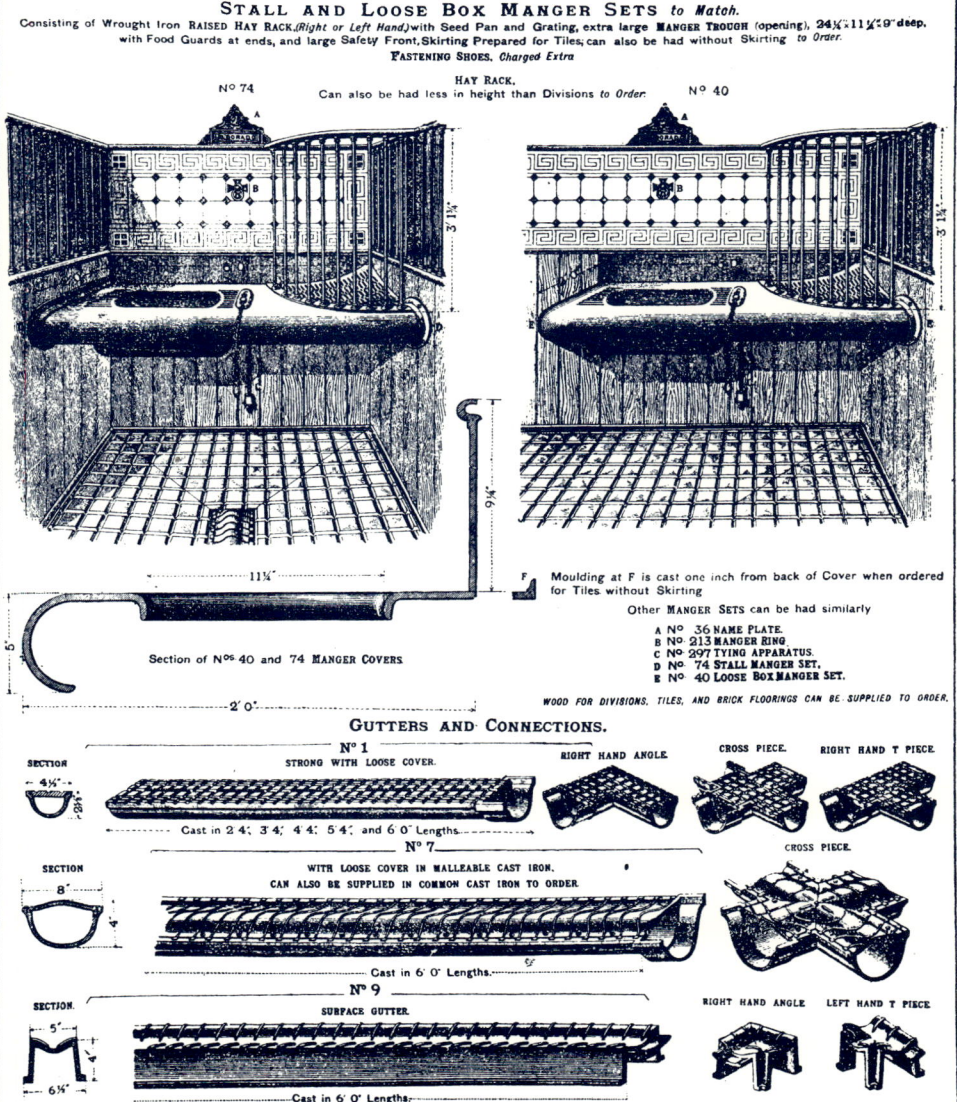

HAY RACK.
Can also be had less in height than Divisions to Order.

Moulding at F is cast one inch from back of Cover when ordered for Tiles without Skirting

Other MANGER SETS can be had similarly
A Nº 36 NAME PLATE.
B Nº 213 MANGER RING
C Nº 297 TYING APPARATUS.
D Nº 74 STALL MANGER SET.
E Nº 40 LOOSE BOX MANGER SET.

WOOD FOR DIVISIONS, TILES, AND BRICK FLOORINGS CAN BE SUPPLIED TO ORDER.

GUTTERS AND CONNECTIONS.

Nº 1 — STRONG WITH LOOSE COVER. Cast in 2' 4", 3' 4", 4' 4", 5' 4", and 6' 0" Lengths.

Nº 7 — WITH LOOSE COVER IN MALLEABLE CAST IRON. CAN ALSO BE SUPPLIED IN COMMON CAST IRON TO ORDER. Cast in 6' 0" Lengths.

Nº 9 — SURFACE GUTTER. Cast in 6' 0" Lengths.

For Prices see separate List.

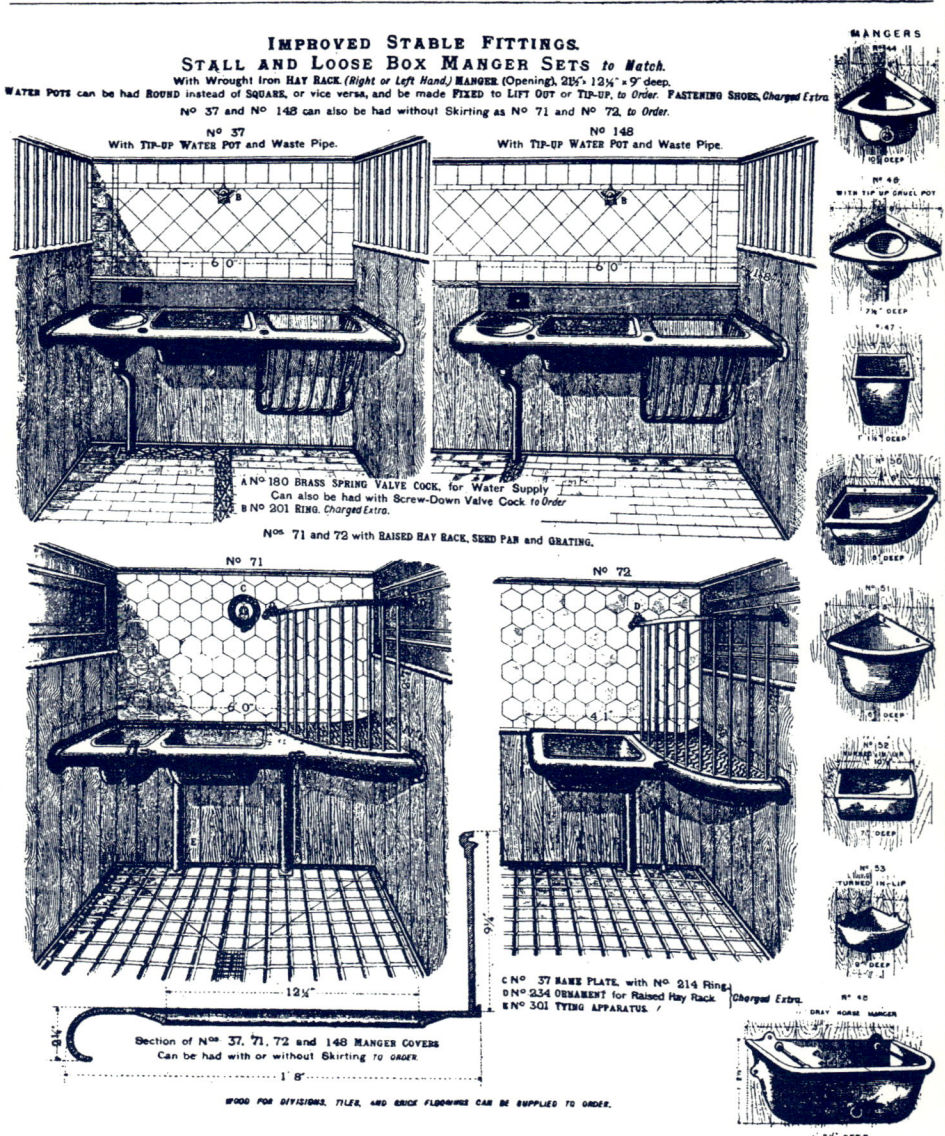

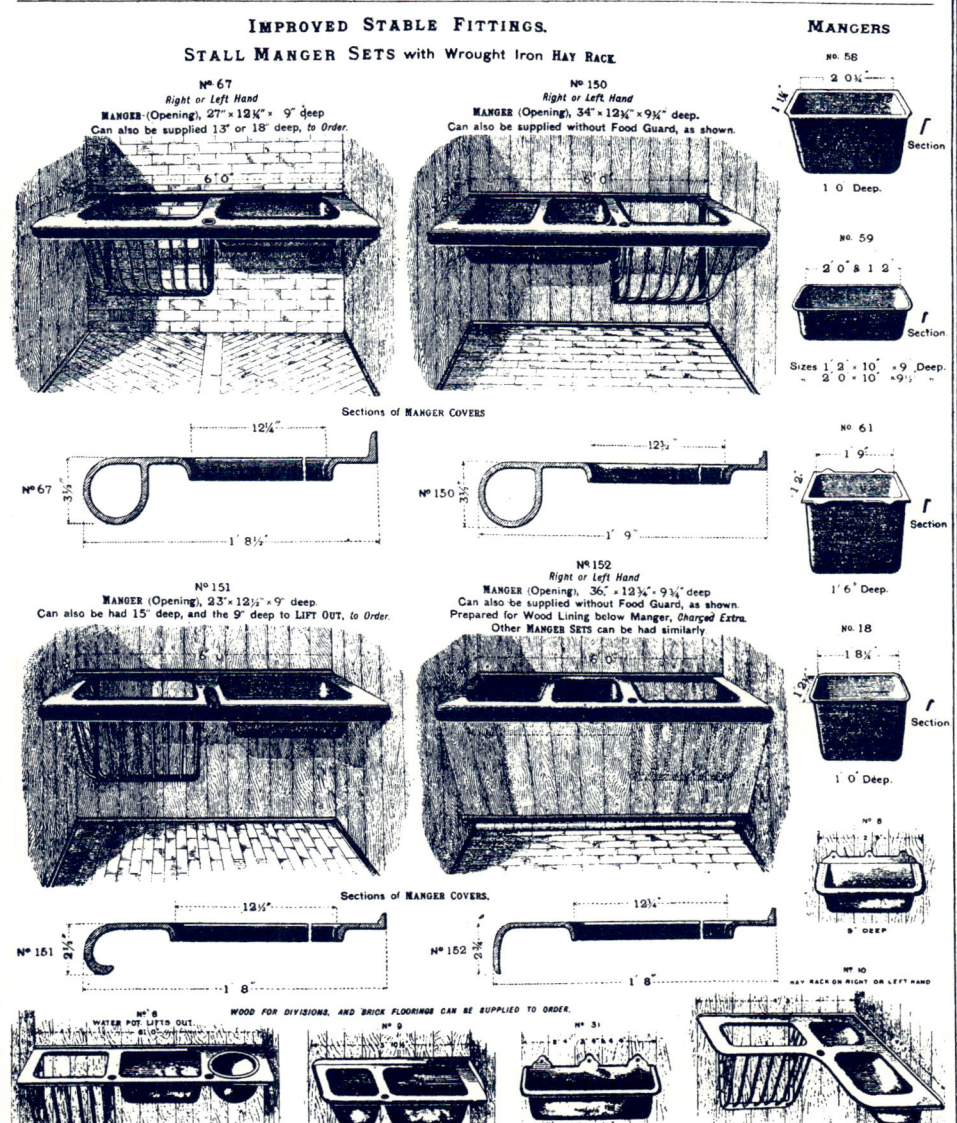

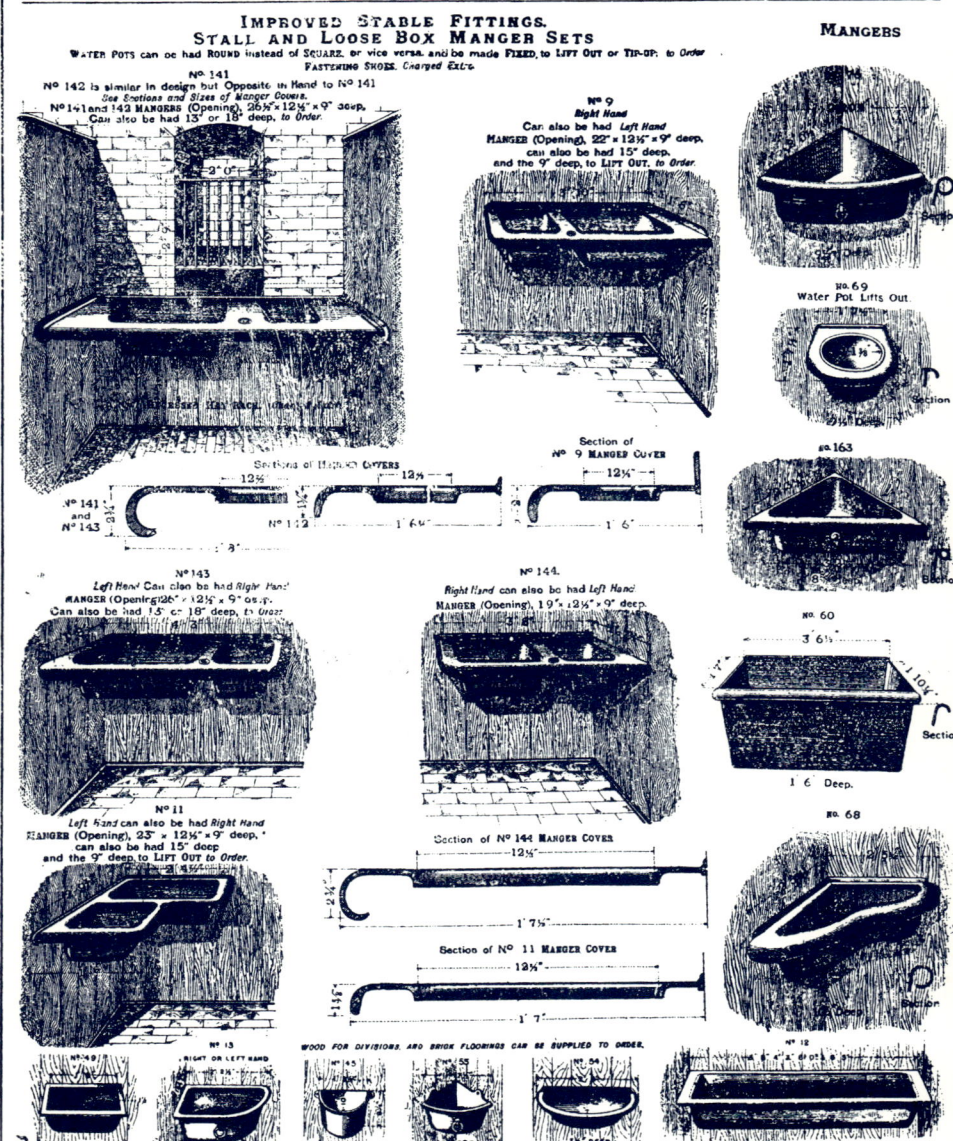

For Prices see separate List.

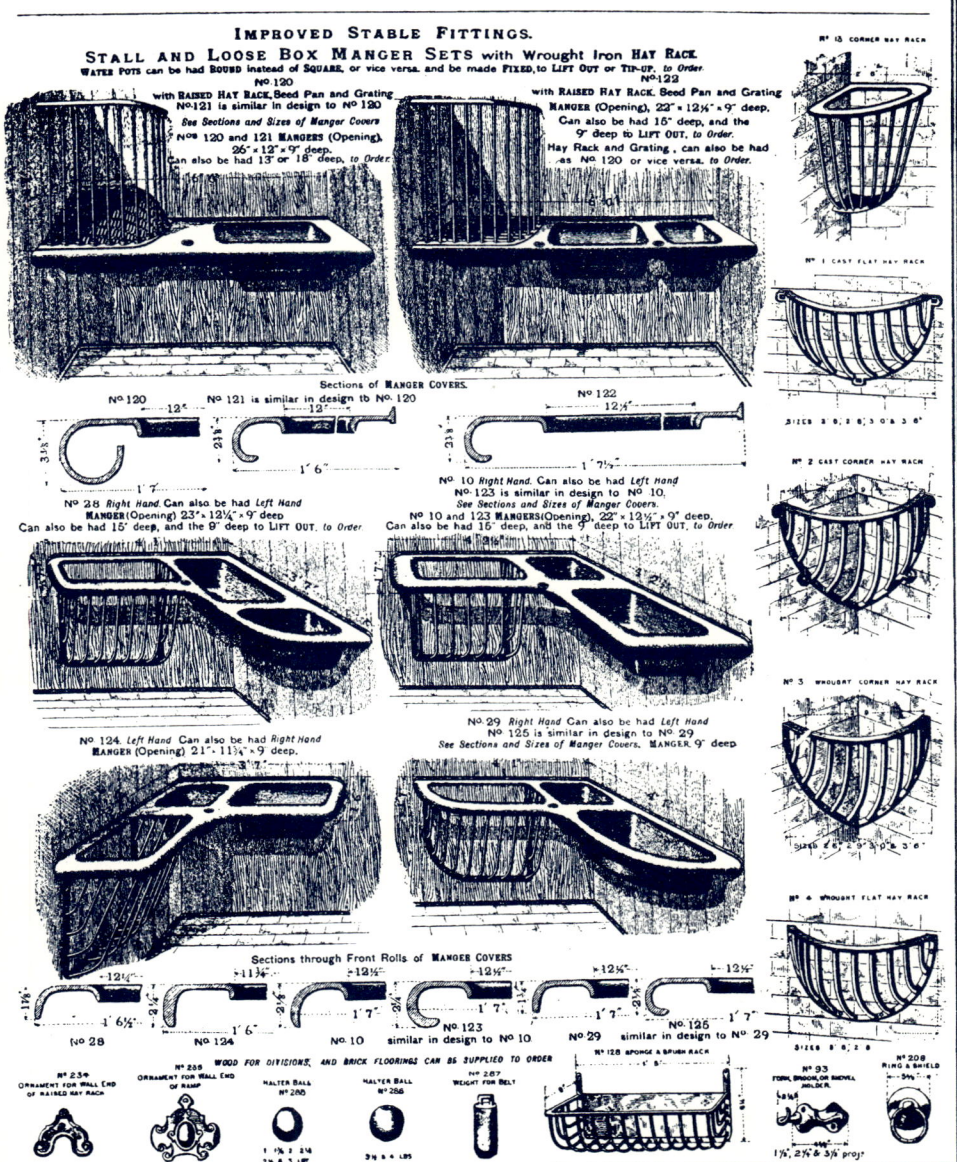

BOULTON & PAUL, MANUFACTURERS.

IMPROVED STABLE FITTINGS.
STALL MANGERS.
FASTENING SHOES, RINGS, FOOD GUARDS, &c., *Charged Extra.*

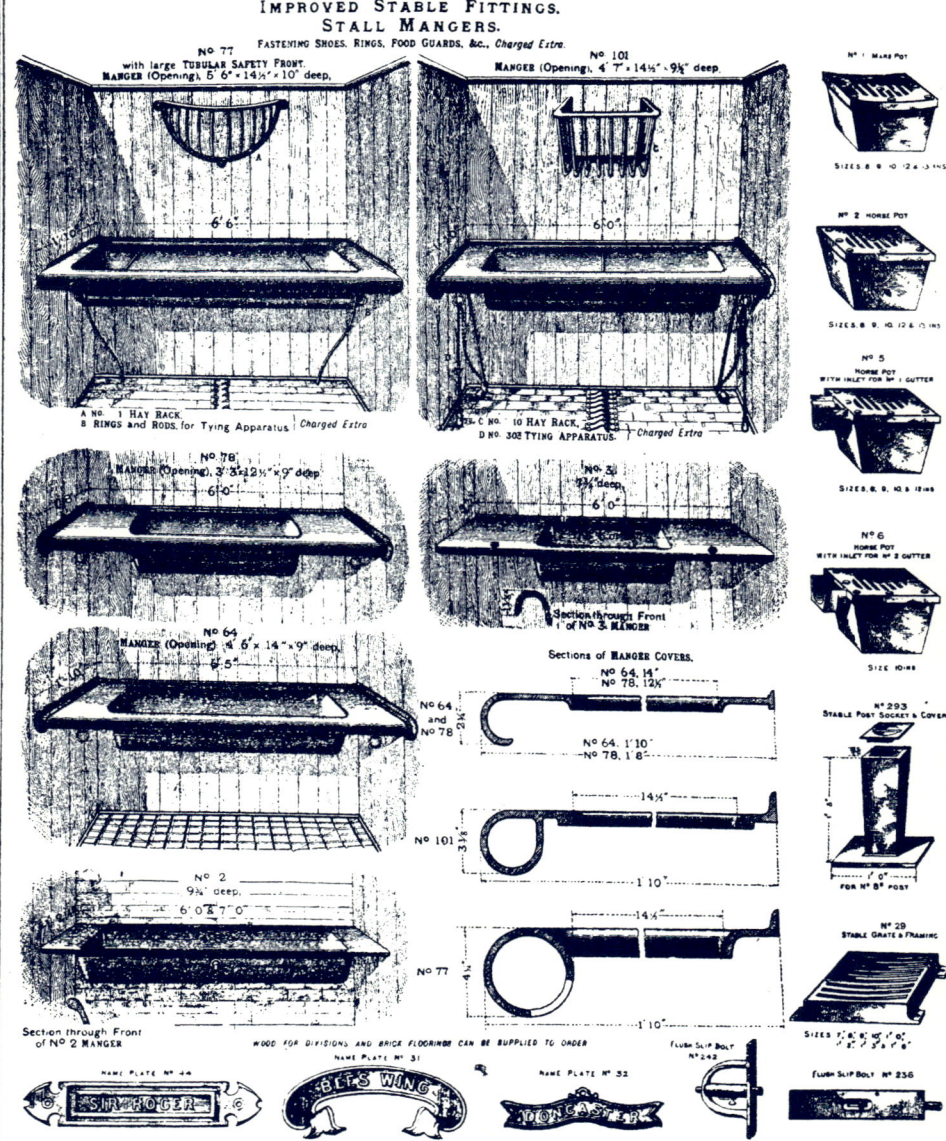

For Prices see separate List.

SECTION
OF
ROSERIES
AND AN
ARRANGEMENT
OF
GARDEN REQUISITES
AND
HORTICULTURAL IMPLEMENTS.

Gentlemen waited upon and Sites surveyed

BOULTON & PAUL, MANUFACTURERS.

No. 10. ORNAMENTAL COVERED WAY

For Ornamental Gardens, Public Parks, &c.

WITH DOMED CORNERS.

As supplied to J. McConnell, Esq., Belfast.

No. 11. GARDEN ARCHES.

No. 12. ROSE BOWER.

No. 13. IRON POSTS AND CHAINS

Special Designs prepared to suit any situation.

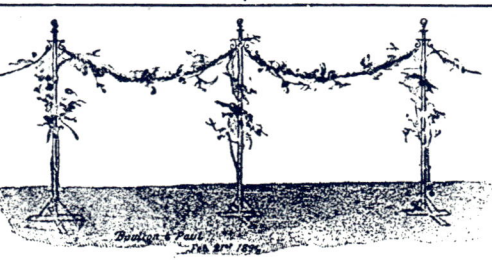

FOR TRAINING CREEPERS.

As supplied to C. W. Dyson Perrins, Esq.

ROSE LANE WORKS, NORWICH. 191

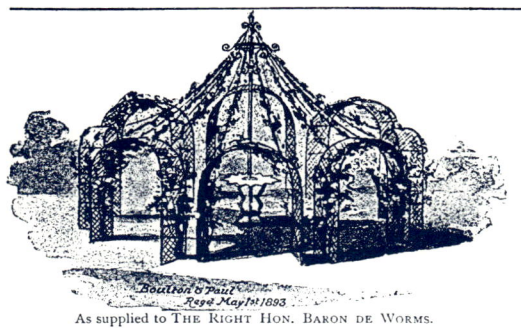

No. 14. ARRANGEMENT OF GARDEN ARCHES.

With Centre Pole, having Wires extending from same to Arches, for training Roses and other climbing plants.

As supplied to THE RIGHT HON. BARON DE WORMS.

TESTIMONIAL.

From G. B. BAMBRIDGE, ESQ., Morpeth.
I have pleasure in stating that the work entrusted to your firm has be n carried out to my entire satisfaction.

No. 15. ROSE BOWER OF THE DOME PATTERN WITH ARCHES.

THIS would make an ornamental feature in centre of garden, and can be made to any size.

THE Bower can also be modified to suit any situation.

Special Estimates on application.

No. 16. DOUBLE ARCHES & CIRCULAR ENCLOSURE.

For Junctions of Paths.

Estimates free on application.

TESTIMONIAL.
From HIS GRACE
THE DUKE OF ABERCORN,
Barons Court.
I am well satisfied with your Wire Work, which is very strong, and I think will be durable.

Special Estimates and Designs given upon receipt of particulars of requirements.

BOULTON & PAUL, MANUFACTURERS.

No. 17.
GARDEN ARCHES,

With Centre Pole, Wires from same to Arches, for training Roses and other climbing plants.

TESTIMONIAL.

From B. W. COOPER, ESQ., Bury.

I enclose cheque for amount of your account, and have great pleasure in expressing my satisfaction at the work you have done.

No. 18.
ROSE BOWER,

Dome Pattern, with Arched Entrances and Side Wings.

CAN be made any size, and to suit any situation.

Estimates on application.

TESTIMONIAL.

From C LEACH, ESQ., Fernhurst.

I wish to say I am perfectly satisfied with the quality, style, and workmanship of the Iron and Wire Work as supplied and erected by you.

No. 19.
OCTAGONAL ROSE BOWER.

As erected for THE RIGHT HON. EARL CADOGAN, K.G., Culford.

No. 20.
FOUR-ARCHED BOWER.

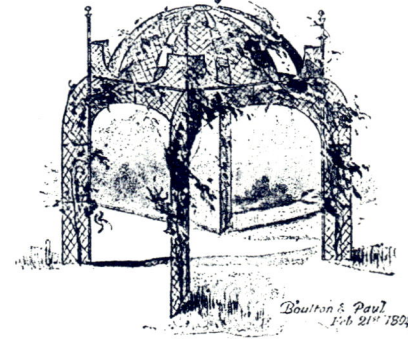

As erected for D. FOX TARRETT, Ellary.

Special Estimates and Designs given upon receipt of particulars of requirements.

ROSE LANE WORKS, NORWICH. 193

ORNAMENTAL IRON ROSE BOWERS.

No. 21. SUITABLE FOR LAWN CENTRES. | **No. 22.** COSY CORNER FOR GARDENS.

No. 23.
ORNAMENTAL COVERED WAY
WITH DOME AND CIRCULAR ENCLOSURE.

Can be modified to suit any situation.

As erected for The Rt. Honourable Lord Battersea, Overstrand.

TESTIMONIAL.

From R. H. M. PRAED, ESQ., Mickleham.

I thoroughly approve of the Fence and style of erecting.

No. 24. ROSE BOWER.
DOME PATTERN FOR CENTRES.

No. 25. CIRCULAR ENCLOSURE
WITH ARCHWAYS FOR JUNCTIONS.

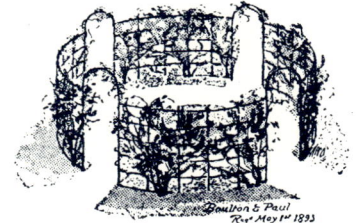

Special Estimates and Designs given upon receipt of particulars of requirements.

BOULTON & PAUL, MANUFACTURERS,

No. 11a. ORNAMENTAL ARCH.

7 ft. high headway, 5 ft. span, 3 ft. deep, with seats as shown, **£5 0 0** each.
Smaller size, without seats, 4 ft. span, 2 ft. deep, **£3 10 0** each.

No. 26. COSY ARBOUR.

6 ft. 6 in. long by 3 ft. wide, covered with 4-in. mesh wire lattice, **£6 0 0**

No. 65a. GARDEN SEAT.

Price, 6 ft. long, **16/-**

No. 27. ROSE SCREEN.

Highly ornamental, suitable for rose and other training plants, made to any size, and to suit any situation.

ORNAMENTAL SPAN-ROOF ENTRANCE PORCH.

MADE with wrought-iron frame, painted green, covered with 4-in mesh galvanized wire lattice, roof covered with sheet iron, having cast-iron cresting on ridge.

Size, 10 ft. high, 5 ft. wide, by 2 ft. deep outside.

Cash Price, Carriage Paid,
£2 19 6

ORNAMENTAL LEAN-TO ENTRANCE PORCH.

MADE with wrought-iron frame, painted green, covered with 4-in. mesh galvanized wire lattice, roof covered with sheet iron.

Size, 10 ft. high at back, 5 ft. wide outside, 2 ft. deep

Cash Price, Carriage Paid,
£2 13 0

BOULTON & PAUL, MANUFACTURERS.

NEW PERMANENT METHOD OF WIRING IN FRUIT GARDENS.

For protecting buds and bloom from frost and birds, seed in the ground, and fruit whilst ripening. By a simple method of straining wires an enclosure of any size can be wired in with bird-proof netting, wire netting on the sides, and string netting on top.

Enquirers should send plan showing measurements and shape of space to be enclosed, with positions of doorways required. *Special Estimates given.*

No. 102. CONTINUOUS WROUGHT-IRON ESPALIER.

This Fencing is complete in itself, requiring no stone blocks; it is readily fixed by a handy labourer, and can be adapted to any line.

Cash Prices.

4 ft. high with 5 bars, standards 3 ft. apart ... per yard 2/6
5 ft.   ,,   7   ,,   ,, ...   ,,   3/-
6 ft.   ,,   8   ,,   ,, ...   ,,   4/-

Cast-iron Terminal Pillars extra.

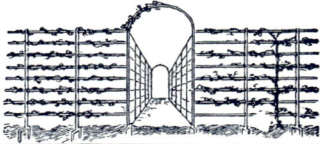

No. 103. CONTINUOUS COVERED-WAY ESPALIER.

For training fruit trees or climbing plants, 8 ft. high, 5 ft. wide. Painted green or any colour.

Price 9/6 per yard run.

Made in 6 ft. lengths, with plate feet, jointed in the centre and fitted together with small bolts and nuts.

Estimates given for any work of this kind upon receipt of specification, giving length, width, and height of arch.

TESTIMONIAL.   Coorheen House, Loughrea.

Dowager Lady Clancarty writes to thank Messrs. Boulton and Paul, and to say the wire and screws arrived all right, and the Espalier is up and looks well.

GARDEN ARCHES.

No. 1.
Covered with Galvanized Wire Netting.
7 ft. high,
4 ft. span,
2 ft. wide.
Price 7/6 painted frames.

No. 3.
All wrought-iron.
7 ft. high,
4 ft. span,
1 ft. 6 in. wide.
Price 9/6 painted.

No. 4.
All wrought-iron.
7 ft. high,
4 ft. span,
1 ft. 6 in. wide.
Price 12/- painted.

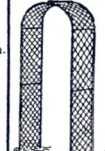

No. 5.
7 ft. high,
4 ft. span,
1 ft. wide.
Price.
Painted ... 5/- ea.
Galvanized 6/6 ,,

Carriage Paid on all Orders above 40/- value to the principal Railway Stations in England and Wales.

ORNAMENTAL ENTRANCE ARCH,
For Covered Way Espalier.

No. 104. COVERED WAY ESPALIER.

Cash Price, for Front Arch only, £3 0 0 each.

5 ft. span, 8 ft. high, constructed with flat iron arch bars, and No. 13 gauge wires spaced about 9 in. apart, having T iron straining bar at top.

Price 3/- per yard.

Terminal Straining Arches fitted with raidisseurs, **35/-** each.

No. 9. Standards & Arch Bars,
For Training Creepers.

6 ft. high by 6 ft. wide.
Standards ... **3/-** each.
Arch Bars **2/6** ,,
Painted green.

No. 7.

No. 6.

GARDEN ARCHES.

No. 6.
All wrought-iron with strong frame.
8 ft. high by 6 ft. span, 2 ft. wide.
Painted, **17/6** each.

No 7.
Highly ornamental. Strong wrought-iron frame, covered with wire lattice.
8 ft. high, 4 ft. 6 in. span, 2 ft. wide.
21/6 each.

No. 100. IMPROVED ESPALIER FENCING AND WALL TRAINERS.
FOR TRAINING FRUIT TREES ON THE FRENCH SYSTEM.

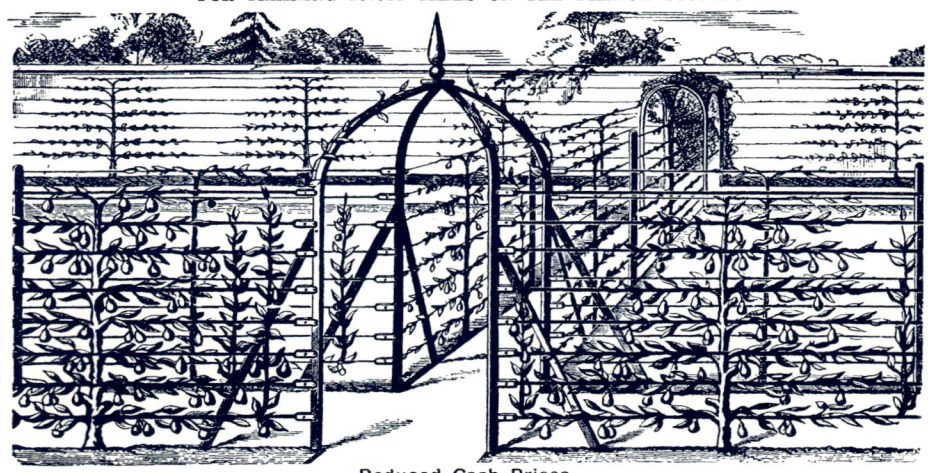

Reduced Cash Prices.

For Pears, Apples, etc. 4 ft. high, with standards 10 ft. apart, and 6 lines of galvanized wire	per yard	6d.
Terminal Posts and Raidisseurs	each	8/-
For Pears, Apples, etc. 5 ft. high, with standards 10 ft. apart, and 8 lines of galvanized wire	per yard	7d.
Terminal Posts and Raidisseurs	each	8/9
For Pears, Apples, etc. 6 ft. high, with standards 10 ft. apart, and 9 lines of galvanized wire	per yard	8d.
Terminal Posts and Raidisseurs	each	11/-
For Raspberries, Gooseberries, or Currants, with standards 4 ft. high, and 4 lines of wire ...	per yard	5d.
Terminal Posts and Raidisseurs	each	8/-

MATERIALS REQUIRED.

No. 104. T IRON TERMINAL OR ANGLE POSTS.

With self-fixing bases. Two required for each length.

Cash Prices.

	Painted.	Galvd.
4 ft. high, each	7/-	11/6
5 ft. ,, ,,	8/6	14/-
6 ft. ,, ,,	10/-	15/-

No. 105. INTERMEDIATE STANDARDS.

With anchor feet. Generally placed 10 ft. apart.

Cash Prices.

	Painted.	Galvd.
4 ft. high, each	1/1	1/9
5 ft. ,, ,,	1/3	2/-
6 ft. ,, ,,	1/6	2/6

No. 106. DOUBLE STRAINING POST FOR CORNERS.

Cash Prices.

	Painted.	Galvd.
4 ft. high, each	10/6	14/3
5 ft. ,, ,,	12/-	16/6
6 ft. ,, ,,	12/9	18/3

GALVANIZED RAIDISSEURS,

For Straining the Wires. One required for each Wire.
Price 3/- per doz.
Stronger pattern per doz. 5/-
Ex. strong for heavy wire per doz. 8/6
Wrought-iron Key for ditto ... 9d.

BEST GALVANIZED WIRE.

No. 13. Suitable for the higher kind of Trainer ... per 100 yards 1/6

No. 14. Suitable for the lower kind of Trainer ... per 100 yards 1/3

Estimates given for any quantity on receipt of particulars.

Carriage Paid on all Orders above 40/- value to the principal Railway Stations in England and Wales.

WIRING GARDEN WALLS.

THE arrangement is so simple, that it can be applied to any walls by inexperienced hands, and in very much less time than the old system.   **All the Fittings are Galvanized.**

Glass Lights and Brackets can be supplied as illustration.   Estimates on application.

MATERIALS REQUIRED.

GALVANIZED WROUGHT-IRON EYES.
For Guiding the Wires on the Wall.
Spaced about 10 ft. apart.

No. 110. 4 in. long . 5d per doz.

No. 112. 2 in. long . 5d. per doz.
3¼ in. ,, ... 9d. ,,

No. 108. GALVANIZED RAIDISSEURS.
For Straining the Wires.

One required for each Wire.
Price 3/- per doz., used for wire up to No. 13 gauge.
Larger size, 5/- per doz.   ,,   ,, No. 10 ,,
Extra large, 8/6 per doz.   ,,   ,, No. 6 & 8 ,,
Keys for ditto, 9d each.

No. 109. GALV. TERMINAL HOLDFASTS

Two required for each Wire.   Price 2/- per doz.

GALV. WROT.-IRON TRAINING HOOKS & EYES

No. 1.   4 in. long . 1/- per doz.
6 in. ,, .. 1/3 ,,
9 in. ,, .. 1/6 ,,
12 in. ,, ... 2/- ,,

No. 111.
STRAINING BOLT AND HOLDFAST.
All Galvanized.

THIS Strainer is preferred by some as neater than the French Raidisseur.

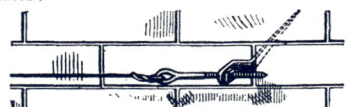

Price, including Holdfast, 4/. per doz.
One required for each wire
Keys for turning the nuts, 4d. each.

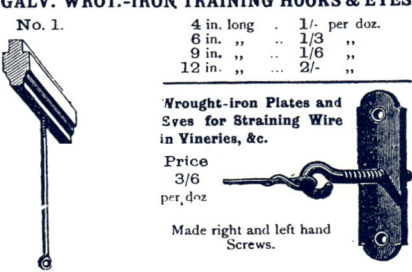

Wrought-iron Plates and Eyes for Straining Wire in Vineries, &c.
Price 3/6 per doz
Made right and left hand Screws.

BEST QUALITY GALVANIZED WIRE.

No. 11 .. 2/6 per 100 yards
No. 12 .. 2/3 ,, ,,
*No. 13 .. 1/6 ,, ,,
*No. 14 ... 1/3 ,, ,,

* *These sizes are recommended as the best for Walls.*

GALVANIZED RODS.
Galvanized Rods. ¼ in., 1½d. per foot.

Carriage Paid on all Orders above **40/-** value to the principal Railway Stations in England and Wales.

ROSE LANE WORKS. NORWICH.

FOSTER'S FRUIT, SEED, AND PLANT PROTECTOR.

Used and recommended by MAJOR FOSTER, Dene Court, Bishop's Lydeard, Somerset.

Method of Fixing.—Lay the strips in position, then drive pickets firmly into the ground at the junction of the strips; adjust the strips on the heads of the pickets, and secure them in position by lightly tapping in the nails. Lay the net, allowing plenty of slack, along the centre of the border, resting on the cross strips. Draw net out on either side, straining it over the projecting nails, and allowing the curtain to rest upon the ground. Should the border be much wider than 7 ft., a double line of framing should be used, the nets to cover same being lightly laced together with a netting needle and twine. To protect narrow borders, the cross strips may be laid diagonally, with a lath of the required length fixed at each end of the line of framing.

Gathering Fruit.—In picking strawberries, carefully lift the net off the nails, and lay it in its original position before being spread. In gathering gooseberries or currants the curtains of the net should be raised and deposited along the strips. For raspberry picking, one curtain only need be lifted, there being sufficient head-room to permit of the fruit being picked under the net. Seed beds require the same arrangements as strawberry borders.

Various Applications.—Lawns, tennis grounds, and cricket pitches, either when fresh sown, or when renovating mixture has been applied, are insured absolute immunity from birds by the use of the Protector. Moreover, the net, by affording protection from spring frosts, enables the grass seeds to be sown a full month earlier than usual, thus assuring by the summer a fine and superior turf.

Onion, cabbage, carrot, parsnip, etc., when maturing seed, are most effectually secured by the framing and net recommended for strawberries.

Cash Prices.

Square deal pickets, with hole in top to take point of nail, with necessary nails.

For Strawberry Borders, Seed Beds, and Plants, 2 ft. long ... ... per doz.	1/-
For Gooseberries and Currants, 4 ft. long .. .. ,,	1/6
For Raspberries, 6 ft. long ... ... ,,	2/-
Deal strips 7 ft. long, 2 in. wide, ½ in. thick, pierced at each end ,,	1/8

Net—the ordinary Herring Net, 4 yards wide.

We do not supply Second-hand Netting, but can quote Special Prices for New Garden Netting of the best make.

From MAJOR FOSTER, Dene Court, Bishop's Lydeard.

"I have used the Frames in question in the gardens here for many years for securing strawberries and bush fruits from birds, and protecting seed beds, etc. They are most effective, handy, and useful. The netting lasts five times as long as when used in the ordinary way."

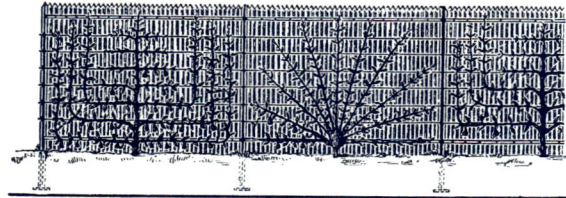

CORRUGATED IRON WALL FOR TRAINING FRUIT TREES.

As used by MR. COTTERELL, Derry Ormond Estate.

THIS Fence will be found very useful for dividing or enclosing gardens, and will be found much less expensive than a brick wall. Fruit Trees may be trained on both sides.

Prices quoted on application.

STAND FOR STORING FRUIT

FRAME made of iron with half-round spline trays.

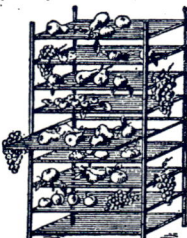

Wire Holders for Pickle Bottles for Preserving Grapes fresh during the whole year.

Price 3/- per doz.

Size 2 ft. 2 in. wide, 1 ft. 6 in. deep, 4 ft. 8 in. high.
Price ... 16/6
Any size made to order.

GALVANIZED WIRE STRAWBERRY PROTECTORS.

No. 1 Pattern.

15 in. diameter, 5/- per doz., or 40/- per 100.

No. 2 Pattern.

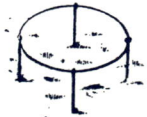

7 in. diameter, 1/- per doz., or 8/- per 100.

Carriage Paid on all orders above 40/- value to the principal Railway Stations in England and Wales.

ROSE LANE WORKS, NORWICH. 198A

WIRE LATTICE HURDLE
For Training Plants.

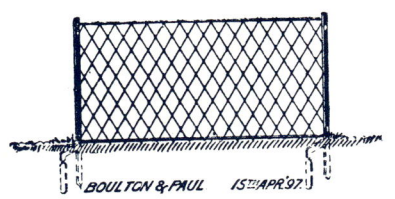

Made with wrought frames covered with 6 in. mesh wire lattice. Size, 6 ft. by 3 ft.
Price **4/6** each.

SUPERIOR SQUARE WOOD TRELLIS.

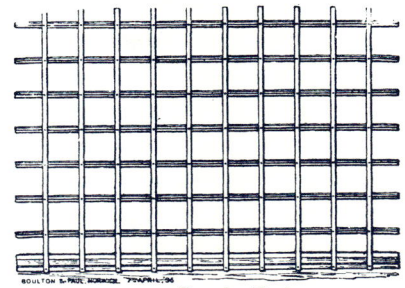

Cash Prices.
Painted two coats.
6 in mesh, **6d.** per ft. super.
9 in. mesh, **5d.** ,,
12 in. mesh, **4d.** ,,
With top and bottom Rails.
Posts for fixing into ground, 3 in. by 4 in., **3d.** per ft run

GALVANIZED SANITARY BASKETS

Made of 1 in. mesh, 13 gauge, corrugated lattice.

Cash Prices.
No. 9. Semicircular - **5/-**
No. 10. Rectangular - **5/6**

No. 9.

No. 10.
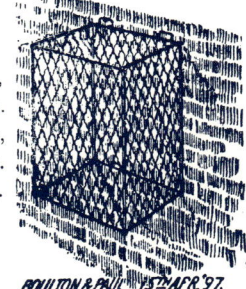

Size, 18 in. wide, 28 in. high.

NEW WROUGHT-IRON PORTABLE FLOWER STANDS.

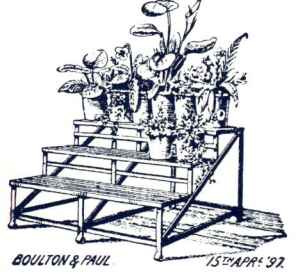

No. 25. Flat Stage.
4 ft. long, 3 ft. high, **33/-**

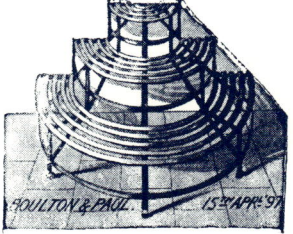

No. 26. Semicircle.
4 ft. wide, 3 ft. high, **38/-**

No. 27. Quarter Circle.
2 ft 6 in. wide, 3 ft. high, **28/6**

BOULTON & PAUL, MANUFACTURERS.

DIAMOND WOOD TRELLIS FOR SCREENS, FENCING, TRAINING PLANTS.

CAN be made in sizes to suit any purpose, either for fixing on walls, or for fastening to posts to train creepers to form light fences. The splines are 1 in. by ¼ in., unpainted. Stock sizes, when open, 4 ft. by 12 ft., and 6 ft. by 12 ft.

Price 1½d per square foot.

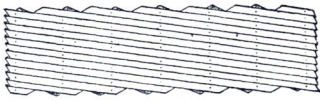

CLOSED FOR PACKING.

Special Estimates given for quantities.

DIAMOND WIRE TRELLIS FOR TRAINING PLANTS.

No. 1.

Cash Prices.

Stock sizes—5-in. Mesh, Light Quality.

| 6 ft. by 3 ft. | each 2/6 | 6 ft. by 5 ft. | each 4/6 |
| 6 ft. by 4 ft. | „ 3/6 | 6 ft. by 6 ft. | „ 5/6 |

Made any other size to order at the following prices per square foot.

Mesh.	ADAPTED FOR	Light Quality. Gauge.		Medium Quality. Gauge.		Strong Quality. Gauge.	
1½ in.		15	6d.	14	7d.	13	7½d.
2 in.	Protecting Windows.	14	4½d.	13	5½d.	12	6d.
4 in.	Plants in Conservatories, on House Walls, etc.	12	3d.	11	4d.	10	4½d.
5 in.		12	3d.	11	4d.	10	4½d.
6 in.		12	2½d.	11	3½d.	9	4d.

Prices of other Meshes on application.

STRAIGHT WIRE LATTICE FOR PROTECTING WINDOWS AND SKYLIGHTS.

No 2.

Cash Prices.

No. 2. Straight Wire Lattice per square foot.

Mesh.	ADAPTED FOR	Light Quality. Gauge.		Medium Quality. Gauge.		Strong Quality. Gauge.	
⅜ in.	Protecting Church and other Windows, Skylights, Aviaries, etc.	15	5d.	14	6d.	13	7d.
½ in.		15	4½d.	14	5d.	13	6d.
1 in.		14	4d.	13	4½d.	12	5d.
1½ in.	Training Plants on Walls, in Conservatories, etc.	13	3½d.	12	4d.	11	5d.

Prices of other Meshes on application.

Carriage Paid on all Orders above 40/- value to the principal Railway Stations in England and Wales.

IMPROVED PEA TRAINERS TO BE USED INSTEAD OF PEASTICKS

No. 1. With straight feet. Hurdles 4 ft. long.

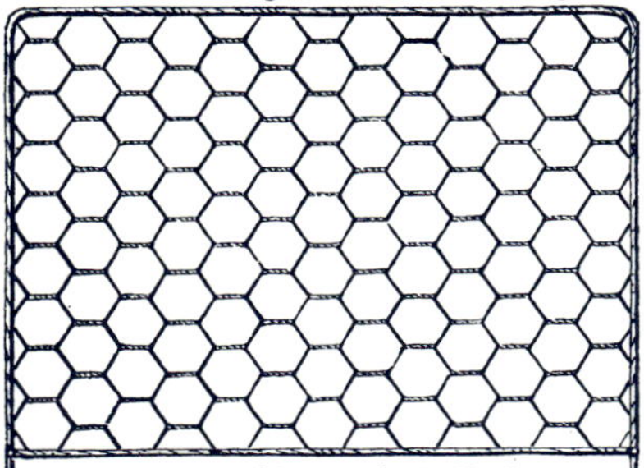

THE advantages of these are, that the Hurdles are complete in themselves, and are readily put down. Can be used for single or double rows. They can also be used for Poultry enclosures, or for any other purpose when the peas are done with.

No. 1. Cash Prices, per Hurdle.

3 ft. high, 1/3; 4 ft. high, 1/6; 5 ft. high, 2/6; 6 ft. high, 3/-

ORNAMENTAL RASPBERRY OR FRUIT BUSH TRAINER.

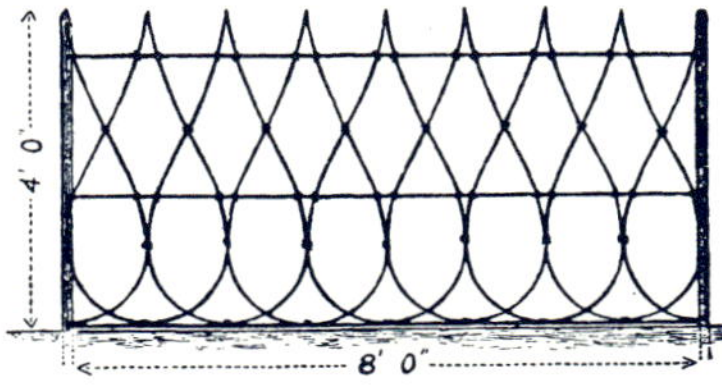

THESE Ornamental Hurdles are specially suitable for training Raspberry or any kind of Fruit Bushes. The rods can be trained perfectly straight or in an ornamental way according to taste, the design of the Hurdle giving a neat appearance in winter They are usually made in 8 ft. lengths, 4 ft. high.

Cash Price, 6/6 per Hurdle.

GALVANIZED PEA AND SEED GUARD.

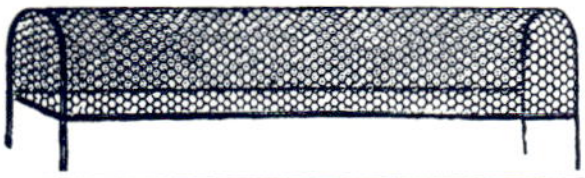

3 ft. long, 6 in. wide, two end pieces to the dozen lengths. Made of ¾-in diamond mesh netting, and galvanized. Cash Price, 4/- per doz.

Mouse-proof Guards, made of ⅜-in. mesh netting. Cash Price, 7/9 per doz.

From CHARLES W SWARBECK, Esq., Thirsk.
The Pea Guards have given entire satisfaction.

STRAWBERRY OR PLANT PROTECTOR.

For protecting Strawberries and Seed Beds from birds, also for protecting Lettuces in winter.

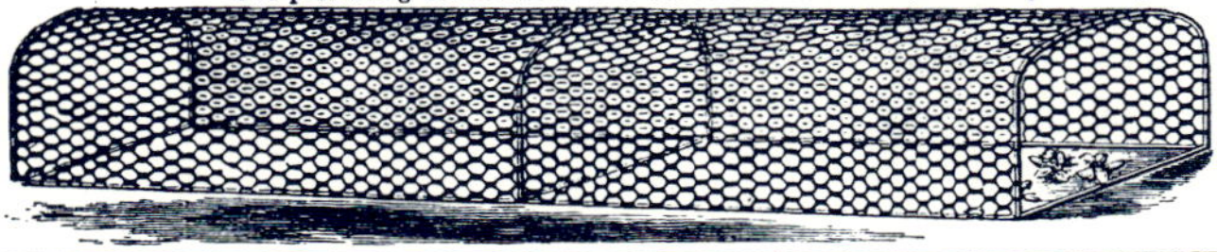

MADE of diamond mesh netting, with elliptic top, galvanized, made of ¾-in. netting.

Cash Prices.

Long.	Wide.	High.	Each.
6 ft.	18 in.	12 in.	1/6
18 in.	18 in.	12 in.	8d.

Ends, 3d. each.

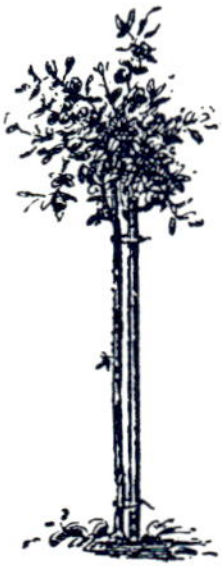

No. 1. No. 3.

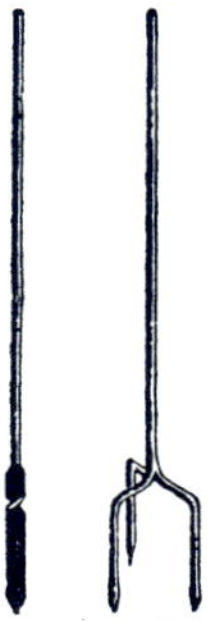

WROUGHT-IRON GARDEN STAKES.

USED for supporting Dahlias, Rose Trees, &c. Much superior to wood, both in appearance and durability.

Cash Prices per dozen.

Height above ground.	Diameter.	GARDEN STAKES. No. 1. Painted.		No. 1. Galvanized.		No. 3. Painted.		No. 3. Galvanized.	
		s.	d.	s.	d.	s.	d.	s.	d.
2½ ft.	$\frac{6}{16}$-in.	3	6	5	0	—		—	
3 ft.	⅜-in.	4	2	5	6	—		—	
3½ ft.	⅜-in.	4	6	6	3	All ⅝-in.		diameter.	
4 ft.	½-in.	5	3	7	3	15	6	23	0
4½ ft.	½-in.	6	6	9	6	16	3	24	6
5 ft.	½-in.	7	3	11	0	17	3	26	0
5½ ft.	⅝-in.	11	0	17	0	18	0	28	0
6 ft.	⅝-in.	12	0	18	9	19	0	29	6

Carriage Paid on all Orders above 40/- value to the principal Railway Stations in England and Wales.

BOULTON & PAUL, MANUFACTURERS.

IMPROVED GARDEN ROLLERS.

For Tennis Courts, Lawns, Drives, and Gravel Paths, Fitted with Counterbalanced Weights.

These Rollers are superior to most in the Market, being **heavy** and **well finished**. The cylinders are bored and faced, and spindles turned in the lathe, which adds considerably to the value of the Double Cylinder Rollers, friction and noise being thus reduced to a minimum.

DOUBLE CYLINDER ROLLER WITH BALANCE HANDLE

The Cylinders are cast in two parts, and have rounded edges, which admit of the Rollers being turned sharply with the greatest ease, and without injury to the surface of the ground. Painted green and black, with varnished wood handles.

Reduced Cash Prices.

Size			Approximate Weight—cwt. qrs. lbs.			Price		
18 in. long by 16 in. diameter			...	2	0	0	...	£1 18 6
20 in. ,, 18 in. ,,			...	2	1	0	...	2 3 6
22 in. ,, 20 in. ,,			...	3	0	0	...	2 11 6
24 in. ,, 22 in. ,,			...	3	2	0	...	3 2 6
26 in. ,, 24 in. ,,			...	4	2	0	...	3 10 6
28 in. ,, 26 in. ,,			...	5	1	6	...	4 12 6
30 in. ,, 26 in. ,,			...	5	2	0	...	4 16 9

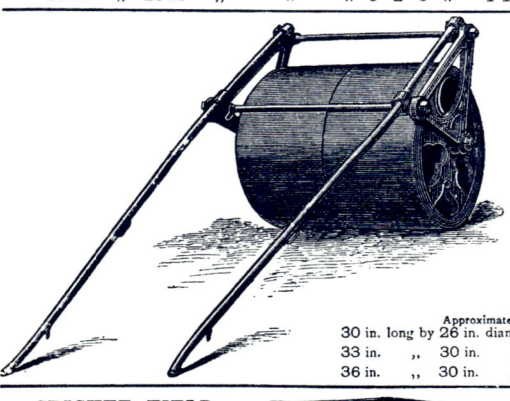

IMPROVED ROLLER WITH BALANCE SHAFTS.

For Parks, Drives, Cricket Fields, or Bowling Greens.

Constructed as the Hand Rollers. Self-acting Scrapers can be fitted to these Rollers at a slight extra charge.

Cash Prices

	Approximate Weight—cwt. qrs. lbs.					
30 in. long by 26 in. diameter ...	5	2	16	...	£5 15 0	for **Donkey**
33 in. ,, 30 in. ,,	..	6	1	0	..	6 10 0 for **Pony**
36 in. ,, 30 in. ,,	...	6	3	8	...	7 10 0 for **Cob**

CRICKET FIELD ROLLER.

In Two Parts.

Cash Prices.

Diam.	Length.	Approximate Weight. cwt. qrs. lbs.				Price		
30 in.	by 32 in.	,.	9	2	0	..	£8 11	0
36 in.	,, 42 in.	...	12	2	0	...	12 12	0
36 in.	,, 48 in.	...	16	1	0	...	14 8	0
42 in.	,, 48 in.	...	20	3	0	..	18 0	0

Carriage Paid on all Orders above **40/-** *value to the principal Railway Stations in England and Wales.*

ROSE LANE WORKS, NORWICH. 200A

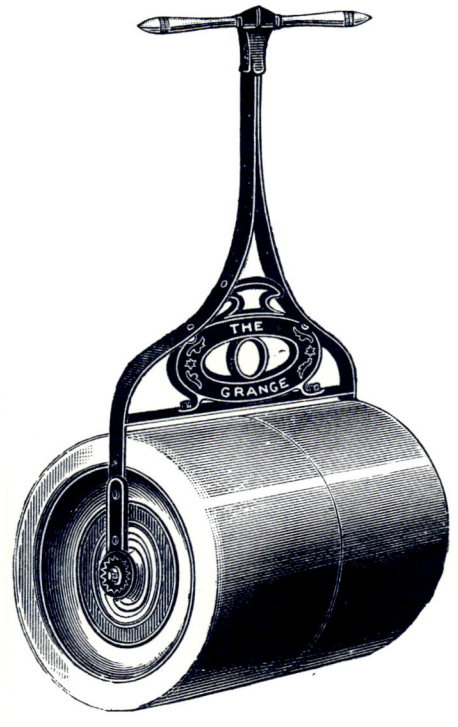

No. 4.
THE WATER BALLAST DOUBLE CYLINDER GARDEN ROLLER.

FITTED with balance handle, painted, and wood cross bar varnished. The joint at centre is turned and faced, the journals bored. The end and inside plate, enclosing water chamber, is cast in a piece with the cylinder, thus preventing all possibility of leakage by doing away with a fitted joint.

Being well fitted they are **very easy** to work, and having rounded ends no marks are made on walks or grass.

Cash Prices, Carriage Paid.

Wide.	Diam.	About Weight empty. cwt. qrs. lbs.			About Weight full. cwt. qrs. lbs.					
21 in. by 21 in.	...	4	3	0	...	5	3	0	... £4	8 6
24 in. ,, 24 in.	...	5	3	18	...	7	2	22	... 5	5 0
27 in. ,, 27 in.	...	7	2	20	...	10	1	2	... 7	10 0
30 in. ,, 30 in.	...	9	1	10	...	13	1	12	... 9	0 0

A Key and Water Filler are supplied with each Roller.

ROSE LANE WORKS, NORWICH.

PATENT IMPROVED LAWN MOWERS FOR 1898.

PATENT "SILENS MESSOR" LAWN MOWERS FOR HAND-POWER

Cash Prices, Carriage Paid, including Grass Box.

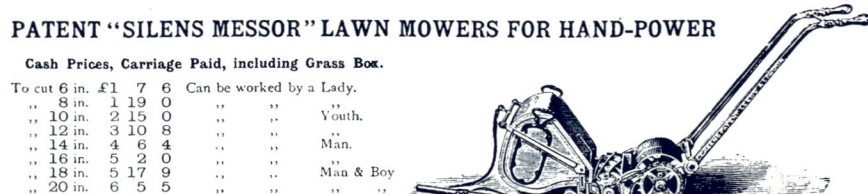

Green's Patent.

To cut 6 in.	£1	7	6	Can be worked by a Lady.		
,, 8 in.	1	19	0	,,	,,	,,
,, 10 in.	2	15	0	,,	,,	Youth.
,, 12 in.	3	10	8	,,	,,	,,
,, 14 in.	4	6	4	,,	,,	Man.
,, 16 in.	5	2	0	,,	,,	,,
,, 18 in.	5	17	9	,,	,,	Man & Boy
,, 20 in.	6	5	5	,,	,,	,, ,,
,, 22 in.	6	13	1	,,	,,	,, ,,
,, 24 in.	7	1	4	,,	,,	,, ,,

The 20, 22, and 24-in. can be fitted with stronger framework and whippletree, suitable to be worked by a Donkey, at 30/- extra. Packing Cases are charged at the following low rates, viz. : for the 6, 8, and 10-in. Machine 3/-, 12 and 14-in. 4/-, 16-in. 5/-, 18 and 20 in. 6/-, 22 and 24-in. 7/-; but if specially ordered the Machines can be sent on small open platforms for which no charge is made. Packing Cases for Donkey Machines, 20 and 22-in., 7/- ; 24-in., 8/-.

Buyers are recommended to purchase the cases in which to put the Machines away when not in use, to prevent them from getting damaged ; if returned, two-thirds will be allowed.

RANSOME'S "NEW AUTOMATON" LAWN MOWER.

Cash Prices, Carriage Paid, including Grass Box.

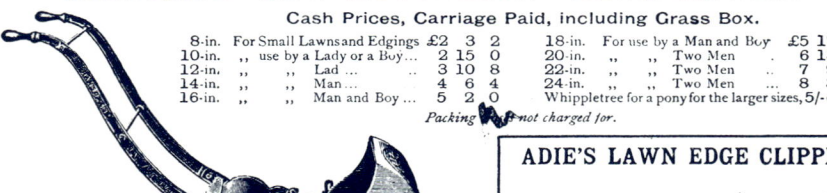

8-in.	For Small Lawns and Edgings	£2	3	2	18-in.	For use by a Man and Boy		£5	17	9
10-in.	,, use by a Lady or a Boy ...	2	15	0	20-in.	,, ,, Two Men	.	6	13	5
12-in.	,, ,, Lad ...	3	10	8	22-in.	,, ,, Two Men	...	7	9	2
14-in.	,, ,, Man ...	4	6	4	24-in.	,, ,, Two Men	...	8	3	10
16-in.	,, ,, Man and Boy ...	5	2	0	Whippletree for a pony for the larger sizes, 5/- extra.					

Packing not charged for.

RANSOME'S "ANGLO PARIS" LAWN MOWER.

Cash Prices, Carriage Paid, without Boxes.

6-in.	£1	0	3
8-in.	1	3	7
10-in.	1	6	11
12-in.	1	13	8
14-in.	2	0	4
16-in.	2	7	2

Grass Boxes, 6-in., 8-in., and 10-in., 3/4, 12-in., 14in., and 16-in., 5/1 each extra.

ADIE'S LAWN EDGE CLIPPER.

Cash Price, 10/- each.

Carriage Paid with other Goods amounting to 40/- value.

BOULTON & PAUL, MANUFACTURERS.

IMPROVED SWING WATER OR LIQUID MANURE BARROWS WITH GALVANIZED STEEL TANKS.

THESE BARROWS are very strong, the frames are made of bulb-iron, and the **wheels of steel**, painted; the tanks are also made of **steel**, galvanized after made. They are well suited for Horticultural purposes, and for conveying liquids of all kinds. Much time can be saved and a large quantity of liquid carted by using two tanks with one frame, one being filled while the other is taken away. The tanks are detached from the frame by raising the handles, and the water may be tilted or dipped out.

FOR GARDEN, FARM, AND STABLE.

REGISTERED DESIGN

TESTIMONIALS.

From
HERBERT WOODS, ESQ.,
Green Lodge, Newport Pagnell.

The Barrow arrived to-day, and I am very pleased with it.

From
MESSRS. WARD & CLARKE
2 Lancaster Street, W.C.

The Water Barrow is a thorough good article, and has given great satisfaction

From
MR. JAMES WATT,
The Gardens, Angmering, Sussex.

I approve of the Swing Water Barrow, it being a well made and strong frame. I shall be pleased to show the Barrow to any one who may require one.

TESTIMONIAL.

From A. GIBSON, Gardener to
F. F. Burnaby Atkin, Esq., The Gardens, Halstead Place, Sevenoaks.

The Water Barrow has arrived safely, and I think it a very useful affair.

Reduced Cash Prices.

12-Gallon size.
With one tank £1 0 0
,, two tanks ... 1 6 6
Lids 0 2 6

18-Gallon size.
Barrow, with steel wheels and one 18-gallon steel tank £1 5 0
,, ,, two 18-gallon steel tanks 1 15 0
Galvanized lids for tanks .. .. each 0 3 6
Height of wheels 22 in., tyres 2¼ in. wide. Extreme width 2 ft. 3½ in.

30-Gallon size.
Barrow, with steel wheels and one 30-gallon steel tank £1 12 0
,, ,, two 30-gallon steel tanks 2 5 0
Galvanized lids for tanks . .. . each 0 4 0
Height of wheels 24 in., tyres 2½ in. wide. Extreme width 2 ft. 7 in.

TESTIMONIAL.

From THOMAS JOHN, ESQ., Prendergast House, Haverfordwest.
The Barrow is very strong, admirably designed, and reasonable in price. I have enclosed cheque for amount due.

VALVE AND COPPER SPREADER.

For Watering Lawns, &c., fitted to tank . . . each 12/-

No. 85.
POWERFUL GARDEN AND CONSERVATORY ENGINE

THROWS a continuous stream a distance of about 50 feet, and is specially adapted for use in connection with the Swing Water Barrow described in this Catalogue, or can be placed in an ordinary pail.

The working parts and mountings are brass.

Price, £1 15 0

PORTABLE SWING CANS FOR GARDEN, KITCHEN, & PIGGERY.
Specially recommended for Workhouses, Hospitals, Asylums, and Large Establishments.

For Prices see above

Carriage Paid on all Orders above 40/- value to the principal Railway Stations in England and Wales.

IMPROVED SWING WATER OR LIQUID MANURE BARROWS WITH OAK TUBS.

This Barrow is made very strong, and is used for conveying and distributing water, pig's wash, liquid manure, &c. A great saving of labour is effected by using swing water barrows, and no Garden, Farm, Stable, or Kitchen Yard ought to be without one. For watering lawns or distributing liquid manure, a valve and copper spreader can be fitted to the tub; when the spreader is not required, the valve answers for drawing off the liquid into water pots, pails, &c. For conveying water for household drinking purposes, a plain oak tub can be supplied at an extra charge **The wheels are made of steel, and the frames of bulb-iron.**

Reduced Cash Prices, Carriage Paid

	£ s. d.
Barrow with one 36-gallon tub	£2 0 0
Barrow with two 36-gallon tubs	2 12 6
Valve and copper spreader, fitted to bottom of tub extra	0 10 0
Wood lid, with handle for above each	0 5 0
Garden Engine and fittings as preceding page	2 5 0

Height of wheels 27 in.; tyres 2½ in. wide
Extreme width 3 ft. 3 in.

TESTIMONIAL.

From W. H. BOND, Esq., Foyom Court

Mr. W. H. Bond begs to enclose cheque in payment of the 36-gallon Water Barrow which has reached him in safety, and which seems to be a thoroughly well made article.

No. 1. IMPROVED WATER CART FOR A DONKEY OR PONY.

A DONKEY can work it easily, and if required to travel long distances over rough ground it is better than the hand barrows. The tub can be detached from the carriage by simply raising the shafts. Constructed to same specification as the hand barrows.

Cash Prices, Carriage Paid.

To hold 36 gallons, suitable for a Donkey, extreme width 3 ft. 2 in. Wheels 27 in. high, tyres 2½ in. wide.

Cart on **steel wheels** with one tub	£3 0 0	Wood lid for handle for above	5/-
Cart on **steel wheels** with two tubs	3 15 0	Valve and copper spreader, fitted to bottom of tub	10/-

Carriage Paid on all Orders above 40/- value to the principal Railway Stations in England and Wales.

SWING WATER OR MANURE CART SUITABLE FOR DONKEY OR PONY.

No. 2.

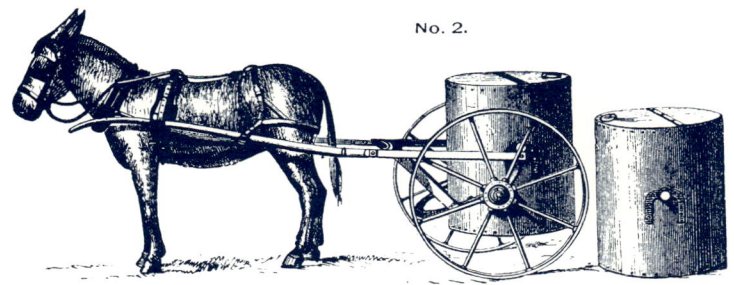

Adapted for the Removal of Night Soil. Specially suited for the Colonies.

This Implement is adapted for conveying water, liquid manure, &c. The Tanks are strong and well galvanized; with close lids fitted to them; the carriages are of wrought-iron, and the wheels of **steel**; the shafts are tubular and easily detached, and by merely raising them the tanks can be set on the ground, and detached from the carriages. With two tanks to one carriage, a large quantity of liquid may be carted in a short time, one tank being filled while the other is taken away. The tanks are also useful as drinking troughs for cattle, and for many other purposes.

A valve and spreader can be attached for distributing water or liquid manure.

Cash Prices, Carriage Paid.

To hold 60 gallons, suitable for a Donkey, extreme width 3 ft. 5 in. Wheels 2 ft. 9 in. high, tyres 2½ in. wide.
Cart, on **steel wheels**, with one tank £4 10 0
Cart, on **steel wheels**, with two tanks 6 10 0
Valve and spreader, fitted to tank ... ... extra 1 0 0

No. 3. SWING WATER OR LIQUID MANURE CART.

In climbing or descending a hill no weight is upon the horse, the Tanks always keeping in a vertical position. Similar to the preceding one, but of stronger make. Suitable for a Cob or Carriage Horse.

Cash Prices, Carriage Paid.

To hold 100 gallons, extreme width 4 ft. 6 in. Wheels 3 ft. 4 in. high, tyres 3 in. wide.
Cart, on **steel wheels**, with one tank £7 10 0
Cart, on **steel wheels**, with two tanks 11 10 0
Valve and spreader ... ... extra 1 0 0

We have received many Testimonials for our Water Carts; the following is for one of the above pattern:—

From J. J. WOOD, Esq., Maidenhead.

SIRS,—I have to inform you the Cart arrived here this morning quite safely, and I am very pleased with it, and feel sure it has only to be seen to be much more widely circulated. It is just the thing that has been much wanted in our neighbourhood, and I shall be glad at all times to bring it to notice amongst my friends.

Carriage Paid on all Orders above 40/- *value to the principal Railway Stations in England and Wales.*

ROSE LANE WORKS, NORWICH. 205

No. 8. HAND WATER CART.

THIS Cart will commend itself to purchasers for many purposes, such as watering lawns or gravel walks, distributing liquid manure, &c., and being fitted with a valve, water may be drawn from it by detaching the spreader. Made of wrought-iron and with **steel wheels**; will stand very rough usage.

Cash Price, Carriage Paid.

To hold 50 gallons £5 10 0
Spreaders, each 0 15 0
Extreme width 3 ft. 10 in., wheels 2 ft. 9 in. high, tyres 2½ in. wide.
This Cart can be fitted with **Shafts for Donkey** at 10/- extra.

No. 9. WATER CART FOR LIGHT HORSE.

For Water, Liquid Manure, &c.

THE Carriage is of wrought-iron, with tubular iron shafts; the **wheels of steel**, and the tank of wrought plates, well riveted; fitted with stop valve, to which either a clear water or a manure spreader can be attached.

Cash Price, Carriage Paid.
PAINTED.

To hold 100 gallons ... £8 10 0
Spreader for Clear Water extra 1 0 0
Spreader for Liquid Manure ,, 1 0 0
Extreme width 4 ft. 6 in., wheels 3 ft. 4 in. high, tyres 3 in. wide.

To hold 150 gallons, £13 10 0

No. 1 PATTERN PUMP,
For Filling or Pumping from Water Cart.

Cash Price
22/- extra.
Indiarubber Suction Hose, 2/3 per ft.

From E. H. BELLAIRS, ESQ., Christchurch.
I have two of your No. 9 Water Carts in constant use, and find them among my best farm investments.

No. 10. IMPROVED STREET WATERING CART.

THIS Cart is made entirely of wrought-iron with cranked axle fitted underneath the centre, which renders it self-balancing, thereby making the draught equal to one horse, and is well adapted for hilly roads.

This Cart is made either for street watering or spreading liquid manure. It has two stays behind to prevent tipping, and is mounted on improved wheels with broad tyres and **steel spokes**.

Cash Prices, Carriage Paid.

To hold 200 gallons ... £24 0 0
Distributor for Clear Water extra 1 0 0
Distributor for Liquid Manure ,, 1 0 0
Pump with 10 ft. Rubber Suction Hose extra 3 0 0
Extreme width 6 ft. 4 in., wheels 4 ft. 6 in. high, tyres 3½ in. wide.

Carriage Paid on all Orders above 40/- *value to the principal Railway Stations in England and Wales.*

BOULTON & PAUL, MANUFACTURERS.

No. 4. IMPROVED WATER OR LIQUID MANURE CART.
FOR AGRICULTURAL PURPOSES.

A GREAT improvement on the ordinary Agricultural Manure Cart. It is constructed almost entirely of wrought-iron; the shafts and lids are arranged to turn out of the way when the Cart is left for cattle to drink from, and the valve is made to screw down, which effectually prevents leakage when travelling over rough ground; the Spreader is readily detached, that sheep troughs or pails may be filled by opening the valve; the wheels are of wrought-iron, with wide tyres, and jolting is avoided by indiarubber springs. For conveying and distributing liquid manure it is invaluable.

This Cart is confidently recommended to Agriculturists and others, as the strongest and most convenient implement yet introduced. No farm should be without one. One horse is sufficient to work the 200-gallon Cart.

Cash Prices, Carriage Paid.

No. 4.	140-gallon Cart, 4 ft. wide, painted tank £10 0 0	galvanized tank	£12 10 0
No. 6	200-gallon Cart, 4 ft. 6 in. wide ,, 12 10 0	,, ,,	14 0 0
No. 21.	Portable Pump, fixed as illustrated } Indiarubber Suction Pipe, 10 ft. long	... ... extra	3 0 0
	Spreader for either size cart	... ...	0 15 0

The Pump and Spreader can be detached from the Cart without trouble or loss of time.

No. 11. WROUGHT-IRON TANK ON WHEELS.

Extreme width 2 ft. 6 in. To hold 50 gallons.

Cash Price, Carriage Paid .. £3 5 0
If galvanized body .. extra 0 15 0
Valve and Spreader fitted ,, 0 17 6
Tank 3 ft. 4 in. long, 1 ft. 9 in. wide, 1 ft. 6 in. deep

No. 7. WATER OR LIQUID MANURE CART.

Extreme width 2 ft 10 in.

To hold 75 Gallons.

Cash Price, Carriage Paid .. £6 0 0
No. 1 Pattern Pump, fitted with India rubber Suction Pipe, 10 ft. long, for clear water . . extra 2 10 0
No. 21 Pump, with 10 ft. Suction Hose, for liquid manure .. extra 3 0 0
Spreader for Cart . ,, 0 15 0

Carriage Paid on all Orders above 40/- value to the principal Railway Stations in England and Wales

No. 5a. WROUGHT-IRON ROAD WATER CART.

To carry 200 Gallons.

THE tank is constructed of strong wrought-iron plates, well painted inside and outside, which prevents corrosion, and fitted with a simple outlet valve, which by means of a lever can be opened and shut by the man driving. Improved distributor for spreading. The Cart is mounted on improved wheels with **steel spokes**, broad tyres, and strong springs.

Cash Price, Carriage Paid, with Water Distributor	£24 0 0
Extra charge for 4-in. detachable Pump (No. 2, page 209)	1 15 6
Rubber Suction Hose, for pump, 10 ft. long, with Union and Strainer	1 10 0

No. 5. WROUGHT-IRON CART FOR LIQUID MANURE.

A smaller size to hold **150** gallons for light horse.

To carry 200 Gallons.

Cash Price £15 0 0

THIS Cart is constructed as No. 5a, and extensively used by Agriculturists and others for distributing liquid manure; the distributor is of an improved make, through which the liquid manure passes freely, and can be easily detached when required. The pump can be readily taken off, and is fitted with improved bucket and valve which cannot clog.

Cash Price, Carriage Paid	£19 15 0
Extra charge for 4-in. detachable Pump (No 2, page 209)	1 15 6
Rubber Suction Hose, for pump, 10 ft. long, with Union and Strainer	1 10 0
Manure Distributor	1 5 0

TESTIMONIAL.

From T RANDALL, ESQ., Holme Lacy.

I am very pleased with the Water Cart which arrived last night, and also with the Keeper's Hut which you supplied us with a short time ago.

Carriage Paid on all Orders above **40/-** *value to the principal Railway Stations in England and Wales*

PORTABLE PUMPS FOR GENERAL PURPOSES.

No. 21.

No. 20.

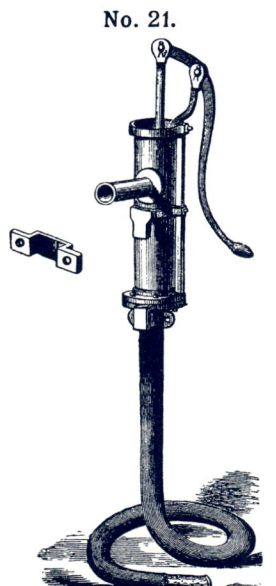

The barrels of these Pumps are made of galvanized iron, of 4½-in. bore, and the valves are constructed so that they cannot clog. They are well adapted for use in connection with the Water or Liquid Manure Barrows and Carts described in this Catalogue.

No. 20.

Fixed on Iron Stand. Can be raised or lowered. Easily carried by one man, the legs folding together.

Cash Price.

With stand
10 ft. Indiarubber Suction Pipe } £3 0 0
with Union and Strainer

No. 21.

For Bolting to a Post, Water Cart, or Tub.

Cash Price.

Including two brackets and bolts
10 ft. Indiarubber Suction Pipe } £3 0 0
with Union and Strainer

Shown fitted to Water-cart on page 206.
Spanner sent with each Pump.

Longer lengths of Suction Pipe can be had, if required, at 2/3 per ft. extra.
Brass Unions, for connecting extra length of suction pipe, 6/- each.

No. 22. FORCE AND LIFT PUMP ON BARROW.

For pumping thick liquids, sewage, or clear water into water-carts, cisterns, &c. Will draw water from a depth of 25 ft., or raise it 30 ft. Made with 4-in. bored cast-iron barrel. All working parts well finished, and not liable to get out of order.

Cash Price, Carriage Paid.

With Connections and Strainer, but without Hose £4 0 0
Suction or Delivery Hose   ..   ..   per ft   0 2 3

Extreme width outside wheels, 22 in

Carriage Paid on all Orders above 40/- value to the principal Railway Stations in England and Wales

ROSE LANE WORKS, NORWICH.

IMPROVED CAST-IRON PUMPS.
FOR LIQUID MANURE OR WATER.

WITH wrought-iron handle and rod, brass union for rubber hose, or screwed flange for iron pipe. The top is reversible, so that the handle can be placed at any angle to the spout. Used for cisterns and shallow wells not exceeding 25 ft. in depth.

No. 1.

For bolting to plank or wall. Used for shallow wells, cisterns, and small water carts.

Cash Prices.

3-in. barrel ... 17/6
4-in. ,, ... 22/-

No. 2.

For water carts or cisterns. This pump is supplied with brackets, so can be readily detached when required.

Cash Prices.

3-in. barrel ... 25/-
4-in. ,, ... 35/6

No. 3.

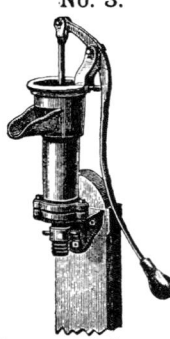

FIXED to plank, for emptying cesspools, filling water carts, etc.

Cash Prices.

3-in. barrel ... 23/6
4-in. ,, ... 35/9

No. 4.

For greenhouse or scullery, supplied with brackets for fixing to wall, for pumping from soft water tanks. With brass barrel.

Cash Price.

2-in. barrel ... 23/9

No. 5. COTTAGE OR CONSERVATORY PUMP.

For Liquid Manure or Water.

WITH closed top, wrought-iron lever, sling, and rod, base plate for bolting to timber or stone work, fitted with tail pipe for either lead or screwed iron suction pipe.

Cash Prices.

3-in. working barrel,
  1¼-in. bore suction £1 12 6
4-in. working barrel,
  2-in. bore suction 2 0 0
Screwed iron pipe 1¼-in.
  per ft. 0 1 0
Screwed iron pipe 2-in.
  per ft. 0 1 6

		per ft.
Patent Embedded Rubber Suction Hose for 3-in. pumps		1/9
,, ,, ,, ,, 4-in. ,,		2/3
Wrought-iron Suction Pipe for ,, 3-in. ,,		1/-
,, ,, ,, 4-in. ,,		1/6
Wood Planks ,, ,, ... each extra		7/6

No. 10.
PORTABLE FORCE PUMP ON TRIPOD STAND.

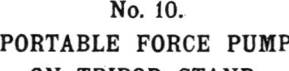

THIS Pump is used for raising water or liquid manure from a pit or cesspool, and lifting it into water carts or cisterns above its own height. The barrel is of cast-iron, bored out accurately, and well fitted.

Cash Prices.

3-in. barrel £2 10 0
4-in. ,, ... 3 0 0
Patent Embedded Rubber Suction and Delivery
  Hose for 3-in. pump ... ... per ft. 1/9
Patent Embedded Rubber Suction and Delivery
  Hose for 4-in. pump ... ... per ft. 2/3

Carriage Paid on all Orders above 40/- *value to the principal Railway Stations in England and Wales.*

BOULTON & PAUL, MANUFACTURERS.

IMPROVED CAST-IRON PUMPS.

A MOST important advantage in our Pumps is that the different parts are connected by *flanges*, *bolts*, and *nuts* which may be readily detached or reversed when required, rendering easy access to the bucket by unscrewing the nuts on the cap, and to the valve by taking off the nuts at the base of the Pump. The handle is reversible, and may be shifted to the right or left hand, or opposite the nose, as desired. Every possible attention and care is exercised in the manufacture and construction of our Pumps, the castings are clean and smooth, and bored out accurately, and fitted together in a most substantial manner.

No. 6. GARDEN PUMP WITH CLOSED TOP.
For Domestic Purposes, Gardens, Stable Yards, &c.

MOUNTED on plank, for raising water or liquid manure from wells not exceeding 25 feet in depth. Wrought-iron handle, sling, and rod, fitted for either lead, iron, or rubber suction pipe.

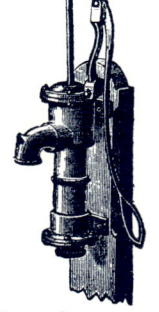

Size of Barrel.	Bore of Suction.	Price. £ s. d.	Wrought Suction Pipe. s. d.	Lead Suction Pipe. s. d.	Patent Embedded Rubber Suction Hose. s. d.
3-in.	1½-in.	1 12 6	1 0 per ft.	0 0 per ft.	1 9 per ft.
4-in.	2-in.	2 0 0	1 6 ,,	0 0 ,,	2 3 ,,

Planks 7/6 each extra.

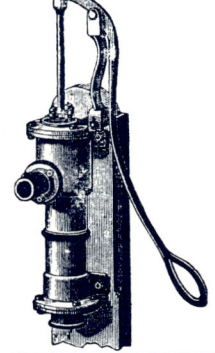

No. 7. LIFT AND FORCE PUMP.
For Dwelling Houses, Greenhouses, Farm Buildings, &c.

MOUNTED on wood plank, for filling supply cisterns from wells not exceeding 25 feet deep, and force to height of 30 feet. With wrought-iron lever, sling, and guide rod: fitted for iron, lead, or rubber suction and delivery hose. Can be used for liquid manure or water.

Size of Barrel.	Size of Suction and Delivery.	Price. £ s. d.	Iron Suction and Delivery Pipe. s. d.	Lead Suction and Delivery Pipe. s. d.	Patent Embedded Rubber Suction and Delivery Hose. s. d.
3-in.	1½-in.	1 15 0	1 0 per ft.	0 0 per ft.	1 9 per ft.
4-in.	2-in.	2 2 6	1 6 ,,	0 0 ,,	2 3 ,,

Planks 7/6 each extra.

No. 8a.

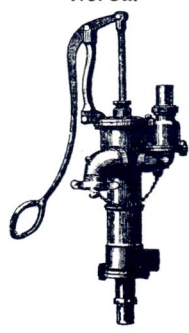

No. 8.
FORCE PUMP WITH RETAINING VALVE.

THIS Pump is used for drawing liquid manure or water from a well or cistern not exceeding 25 feet, and forcing same up into baths, tanks, or other elevated receptacles, to height of 50 feet. With brackets for wood plank or wall. Wrought-iron handle, sling, and guide rod, screwed for wrought-iron or lead suction pipe.

Size of Barrel.	Size of Suction and Delivery.	Price. £ s. d.	Iron Suction and Delivery Pipe. s. d.	Lead Suction and Delivery Pipe. s. d.
3-in.	1½-in.	1 17 6	1 0 per ft.	0 0 per ft.
4-in.	2-in.	2 5 0	1 6 ,,	0 0 ,,

If fitted with Spout and Cap as No. 8a, 7/6 extra.
Planks 7/6 each extra.

Carriage Paid on all Orders above 40/- value, to the principal Railway Stations in England and Wales.

No. 11. CAST-IRON PUMPS FOR DEEP WELLS.

WE here illustrate a Fluted-head Yard Pump of neat design, for drawing water out of wells from 10 feet to 80 feet deep. When the distance from which the water has to be drawn does not exceed 25 feet the pump case itself is bored, which answers as the working barrel, and nothing else is required but a straight pipe and strainer. In cases where the distance exceeds 25 feet, it is requisite to place the working barrel down the well, and to connect the top case with suitable piping as at **A**, through which the bucket rod passes to the barrel, this being fitted with detachable inspection doors, rendering easy access to the valves and brass buckets, which are of improved construction. The barrels are smoothly bored, and fitted in a most substantial manner. The working barrels may be connected with cast-iron pipe as at **B**, sufficiently large enough to allow the bucket to pass entirely out at the top of pump case for re-leathering, thus obviating the descent of the well, saving much labour and expense.

Estimates on Application.

No. 9.

CAST-IRON

STANDARD.

Suitable for
Nos. 5, 7, and 8 Pumps,
instead of
wood or stone bases.

Cash Prices.
For 3-in. Pump, 12/-
For 4-in. ,, 15/-

Carriage Paid on all Orders above 40/- *value to the principal Railway Stations in England and Wales.*

ROSE LANE WORKS, NORWICH. 212A

THE NEW NOZZLE LAWN SPRINKLER.
No. 6a.

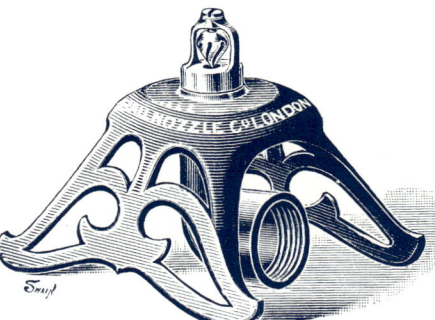

No. 6 Ball Nozzle for high pressure, to fit $\frac{1}{2}$-in. to 1-in. hose with japanned stand, bronze elbow and top.

Cash Price 6/6 each.

These Nozzles do not burst the hose, and never get out of order. No directions required, simply turn on the water, which falls like natural rain, and covers more ground than the ordinary nozzle.

No. 6a Ballcone Nozzle for low pressure, to fit $\frac{1}{2}$-in. to 1-in. hose with japanned stand, &c.

Cash Price 6/6 each.

Illustration showing the Sprinkler at work on the lawn.

BOULTON & PAUL, MANUFACTURERS.

WROUGHT-IRON HOSE REELS.
FOR WATERING GARDENS, LAWNS. &c.

These Hose Reels are indispensable in large gardens where there is a water supply, they preserve the tubing, and when not in use are easily stowed away.

Cash Prices. *Reel painted Green.*

No. 66 Reel	16 in. diameter,	10 in. wide, will carry about 200 ft. of ½-in. tubing	10/6
No. 67 ,,	18 in. ,,	12 in. ,, ,,	200 ft. of ⅝ in. ,,	12/6
No. 68 ,,	20 in. ,,	14 in. ,, ,,	200 ft. of ¾ in. ,,	15/-

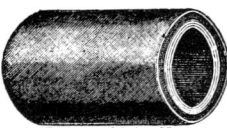
Full size of ½-in. Hose.

Full size of ¾-in. Hose.

BEST RUBBER HOSE.
FOR GARDENS, STABLES, YARDS, &c.
Best Quality.

This is good, sound, clean Hose, that does not suck up wet and hold grit.

Cash Prices.

	per coil of 60 ft.	Additional lengths of Hose Pipe, ½-in. diam. per yard	10d.
½-in. diameter, two-ply	£0 17 0	,, ,, ⅝-in. ,, ,,	1/1
⅝-in. diameter, three-ply	1 2 6	,, ,, ¾-in. ,, ,,	1 4
¾-in. ,, ,,	1 5 6		

ARMOURED GARDEN HOSE.

	per coil of 60 ft.
½-in. diameter, two-ply	£1 5 0
⅝-in. diameter, three-ply	1 11 9
¾-in. ,, ,,	1 16 0

For Prices of Directors and Unions see below.

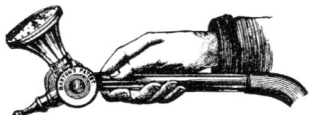

BARTON'S PATENT TWIN WATER DIRECTOR.

This most handy branch pipe, with three-way cock, combines rose, jet and stopcock all in one.

There is nothing else of the kind made so convenient and serviceable for use with garden hose.

For ½-in. hose	...	...	...	4/6 each.
For ⅝-in. ,,	...	...	...	5/- ,,
For ¾-in. ,,	...	...	...	6/- ,,

No. 69d. BRASS UNION.
For Connecting to Tap.

For ½-in. hose	...	...	...	...	1/-
For ⅝-in. ,,	...	...	...	...	1/6
For ¾-in. ,,	...	...	...	...	2/-

No. 69e. BRASS UNION.
For Joining Hose.

For ½-in. hose	...	...	...	...	1/-
For ⅝-in. ,,	...	...	...	...	1/6
For ¾-in. ,,	...	...	...	...	2/-

Nos. 66, 67 & 68 HOSE REELS.

Are fitted with a Catch at the Handles, so that they can be used for Watering Lawns as shown below.

Carriage Paid on all Orders above 40/- value to the principal Railway Stations in England and Wales.

No. 87. THE HAMBURG FIRST PRIZE WATERING MACHINE.
WITH POWERFUL GARDEN ENGINE.

THIS Implement is very complete, and most useful in large gardens. The engine is of entirely new design and improved construction; all the working parts are brass, and the pump barrel being brass-lined throughout, rust or corrosion is impossible. By a simple arrangement of the valve, which is regulated by the wheel shown, the pump will draw water from the barrel or direct from a pond or cistern, which can be delivered through the spreader as shown in illustration (it will easily throw with great force 3½ gallons of water per minute in a continuous stream to a distance of 60 feet). When required, it answers as a water barrow, from which, with the same valve, water-pots and pails can be filled. The barrel is fitted with valve and spreader for watering garden paths and lawns, and special spreader for the distribution of liquid manure for irrigation, &c.

The barrel is of oak, and will hold 36 gallons. The framework is of wrought-iron, and the wheels of steel with broad tyres. The whole is made of best materials, and highly finished.

REGISTERED.

Cash Prices, Carriage Paid.

To hold 36 gallons, fitted with Galvanized Water Spreader, Pump, and 10 ft. length Rubber Suction Pipe, complete .. £5 10 0
If without Pump 4 0 0
Longer lengths of Suction Pipe can be had, if required, at 1/3 per ft
An extra Galvanized Spreader, especially adapted for distributing liquid manure, 5/- extra.
The extreme width, outside wheels, is 3 ft. 2 in.
Width of tyres 2¼ in.

No. 87a.

THE machine is specially constructed for destroying Winter Moths & Caterpillars on Fruit trees with Paris Green or other insect destroyer, as recommended by Mr. C. Lee Campbell, Glewstone Court, Ross. It is fitted with extra powerful pump and two delivery hoses, with jets and sprayers, also dashers for keeping the solution well mixed whilst in use.

Price £8 0 0

This implement is well suited for hop washing.

No. 14. DONKEY BARROW WITH POWERFUL PUMP.

CONSTRUCTED on same principle as the above, but of larger dimensions. Stout plate iron tank to hold 50 gallons, mounted on strong wrought-iron frame and wheels. Very powerful 3-in. Force and Lift Pump attached to tank as shown, fitted with union, strainer, and director.

Cash Price, Carriage Paid, £8 10 0
1½-in. Rubber Suction Hose, extra per ft. 1/9
1-in. Rubber Delivery Hose ,, ,, 1/-

Wheels 2 ft. 9 in. diam., tyres 2½ in. wide, extreme width, 3 ft. 8 in.

No. 86. NEW FIRE ENGINE AND CONSERVATORY PUMP COMBINED.

This has the double advantage of being a serviceable Conservatory Pump, and also a powerful House Fire Engine. Being very compact and highly finished, it is suited to stand in a hall or landing for immediate use, and being narrow it will pass through any doorway, the pivot wheel turning the engine in its own length. Cast-iron wheels, with rubber tyres.

The cistern is made of galvanized wrought-iron, to contain 30 gallons, and the pump is of improved construction; all the working parts are brass, very simple, and easily got at. It will throw a continuous stream of water a distance of about 50 feet.

Cash Price, Carriage Paid.

Painted vermilion and black, with lacquered brass fittings,
12 ft. Rubber Delivery Hose and Jet £6 5 0

Outlet Valve with Union and 12 ft. of Suction Hose can be attached at an additional cost of **£1 17 6**

Size :—2 ft. 9 in. long, 1 ft. 7 in. extreme width, 2 ft. 9 in. high.

REGISTERED.

No. 86a. GARDEN ENGINE.

Specially constructed for spraying fruit trees with insecticide. Tank of strong plate-iron, galvanized, strongly made, mounted on wheels with wide tyres and scrapers, admirably suited for gravel paths, &c. Pivot wheel in front, turning engine in its own length. Fitted with two lengths of delivery hose, stop taps and patent sprayers, also outlet cap for emptying. The Engine is invaluable to fruit growers, and has no equal in the market. See Testimonial below.

TESTIMONIAL.

From F. NIXON, Fruit Grower, Gt. Eversden, near Cambridge.

The Engine is the best that has yet been brought out for a fruit grower's purposes. I have never seen anything so well adapted for spraying fruit trees, gooseberry and raspberry bushes, &c. It is the best adapted for the purpose of any in the market.

Cash Price, Carriage Paid.

Painted, with brass fittings, 24 ft. Rubber Delivery Hose and Jets, etc., complete £9 15 0

No. 27. FIRE BUCKET.

Close Riveted.

Diameter at top 10 in., depth 12 in.

Cash Price, 30/- per doz. galvanized.

Painting outside red (two coats) ⎫ 1/- each
Lettering Fire ... ... ⎬ extra.
Brackets .. .. per doz. 6/-

ROSE LANE WORKS, NORWICH. 215

FORCE PUMPS ON WHEELS OR STANDS.

REGISTERED, NO. 97531.

No. 90. FORCE PUMP ON WHEELS.

THREE-INCH bore Force and Lift Pump, combining lightness, strength, and simplicity. Offered at a price much below any other of equal capacity.

In the garden it will be found invaluable, as water can be drawn from a moat or well 25 ft. deep, and forced in a continuous stream a distance of 55 ft. from the spreader, and about 47 ft. high.

It also answers as a powerful **Fire Engine**; being fitted with high wheels with tested spokes attached to an extra strong hard wood frame. It is well adapted for running at high speed over rough stony roads without fear of injury.

It is a first-class Pump for filling supply cisterns for farm and other buildings, and is fitted with 1½-in. wire embedded suction hose, union and strainer, also 1-in. delivery hose, union and brass director.

Cash Price, Carriage Paid.

Pump only	£4 0 0
10 ft. Suction Hose and Strainer	1 0 0
10 ft. Delivery Hose with Director ..	0 10 0

Extreme width 27 in.

No. 88. CONSERVATORY PUMP.

New Implement.—This Pump has been specially designed to meet a want long felt in Conservatories, where this light portable pump, which can be carried about, is almost indispensable. It has a brass-lined barrel, and all the working parts are of brass. It is mounted on a hollow rigid stand, and complete with 10 ft. of suction hose, union, and strainer; also delivery hose, union, and spreader. It occupies a floor space of only 16 in. by 13 in., and weighs only 32 pounds.

No. 89. PUMP ON WHEELS.

New Implement.—This Pump is in every respect the same construction as No. 88, but mounted on wheels to facilitate its being easily moved by a lady, and, owing to its small dimensions it can be wheeled through a narrow doorway, or on the borders of ornamental flower-beds. It is recommended in connection with our Swing Water Barrows.

REGISTERED, NO. 97532.

Cash Price, Carriage Paid.

With 2 ft. of ⅞-in. Rubber Delivery Hose, and 10 ft. of 1-in. Wire Embedded Suction Hose .. £2 7 6

No. 13. WATER TANK,
with Powerful Force Pump attached.

Cash Price, Carriage Paid.

With Tank, Pump, and Director ... £6 0 0
1-in. Rubber Delivery Hose, per ft. 1/-
If Tank galvanized, 15/- extra.

REGISTERED, NO. 97532.

Cash Price, Carriage Paid.

With 2 ft. of Delivery Hose, and 10 ft. of 1-in. Suction Hose ... £2 15 0

BOULTON & PAUL, MANUFACTURERS,

Awarded the Only Prize by the Royal Agricultural Society, at the Cambridge Show, 1894, for Machine for Spraying Fruit Trees, Bushes, &c.

No. 13. WATER TANK WITH POWERFUL PUMP.

THIS Machine was the one entered for competition at the Cambridge Royal Show. The Pump has two deliveries with stop-tap to each, and is provided with two 6-ft. lengths of $\tfrac{5}{8}$-in. delivery hose. It can also be fitted with one 1-in. delivery hose, and $1\tfrac{1}{2}$-in. suction hose, for pumping from ponds or wells.

Cash Prices, Carriage Paid.

With Tank, Pump, with two 6-ft. lengths of Delivery Hose attached, (Prize Machine) £7 0 0

With one Director only (as illustrated) £6 0 0

If fitted with Shafts for Donkey, 10/- extra.

If Galvanized Tank, 15/- extra.

REGISTERED COPYRIGHT.

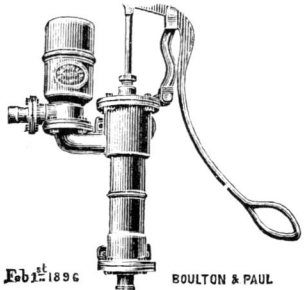

No. 8b. LIFT & FORCE PUMP.

THIS Pump is made similar to the No. 8 Pump shown on page 210), but arranged with air vessel, thus giving a constant flow.

Size of Barrel.	Size of Suction and Delivery.	Price.
		£ s. d.
3-in.	$1\tfrac{1}{2}$-in.	2 5 0
4-in.	2-in.	2 12 6

BOULTON & PAUL, MANUFACTURERS.

No. 83. IMPROVED POWERFUL GARDEN ENGINE.

Stott's Sprayers can be attached to this Engine if required.

The Judges at the great International Horticultural Exhibition held at Manchester tested this Engine very severely, and although all the principal Makers competed, it was declared to be the best, and was awarded the only prize—a Silver Medal.

A Bronze Medal was awarded for this article at the International Exhibition, Hamburg.

The working parts are brass, very simple and easily got at. It has a strong, galvanized iron body, well painted inside and out; wrought-iron wheels with wide tyres, steel spokes, indiarubber delivery pipe and registered spreader; throws a continuous stream.

This Pump will throw water a distance of about 60 feet.

Cash Prices, Carriage Paid.

Width 2 ft. 0 in., to hold 15 gallons ... £3 0 0
  ,,   2 ft. 2 in.   ,,   20   ,,   ...   3 10 0
  ,,   2 ft. 4 in.   ,,   25   ,,   ...   4 0 0

Extreme width—15 galls., 25 in.; 20 galls., 26 in.; 25 galls., 28 in.

No. 84. THE AMATEUR'S GARDEN ENGINE AND WATER BARROW COMBINED.

This Implement will be found very useful in any garden. It is specially suited for Amateur Gardeners, as it can be used for several purposes, viz :—

    1st.   As a Garden Syringe.
    2nd.   The Engine can be taken out and used in a pail.
    3rd.   As a Water Barrow when the Engine is taken out.
    4th.   For watering lawns, paths, etc., an outlet tap and distributor being provided.

The working parts of the Pump are brass, and will work for years without getting out of order. It throws a continuous stream of water, and is fitted with combined jet and spreader, attached by a universal joint.

Cash Prices, Carriage Paid.

15-gallon size £3 10 0
20   ,,   ,, ...   4 0 0
25   ,,   ,, ...   4 10 0

Width same as No. 83.

Rose can be attached if preferred.

TESTIMONIALS.

From Mr. J. LUDSBY, The Gardens, Exton Park.

I am highly pleased with the Garden Engine, it is as near perfection as anything I have seen in that way.

From Mr. CHARLES TYLER, Gardener to the Right Hon. the Earl of Wicklow, The Gardens, Shelton Abbey.

I herewith enclose Post Office Order in payment of account for Garden Engine, which pleases me very well.

Jet & Spreader as used with Engines.

2/6 each.

Price 2/6 extra.

Carriage Paid on all Orders above 40/- *value to the principal Railway Stations in England and Wales.*

No. 3. LIGHT HAND MILK CARRIAGE

IMPROVED PATTERN, REGISTERED.

WROUGHT-IRON frame and steel wheels, with turned axle pins and dust caps on wheels, fitted with improved bearings, name boards, wire crates, and box for books, &c. The Can may be raised or lowered without assistance.

Cash Prices, Carriage Paid.

With 12-gallon Tinned Steel Can and Brass Tap ... £4 5 9
With 16-gallon Tinned Steel Can and Brass Tap ... 5 2 0
Serving Cans extra. Painting Name on Side Boards and Lettering extra.

TESTIMONIAL.

From E. BAKER, ESQ., Chichester.
I received the Milk Carriage quite safe, and it gives great satisfaction.

REGISTERED COPYRIGHT.

No. 4. SINGLE MILK CARRIAGE.

FRAME made entirely of wrought-iron, wheels fitted with steel spokes and dust caps, tinned steel cans, with ventilated lids. Fitted with Rails for Serving Cans only.

Cash Prices, Carriage Paid.

With 12-gallon Tinned Steel Can and Brass Tap ... £3 10 0
With 16-gallon Tinned Steel Can and Brass Tap ... 4 5 0
Serving Cans extra.

TESTIMONIAL.

From MR. W. MARSHALL, Worcester.
I received the Hand Milk Carriage quite safe, which gives great satisfaction.

REGISTERED COPYRIGHT.

No. 8. SINGLE MILK CARRIAGE.

FRAME made entirely of wrought-iron, wheels fitted with steel spokes and dust caps, tinned steel cans, with ventilated lids. Fitted with Name Board, Wire Crate, and Butter Box.

Cash Prices, Carriage Paid.

With 12-gallon Tinned Steel Can and Brass Tap ... £4 5 0
With 16-gallon Tinned Steel Can and Brass Tap ... 5 0 0
Painting Name and Lettering extra.

TESTIMONIAL.

From MR. EDWARD PEARSON, Scarborough.
The Can and Cart arrived all right to-day. I noticed several improvements since the last one I got, which are very good.

REGISTERED COPYRIGHT.

Illustrations of Milk Carts for Horse or Pony sent on application.

218 BOULTON & PAUL, MANUFACTURERS.

WROUGHT-IRON WHEELBARROWS.

No. 2. GARDEN BARROW.

EXTRA strong Barrow, with large body. For Garden and Stable use.

Cash Price.
Painted 21/-
Body galvanized ... ... extra 4/-

No. 7. WROUGHT-IRON BARROW.

For Ashes, Coals, Stable, &c.

Cash Price, £1 5 0

TESTIMONIAL.

From W. TAYLOR BIRCHENOUGH, ESQ., Gawsworth New Hall.
The Wrought-iron Ashes Barrow you made me the other day is very good. Please make me another exactly like it.

No. 6. WROUGHT-IRON ASHES OR OFFAL BARROW.

Garden Barrow and Cinder Sifter Combined.

The top can be taken off, making an excellent Leaf and Garden Barrow.

Size of Body—2 ft. 7 in. long, 2 ft. wide, 1 ft. 5 in. deep.

Cash Price.
Painted 30/-
Body galvanized ... ... extra 6/-
If fitted with Registered Cinder Sifter ... ,, 4/6

REGISTERED, NO. 7655.

NEW CINDER SIFTER.

REGISTERED, NO. 120,537.

CAN be used in connection with any ordinary Wheelbarrow, Offal Barrow, or Ash Pit. No fixing required. Any one can use it.

Price 4/- each.

WROUGHT-IRON GREASING JACK.

FOR CARRIAGES, &C.

Prices.
No. 1.	No. 2.	No. 3.
4/6	5/-	6/-

No. 5. WROUGHT-IRON FODDER OR ASHES BARROW.

MADE entirely of wrought-iron. Suited for carrying leaves, and for carting and conveying food for cattle.

Size of Body—2 ft. 6 in. long, 1 ft. 10 in. wide, 1 ft. 2 in. deep.

Cash Price.
Painted 35/-
Body galvanized ... ... extra 5/6

WROUGHT-IRON HAND SACK BARROW.

Price 6/6 each.

Carriage Paid on all Orders above 40/- *value to the principal Railway Stations in England and Wales.*

No. 8. THE GARDENER'S BARROW.

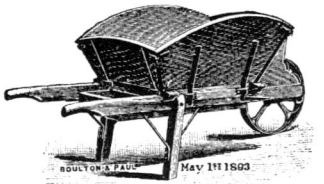

With sides and end removable, as shown, for conveying shrubs, trees, luggage, &c.

Cash Price complete, 45/-

The Barrow is also fitted with small drawer for gardener's small tools, as shown in Illustration.

No. 4. STRONG GARDEN WHEELBARROW.

IMPROVED COMBINATION CINDER SIFTER AND PORTABLE DUST BIN.

As Recommended by the Leading Sanitary Authorities and Medical Profession.

Cash Price.

Ash frame, deal body, wrought-iron wheel,
painted £1 10 0
Larger size £2 2 0

Shifting Boards for carrying leaves, etc., fitted to barrow, 9/- extra.

By using these Dust Bins a great saving of fuel is made, and the objectionable nuisances of the fixed "dust-holes" are dispensed with.
Size—2 ft. 8 in. long, 1 ft. 9 in. wide, 2 ft. 1 in. deep.

Cash Prices,

Body of wood, painted (as illustrated) £2 5 0
If made of Galvanized Iron 2 15 0

No. 12. WROUGHT-IRON TANK ON WHEELS,

for Ashes or House and Garden Refuse.

A useful article, specially suited for Hospitals, or large Establishments.

Extreme width, 2 ft. 6 in.

Tank, 3 ft. 4 in. long, 1 ft. 9 in. wide, 1 ft. 6 in. deep, painted.

Cash Price, Carriage Paid £3 7 6
If galvanized body ... extra 0 15 0

Carriage Paid on all Orders above 40/- value to the principal Railway Stations in England and Wales.

BOULTON & PAUL, MANUFACTURERS.

GALVANIZED WROT.-IRON SANITARY TRUCK.
No. 5.

24 in. by 18 in. at top } 18 in. deep.
,, ,, at bottom }
Cash Price 22/6.
If without wheels, 7/6 less.

GALVANIZED WROUGHT-IRON OFFAL PANS,
No. 1.   with Covers.   No. 2.

Cash Prices.
14 in. diam., 21 in. depth each 6/-
16   ,,   24   ,,   ,, 8/6
18   ,,   27   ,,   ,, 11/6

Cash Prices.
14 in. square, 21 in. depth each 9/-
16   ,,   24   ,,   ,, 11/6
18   ,,   27   ,,   ,, 14/6

GALVANIZED SANITARY PANS,
For Ashes and House Refuse, &c.

No. 7.
17 in. diam. at top,
by 17 in. deep.
Cash Price 5/6.

No. 8.
18 in. diam. by 18 in. deep.
Cash Price 5/6.

No. 50. SANITARY IRON DUST BINS.
Cash Prices.

No.	Width.	Depth Front to Back.	Height at Back.	Height at Front.	First Quality.	
					Galvanized.	Painted.
	ft. in.	ft. in.	ft. in.	ft. in.	£ s. d.	£ s. d.
1	1 8	1 4	2 1	1 11	1 15 0	1 9 6
3	2 0	1 8	3 1	2 10	2 8 0	1 18 0
4	2 6	1 10	2 8	2 4	2 8 0	1 18 0
6	3 0	2 2	2 10	2 6	2 15 0	2 5 0

SWING SANITARY BARROW.
For the Removal of House Refuse, Milk, Pigs' Wash, Cinder Ashes, &c.

The Tank tips over on the Carriage.

Cash Prices.
18-GALLON SIZE—
Barrow, with one Steel Tank ... £1 5 0
  ,,   ,,   two Steel Tanks ... 1 15 0
Galvanized Lids for Tanks   each 0 3 6
Height of wheels 22 in., 2¼ in. tyres.
Extreme width 2 ft. 2 in.   See also page 203.

Carriage Paid on all Orders above 40/- value to the principal Railway Stations in England and Wales.

No. 4. BOTTLE RACKS FOR WINE CELLARS.

THESE bins are made with wrought-iron lattice divisions, **T** iron bearers, and plate shelves.   Can be made to suit any situation.

Special Estimates on receipt of particulars of requirements.

WROUGHT-IRON STAND FOR BEER CASKS.

THIS Stand is very portable, and easily taken to pieces for stowage if desired.   It is fitted with lever in front for tilting the cask.

Cash Price 7/6

WROUGHT-IRON PORTABLE CELLARETS.

In issuing this List of Bottle Racks, we wish to call attention to the Reduction in Prices, and to point out the great superiority over those known as the French Wine Bins, which, when full of Wine, have frequently been known to collapse, causing great loss to the owner. Our bins having flat bearing bars of wrought-iron with strong frames, can be confidently recommended to carry safely the quantity allotted to the different sizes. They are very compact, occupying much less space than the old wood racks. A bottle can be taken off either shelf without disturbing any other bottles. No sawdust or any other packing is necessary. They are well adapted for export, being so fitted that they can be easily taken to pieces, to pack in a very small compass, and can be fixed by the most inexperienced hands; those fitted with doors and covered tops are almost as secure as lock-up cellars. Special sizes can be made to fit any unoccupied space or recess.

Cash Prices.

	Painted.	To hold.	In a Row.	Width.	Height.	Front to back without doors.	Without doors and top.	With grated doors, padlock, and covered top.
							£ s. d.	£ s. d.
Single Bin	No. 1	6 doz.	6	2 ft. 1 in.	4 ft. 2 in.	1 ft. 0 in.	0 9 0	0 14 6
,, ...	,,	9 ,,	6	2 ft. 1 in.	6 ft. 3 in.	1 ft. 0 in.	0 11 6	1 0 0
,, ...	,,	12 ,,	12	4 ft. 1 in.	4 ft. 2 in.	1 ft. 0 in.	0 15 0	1 5 0
Double Bin, *as illustrated*	No. 3	18 ,,	12	4 ft. 4 in.	3 ft. 5 in.	1 ft. 8 in.	1 0 0	1 15 0
,, ...	,,	24 ,,	12	4 ft. 4 in.	4 ft. 6 in.	1 ft. 8 in.	1 10 0	2 5 0
,, ...	,,	30 ,,	12	4 ft. 4 in.	5 ft. 7 in.	1 ft. 8 in.	1 17 6	2 10 0

Odd sizes calculate without Doors, 1/9 per doz. bottles. Galvanized instead of painted, 1/- per doz. bottles extra. With Doors, 2/3 per doz.

Carriage Paid on all Orders above 40/- *value to the principal Railway Stations in England and Wales.*

BOULTON & PAUL, MANUFACTURERS,

NEW FENCING FOR ENCLOSING TENNIS COURTS.

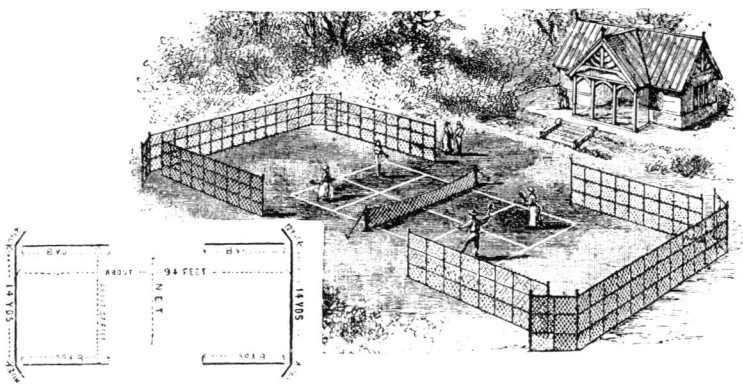

THE above Fencing is constructed with hurdles, each 6 ft. long, having prong-formed feet, pointed so as not to injure the grass, and fitted together with bolts and nuts. The hurdles are covered with 1½ in. mesh netting, galvanized after made, with black varnished frames. Gates can be placed in any part of the fence. The corners are set off at an angle (see plan) for the purpose of throwing the ball inwards, and so arranged to admit the ingress and egress of players without gates. Twelve strong angle iron Terminal Standards with cast heads, to keep the fence rigid, are supplied for the ends of the hurdles. This arrangement requires 68 yards of Fencing and 12 Terminal Standards.

Cash Prices, complete.

6 ft. high	... ...	£7 15 6
7 ft. ,,	... ...	9 7 6

Hurdles only.

6 ft. high, per yard run		2/3
7 ft. ,, ,, ,,		2/6
Angle Iron Corner Pillars 6 ft. high	each	1/6
Angle Iron Corner Pillars 7 ft. high	each	1/9

CHEAPER ARRANGEMENT FOR ENCLOSING TENNIS COURTS.

A cheaper method of enclosing courts with wire netting supported by angle iron standards **A**, which can be driven firmly into the ground without digging. Hook bolts **B** are provided for fastening the netting to the standards. This Fencing is not so neat or durable as the hurdles.

Cash Prices.

Galvanized Wire Netting 6 ft. high, 1½-in. mesh, in 50 yard rolls, per roll **17/7**
Angle Iron Stakes and Hook Bolts (usually spaced 8 ft. apart) per doz. **15/3**

ROSE LANE WORKS, NORWICH. 222A

NEW MOVABLE FENCING FOR CRICKET FIELDS.

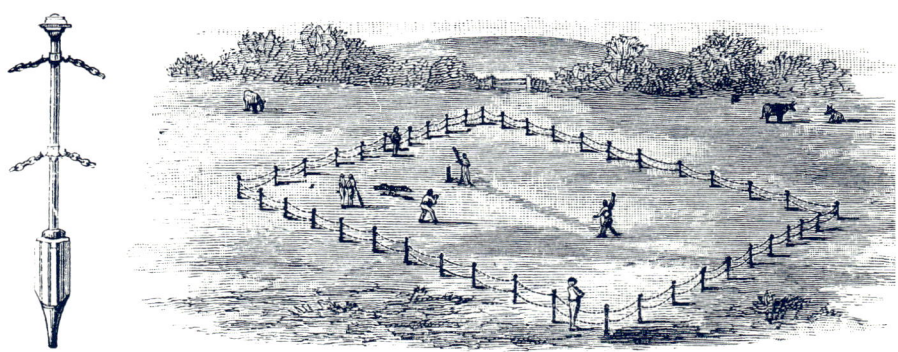

THE above shows an arrangement for protecting the wicket from horses and cattle, at the same time enabling sheep to pass in and out for feeding. The posts are made with detachable cast bases, into which they are "locked" by slightly turning them, and they can be removed and refixed without disturbing the ground.

Cash Price.

3 ft. 6 in. high, with two lines of chains, 3/- per yard.

A cheaper method with angle-iron standards, with hook bolts and three lines of galvanized strand wires, 1/- per yard.

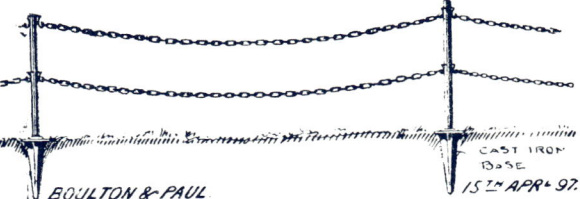

Similar Fencing constructed with T iron standards and detachable cast sockets.

Cash Price.

3 ft. 6 in. high, with two lines of chains, 2/3 per yard.

Standards spaced 12 ft. apart.

Carriage Paid on all Orders above 40 - *value to the principal Railway Stations in England and Wales.*

BOULTON & PAUL, MANUFACTURERS,

LAWN TENNIS NETS.

Cash Prices.

The **"J"**—Best Strong Steam Tarred Net (42 ft. by 3½ ft. high) ... 9/- ⎫
With White Canvas Binding, 2¼ in. deep ... ... 3/6 extra ⎬ Total **16**/-
Strong Copper Cord ... ... 3/6 ,, ⎭

This Net is recommended for all Private Courts, Small Clubs, or Private Houses, &c.

The **"K"**—Extra Best and Extra Strong Steam Tarred Net ... ... **12**/- ⎫
White Canvas Binding ... ... 3/6 extra ⎬ Total **20/6**
Extra Strong Copper Cord ... ... 5/- ,, ⎭

Suitable for the Largest Clubs, Tournaments, and roughest and hardest work.

NOTE.—The Prices of Net include a **Cotton Mounting Line** on each side of the net, **top and bottom.**

THE TOURNAMENT POST.

THIS is an excellent Post for Club or Private use, and the best of its kind. The Posts are superior polished ash, 3 in. diameter, fitted with well-made and highly finished brass furniture.

Price **32**/- per set.

~~The Brookfield Post is made similar to the above, but has tubular iron posts instead of ash.~~
~~Price £0/- per set.~~

Tennis Court Measuring Tapes, 78 ft., in leather case, **5**/- each.

Carriage Paid on all Orders above **40**/- *value to the principal Railway Stations in England and Wales.*

ROSE LANE WORKS, NORWICH. 222C

IRON PILLARS & TOP RODS FOR SUPPORTING STOP NETS.

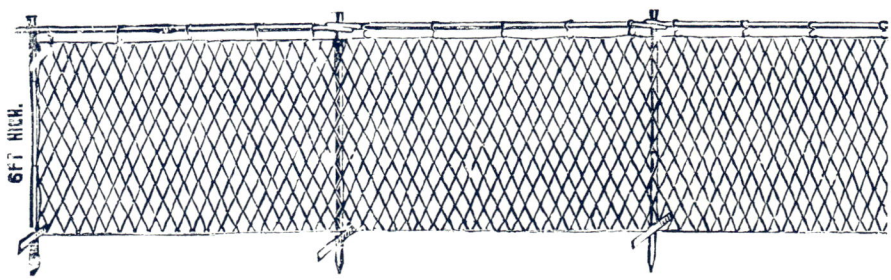

THE Pillars and Top Rods are tubular, and therefore light and strong. They answer either for a Cord or Wire Net, and need no stays. This system of Stop Nets has been some time in use by the Manchester Club of the Northern Counties Association, and by numerous other clubs and private players, with great satisfaction.

Prices, Painted Green.

Pillars, 6 ft. high, **2/-** ; 7 ft., **2 3** ; 8 ft., **2/6** ; 9 ft., **2/9** each. Top Rods, 9 ft., **1/3** each.

Long Staples, to pin bottom of Stop Net to the ground, **6d.** per doz. ; 9 ft. rods, **9d.** each. Strong Steam Tarred String Netting, $4\frac{1}{2}$d. per square yard.

Strong Galvanized Wire Netting, $1\frac{1}{2}$ in. mesh, 6 ft. high, price **20/-** per roll of 50 yards.

THE BOURNE No. 3 TENNIS MARKER.

THIS well known Tennis and Cricket Marker is now made in cast-iron, it is consequently much better, heavier, and more durable than the old pattern, which was made in tin.

Cash Price 10/6 each.

Carriage Paid on all Orders above **40/-** *value to the principal Railway Stations in England and Wales.*

BOULTON & PAUL, MANUFACTURERS,

THE GARDEN TENT FOR TENNIS LAWNS, CRICKET MATCHES, PICNICS, &c.

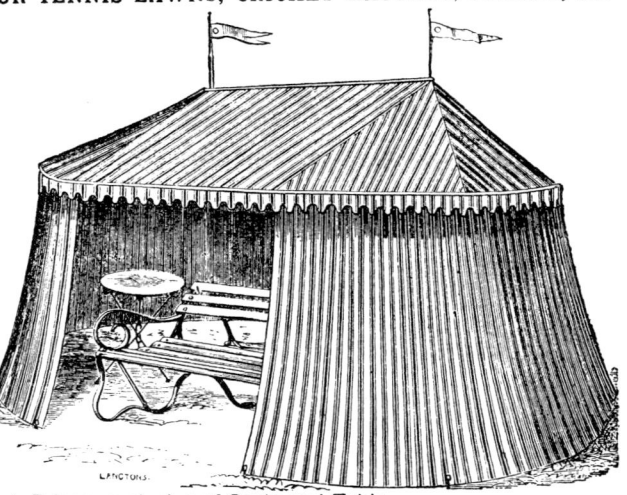

THE frame-work is made of light iron tubing, so fitted that it can be taken to pieces or put together in a few minutes.

All the fittings are iron, the usual upright supports and outside lines are entirely done away with, the interior space being quite clear of all obstruction.

Cash Prices, exclusive of Seats and Table.

No. 1. Length 10 ft., width 8 ft., height to eaves 5 ft. 6 in., height to ridges 8 ft. 6 in. .. £5 5 0
No. 2. ,, 12 ft., ,, 8 ft., ,, 6 ft. 0 in., ,, 9 ft. 0 in. ... 6 15 0

Including Packing Cases, strongly made and well finished with iron hinges and hasps, useful for keeping Tents when not in use. No allowance made for these boxes if returned.

NEW GARDEN SHELTERS. Can be easily moved from place to place.

MADE with strong light deal frame, covered with striped ticking, having seat at each end and flap table in centre. **Cash Price.**
Size, 5 ft. by 2 ft. **£3 10 0**

MADE with strong light deal framing, covered with striped ticking, having seat along the whole of the back. **Cash Price.**
Size, 5 ft. by 2 ft. **£3 0 0**

BOULTON & PAUL, MANUFACTURERS.

No. 7. INDEPENDENT AWNING.

This Awning is for use in connection with any garden chair not exceeding 6 ft. in length. It is portable, and easily fixed into the ground. The Awning is of strong striped ticking; it can be instantly rolled up and protected from the weather by a zinc cover.

Cash Prices, Carriage Paid.

6 ft. wide, neatly Japanned, including box (not returnable), for packing Awning and zinc cover; useful for stowing away in winter £2 7 6
Zinc cover to protect Awning when rolled up ... 7/6 extra.

GARDEN SWING.

No. 48. GARDEN SEAT.

This Seat is very light and comfortable. The wood splines are bolted to the iron-work. Can be put together in a few minutes. Iron-work Japanned green.

Cash Prices.
Pitch Pine, varnished, 5 ft. long 12/9

By using the new lever arrangement, the swing can be kept in motion by the occupants of the chair, or independently outside.

Cash Price, Carriage Paid, £4 10 0

Carriage Paid on all Orders above 40/- value to the principal Railway Stations in England and Wales.

No. 20. GARDEN CHAIR.

MUCH improved in shape, and finished in the best possible manner. For strength and finish this is the cheapest and best chair made. Wood seat and back; ends and elbows wrought-iron.

Cash Prices.

	Length—5 ft.	6 ft.
Pitch Pine, varnished ...	... ... 12/	... 13/-

Iron-work Japanned Green.

No. 23 pattern, with centre support, 3/- extra on above prices.

No. 24. GARDEN SEAT.

THE wood splines are fixed to wrought-iron standards, with galvanized iron bolts; it can be put together in a few minutes, and is well suited for export.

Cash Prices.

	Length—6 ft.	7 ft.
Pitch Pine, varnished	... ... 16/6	... 18/-

Iron-work Japanned Green.

This Chair can be had 5 ft. long, with two standards at 3/- less than the 6 ft.

No. 15. NEW GARDEN CHAIR.

Cash Prices.

	Length—5 ft.	6 ft.	7 ft.
Pitch Pine, varnished	... 15/-	... 16/6	... 18/-

Iron-work Japanned Green.

If without elbows, 2/- less each Chair.

PORTABLE FOOT RESTS.

For use in connection with any of the long Chairs.

Cash Prices.

	Length—4 ft.	5 ft.	6 ft.
Pitch Pine, varnished	... 3/-	... 3/6	... 4/-

Made with wood blocks, and not with iron scrolls as illustrated.

Carriage Paid on all Orders above **40/-** value to the principal Railway Stations in England and Wales.

No. 30. RUSTIC GARDEN CHAIR.

THIS Seat is specially designed for the Park and Public Pleasure Grounds. The iron-work is painted in imitation of oak branches when peeled.

Cash Price.

Pitch Pine, varnished, 6 ft. long £0 15 9

No. 16. THE PARIS GARDEN SEAT.

THIS handsome Seat, with cast-iron rustic supports, is very comfortable and strong, and is suitable for Balconies, Porticos, Terraces, Promenades, Lawns, or any situation where a Garden Seat is required.

Cash Prices.

Pitch Pine, varnished, 6 ft. 6 in. long ... £2 7 6
If fitted with Awning .. .. extra 2 0 0

No. 22. IMPROVED WROUGHT-IRON GARDEN CHAIR.

THE above is made entirely of wrought-iron, and recommended for strength, comfort, and lightness. Can be packed in a small compass for shipment.

Cash Prices.

Length— 5 ft.   6 ft.   7 ft.
Japanned Green, or any plain colour   16/6 ... 18/- ... 21/-
Longer lengths can be had at proportionate prices.

No. 7. WROUGHT-IRON TREE SEAT.

THIS illustration represents a seat $4\frac{1}{2}$ ft. in diameter, to fix round a tree 18 in. in diameter, back 3 ft. high from the ground, made in halves, and fitted with bolts and nuts.

Cash Prices.

As illustrated, painted green £2 10 0
Without elbows and back   2 0 0
Circular foot rest ... 1 10 0

Any size made to order.

No. 7. RUSTIC TABLE.

CAST-IRON legs and wood top, 24 in. in diameter; legs painted in imitation of birch; top grained oak or stained.

Cash Price, 15/-

Carriage Paid on all Orders above 40/- *value to the principal Railway Stations in England and Wales.*

ROSE LANE WORKS, NORWICH.

SEATS FOR PUBLIC PLACES.

FOR PARKS, PROMENADES, PUBLIC GARDENS, OR RAILWAY STATIONS.

No. 10. THE ARBORETUM SEAT.

The seats of this Chair incline backwards, preventing water remaining on them, and rendering the seat very comfortable.

Cash Prices.

	£ s. d.
Strong Seat and Back, grained oak, standards plain colour, highly finished and varnished, 6 ft. long, with two standards	£1 15 0
Ditto, 9 ft. long, with three standards	2 10 0
Ditto, 12 ft. long, with three standards as illustrated	3 0 0

No. 12. THE PROMENADE OR PARK SEAT.

The seat of this Chair inclines backwards, which makes it very comfortable, and allows water to run off, thus keeping the seat always dry, and preventing its decay. Standards are cast-iron, very heavy.

Cash Prices.

	£ s. d.
Seat and Back, grained oak and varnished, standards bronzed green, well finished, 6 ft. long, with two standards	£1 7 6
Ditto, 12 ft. long, with three standards as illustrated	2 5 0

No. 6. WROT.-IRON PROMENADE SEAT.

Cash Prices.

	Length—4 ft.	5 ft.	6 ft.	7 ft.	8 ft.
Painted Green	13/6	14/6	17/-	21/-	25/-

No. 47. CLUB SEAT.

For Balconies, or Public Rooms.

Cash Price.

Pitch Pine, varnished, 6 ft. long ... 15/-

Iron-work Japanned Green.

ROSE LANE WORKS, NORWICH. 227A

No. 63. PUBLIC PARK SEAT

Park Seats 8 ft. long, to seat six persons, made of 3-in. by 1-in. oak battens $\tfrac{3}{4}$-in. apart, on galvanized wrought-iron standards $1\tfrac{1}{2}$ in. by $\tfrac{1}{4}$ in., and four galvanized wrought-iron arms $1\tfrac{1}{4}$ in. by $\tfrac{1}{4}$ in., on oak sleepers 3 in. by 4 in. for bottoms of standards, wood-work unpainted. The necessary galvanized bolts and nuts.

Cash Price   £1   19   6

If with deal splines, and the iron-work painted, on oak sleepers, £1 12 0

No. 64. PARK OR GARDEN SEAT.

Garden Seat 8 ft. long, of 3-in. by 1-in. oak battens, $\tfrac{3}{4}$ in. apart, four galvanized wrought-iron standards $1\tfrac{1}{4}$ in. by $\tfrac{3}{8}$ in., and four galvanized wrought-iron arms $1\tfrac{1}{4}$ in. by $\tfrac{1}{4}$ in., on oak sleepers 3 in. by 4 in. for bottoms of standards. The necessary galvanized bolts and nuts.

Cash Price £2 2 0

If with deal splines, and the iron work painted, on oak sleepers, £1 14 0

No. 22a. WROUGHT-IRON GARDEN SEAT.

Made extra strong, with arched back, two standards and four elbows.

Cash Price.

6 ft. long £1 10 0

BOULTON & PAUL, MANUFACTURERS,

NEW GARDEN SEATS.

No. 65. GARDEN SEAT

THESE Garden Seats are 7 ft. long, made of 3-in. by 1-in. oak battens, and three galvanized wrought-iron standards 1½ in. by ¼ in. Necessary galvanized bolts and nuts.

Cash Price £1 4 6

If with deal splines and iron-work painted, **19/6**

No. 52. LADIES' SINGLE LAWN LOUNGE CHAIR.

Constructed on the same principle as No. 65a.

Cash Price.

Pitch Pine, varnished ... 8/-

Ironwork galvanized.

No. 65a. GARDEN SEAT.

THIS Seat is similar to the No. 65, of a lighter design, but so constructed as to give great durability and strength in the legs.

Cash Price.

Pitch Pine, varnished, 5 ft. long   **14/6**

Ironwork galvanized.

Carriage Paid on all Orders above **40/-** value to the principal Railway Stations in England and Wales.

No. 66. AUTOMATIC FOLDING SEAT AND FOOT REST.

THIS arrangement enables the back to be folded over the seat, thus keeping the seat dry during a shower of rain.

Cash Price.
6 ft. long 32/-

Illustration showing No. 66 Seat folded when not in use.

No. 67. AUTOMATIC FOLDING SEAT.

THIS Seat is exactly similar in construction to the No. 66 Seat, but without foot-rest.

Cash Price.
6 ft. long 25/-

No. 9. OBLONG TABLE.

WITH iron top 4 ft. by 2 ft., and 2 ft. 3 in. high, having wrought-iron standards and stays.

Cash Price ... £1 5 0

VILLA GARDEN CHAIR.

With Cast-iron Rustic Supports, Wood Seat and Back.

The iron-work is painted oak, woodwork olive green, and well varnished.

Cash Prices.

No. 28. 5 ft. long, without arms 12/6
No. 29. 5 ft. long, with arms 14/-

No. 506. SEAT.

This Seat is very strong and comfortable. The wood splines are bolted to the iron-work. Can be put together in a few minutes. Iron-work japanned green.

Cash Prices.

Pitch Pine, varnished, 6 ft. long 17/6

No. 60. THE BRIGHTON SEAT.

As supplied to the Corporation at Brighton.

Specially suited for Esplanades, Piers, and Public Walks. Made with strong cast-iron standards.

Cash Prices.

Red Deal, painted, 6 ft. long 31/-
,, ,, 9 ft. long, with three standards 45/-
,, ,, 12 ft. long ,, ,, 52/-

No. 61.
THE RAILWAY SEAT.

As supplied to the Gt. Eastern Railway Company.

With heavy cast-iron standards, and wood seat and back; suitable for Railway Platforms or Club Grounds, with monograms cast in standards as illustrated.

Cash Prices.

Pitch Pine, varnished, 9 ft. long 30/-
,, 12 ft. long, with three standards 42/-
Monograms cast in standards extra.

ROSE LANE WORKS, NORWICH.

No. 36. TENNIS CHAIR.

Cash Price.
Pitch Pine, varnished ... 5/-
Iron-work Japanned Green.

No. 42. THE LADY'S LOUNGE.

Cash Price.
Pitch Pine, varnished ... 7/6
Iron-work Japanned Green.

No. 50. SINGLE CHAIR.

Cash Price.
Pitch Pine, varnished ... 6/6
Iron-work Japanned Green.

IRON TABLES.

WILL fold and pack in a very small space when not in use. Grained oak and varnished, or japanned green.

Cash Price.
No. 1. 21 in. diameter ... 6/6
No. 2. 25 in. ,, extra strong 7/6

Grained oak and varnished, or japanned green.

Cash Prices.
No. 8. 21 in. diameter .. 6/6
No. 8a. 25. in. ,, . 8/-
Fancy colours extra.
Finished in superior style.

WROUGHT-IRON FOOT REST.

For use in connection with small Garden Chairs.

Cash Price.
Japanned Green .. 1/6

No. 38. TENNIS ARM CHAIR.

Cash Price.
Pitch Pine, varnished ... 7/6
Iron-work Japanned Green.

No. 35. FOLDING CHAIR.

Cash Price.
Pitch Pine, varnished .. 5/-
Iron-work Japanned Green.

No. 51. SINGLE CHAIR.

Cash Price.
Pitch Pine, varnished ... 6/-
Iron-work Japanned Green.

Carriage Paid on all Orders above 40/- *value to the principal Railway Stations in England and Wales.*

BOULTON & PAUL, MANUFACTURERS,

IMPROVED WOOD SACK BARROW.

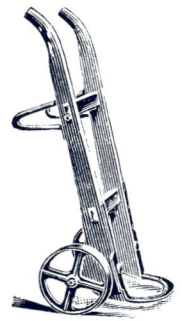

WITH strong ash frame, wrought-iron legs and fittings, and cast-iron wheels.

Cash Price 23/-

PORTABLE DUST HOUSE AND CINDER SIFTER.

MADE of red deal framing, covered with matchboarding.

Size—3 ft. square and 5 ft. high.

Cash Price £4 0 0 complete.

No. 3. PORTABLE HIVE HOUSE.

Strongly made and well painted.

Cash Prices, Carriage Paid.

3 ft. 6 in. long, 2 ft. wide, 4 ft. high, for two Hives £2 10 0
5 ft. 6 in. ., 2 ft. ,, 4 ft. ,, for three Hives 3 10 0

No. 2. MANIPULATING OR BEE HOUSE.

WITH Revolving Window for turning the Bees out, Bench for working or for Standing Hives upon, and Floored

Cash Prices, Carriage Paid

6 ft by 5 ft. £8 0 0
8 ft by 6 ft. 11 0 0

ROSE LANE WORKS, NORWICH. 231

WROUGHT-IRON DOOR AND GARDEN SCRAPERS.
These are specially suited for Greenhouses or Iron Buildings

No. 50.   **No. 51.**

Cash Prices.
30 in. by 12 in., painted 6/- .. galvanized 8/-
30 in. ,, 18 in., ,, 9/- ,, ,, 12/-
Without Brushes 2/- less.

Cash Prices.
With cast-iron Frame and wrought-iron Scraper, with Brushes,
15 in. by 15 in., painted 5/6
Ditto galvanized Scraper and painted Frame 7/6

TRUCK FOR GARDEN OR HOUSE.
For carrying Lawn Mowers, Plants, Flower Pots, Luggage, Linen, &c.

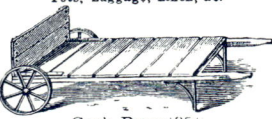

Cash Price, 35/-

TRUCK FOR REMOVING TREES, SHRUBS, &c.

Price on application.

No. 200. IMPROVED FOLDING STEPS.

No. 201. WROUGHT-IRON UMBRELLA STAND.
Specially suited for Greenhouses and Iron Buildings.

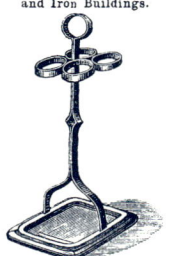

For house and garden. With box at top for dusters, tools, &c. *Made of red deal, varnished.

Cash Prices.
With 5 steps, 4 ft. 0 in. high ... 9/-
With 6 ,, 4 ft. 10 in. ,, ... 10/6
With 7 ,, 5 ft. 6 in. ,, ... 12/6

Cash Prices.
With 4 Rings and cast-iron Pan .. 4/6
If galvanized .. . 6/-
Any size made to order.

CINDER OR GRAVEL SCREEN.
For cinder ashes, gravel, and lime

Wood frames, clamped with iron.

	⅜-in. mesh	½-in. mesh
5 ft. 0 in long, 27 in. wide	19/6	18/6
5 ft. 6 in. ,, 30 in. ,,	21/-	19/6
6 ft. 0 in. ,, 33 in. ,,	22/6	21/-

No. 202. WROUGHT-IRON LINEN POST.
With cast socket base.

Cash Price.
7 ft. above ground 10/-
3 ft. 6 in. high, for enclosing Cricket and Tennis Grounds 7/6

No. 5. GALVANIZED CONICAL COVERS FOR BEE HIVES.

Cash Price.
With Loops for fastening to Straw Skep 2/-

CAST-IRON LINEN POST SOCKET.
With cap, for wood posts.

Price 3/- each

GALVANIZED STRAND ROPE.
For linen lines and back nets for tennis courts.
4/6 per 100 yards

No. 203. IMPROVED PLANT CARRIER.

By the use of this Barrow much time will be saved in moving plants, as a quantity can be carried at a time without risk of damage to foliage or bloom. On **wheels** and spring axle, so that one man can use it.

Cash Price.
Length of body 4 ft., by 2 ft. wide 25/-
If without wheels 20/-

Carriage Paid on all Orders above **40**/- value to the principal Railway Stations in England and Wales.

BOULTON & PAUL, MANUFACTURERS.

CAST-IRON FOUNTAINS & VASES FOR GARDEN & CONSERVATORIES.

Cash Prices.
No. 19 Vase, "The Britannia" ... 12/-
No. 8a. Pedestal ... 7/6

No. 21. FOUNTAIN.

Cash Price.
4 ft. 9 in. high ... £4 10 0
Ground Basins extra.

No. 5. WROUGHT-IRON SPIRAL BORDERING.

Cash Prices.
8 in. wide, galvanized .. per doz. 2/6
12 in. ,,   ,,   ,,   ,,   3/-
15 in. ,,   ,,   ,,   ,,   3/6

No. 7.

GALVANIZED WIRE GARDEN BORDERING
No. 6

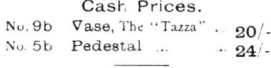

Made any length or shape required
No. 6 12 in. high ... per yard 9d
No. 7 15 in. ,,   ,,   ,,   10d
No. 8 18 in. ,,   ,,   ,,   1/-

Cash Prices.
No. 9b Vase, The "Tazza" .. 20/-
No. 5b Pedestal ... 24/-

Decorative Painting and Packing extra

No. 20. WROUGHT-IRON GARDEN VASE.

Cash Prices without Wood Pedestal.
18 in. diam. 13/- | 30 in. diam 21/-
24 in. ,,   18/- | 36 in. ,, 26/6

Cash Prices.
No. 9a. Vase, The "Tazza" .. 10/-
No. 8a. Pedestal ... 7/6

CAST-IRON GARDEN BORDERING.

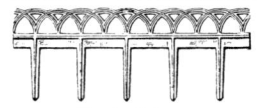

Cash Price.
No. 4 1 ft. lengths, painted, per ft. 5d.

No. 8.

Carriage Paid on all Orders above 40/- value to the principal Railway Stations in England and Wales.

ORNAMENTAL GARDEN BORDERING.

Made in 6 ft. lengths, and from 1 ft. in height. Vertical bars ¼ in. round, about 3 in. apart. Painted one coat. May be curved to any given radius.

No. 31.   No. 32.

No. 33.   No. 34.

12 in. high,   15 in. high,   18 in. high,

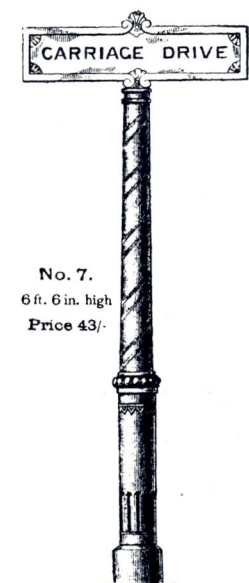

No. 7.
6 ft. 6 in. high
Price 43/-

No. 6.
TO THE CONSERVATORY.
2 ft. 6 in. by 9 in. .. 3/9

No. 72.

HIGH STREET
29 in., 32 in., 38 in., and 43 in. by 6 in. wide.
No. 13.   Prices on application.

BOTANICAL TABLETS.

No. 4.
   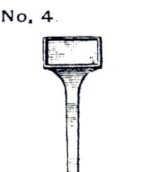
18 in. long ... 4/6   2 ft. 1 in. by 6 in. ... 2/6

LINE REEL.

With 60 yds. Cord 4/6

No. 12.
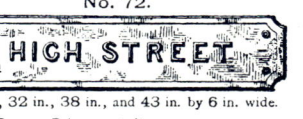
PLEASE KEEP OFF THE GRASS
16 in. high ... 4/6

Carriage Paid on all Orders above 40/- *value to the principal Railway Stations in England and Wales.*

BOULTON & PAUL, MANUFACTURERS.

PATENT BORDERING AND RAILING.
For Fencing in Fountains, Gardens, Rivers, Parks, Lawns, Flower Beds, &c.

This new kind of Bordering is made of hard steel wire; it is prettier and stronger, notwithstanding its low price, than any other at present in the market; made in any height from 6 in. to 80 in., galvanized or painted in any colour (usually green). Although very hard, it can be placed in any shape round flower beds, etc.

Prices per running yard of Pattern No. 1.

Nos. BWG	13		12		11		10	
Height.	Galvd.	Painted.	Galvd.	Painted.	Galvd.	Painted.	Galvd.	Painted.
	s. d.	s. d.	s. d.	s. d.	s. d.	s. d.	s. d.	s. d.
6 in.	0 5½	0 6	—	—	—	—	—	—
8 in.	0 6	0 7	0 7	0 7½	—	—	—	—
10 in.	—	—	0 9	0 9½	0 10	0 10½	0 11½	1 0½
16 in.	—	—	—	—	1 0½	1 1	1 2	1 3
20 in.	—	—	—	—	1 3	1 4½	1 4½	1 5½
24 in.	—	—	—	—	—	—	1 7	1 8

Prices for Higher Sizes on application.

This Bordering is fixed with iron hooks for the heights from 6 in. to 10 in. (Price 1/6 per doz.), and with iron or wooden stakes for heights above 10 in.

NOTICE.—The space inside the bars for the Borderings of No. 13 wire is 1⅜ in., of Nos 12, 11, and 10 wire is 1¾ in., and of Nos. 8 to 3 wire is 2¼ in.

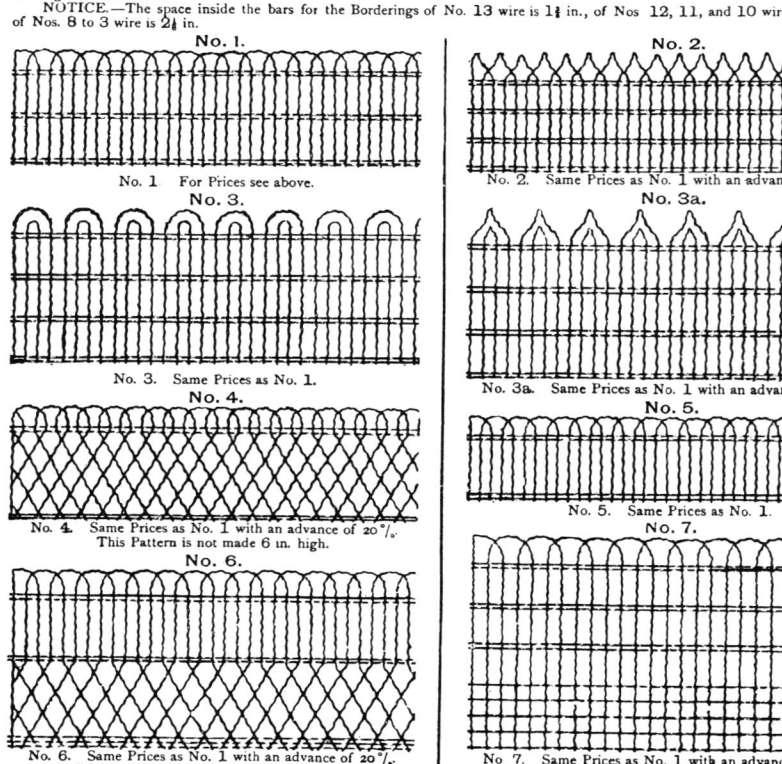

No. 1. For Prices see above.
No. 2. Same Prices as No. 1 with an advance of 20%.
No. 3. Same Prices as No. 1.
No. 3a. Same Prices as No. 1 with an advance of 20%.
No. 4. Same Prices as No. 1 with an advance of 20%. This Pattern is not made 6 in. high.
No. 5. Same Prices as No. 1.
No. 6. Same Prices as No. 1 with an advance of 20%. This Pattern is only made from 20 in. high.
No 7. Same Prices as No. 1 with an advance of 10%. This Pattern is only made from 16 in. high upwards.

ROSE LANE WORKS, NORWICH 235

ORNAMENTAL GAME-PROOF HURDLES.

For Pleasure Grounds, Enclosing Flower Beds, Shrubberies, Ornamental Waters, &c.

No. 23.   No. 24.

Special Estimates on application.

Nos. 23 and 24. Hurdles in 6 ft. lengths (straight or curved), 3 ft. 6 in. high. $\frac{1}{2}$-in. round vertical bars, $1\frac{1}{8}$ in. apart, about 2 ft. from ground, 3 in. apart above. Frames 1 in. by $\frac{3}{8}$ in. Painted one coat. Shorter lengths to order.

Hand Gates to match.   Cast Posts with self-fixing bases.

For Prices see accompanying Price List.

WIRE NETTING PLANT AND TREE GUARDS.

No. 153.   No. 154.

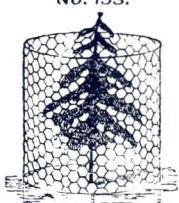 

THESE Guards are made of Galvanized Wire Netting, with prongs for fixing, and for convenience of carriage are sent out flat, but are easily bent to a circular form.

THESE are made with a strong frame, in two parts for the convenience of carriage, and are covered with Galvanized Wire Netting, sufficiently fine to exclude hares and rabbits.

Cash Prices.

18 in. high,	12 in. diameter	...	10/6 per doz.
24 in. ,,	18 in. ,,	...	13/6 ,,
30 in. ,,	24 in. ,,	...	18/6 ,,
18 in. ,,	18 in. ,,	...	12/- ,,
24 in. ,,	24 in. ,,	...	15/6 ,,

Cash Prices.

				3 ft. high
24-in. diameter	...	...	...	each 7/6
30-in. ,,	...	...	...	,, 9/-
36-in. ,,	...	...	...	,, 10/6

Carriage Paid on all Orders above **40/-** *value to the principal Railway Stations in England and Wales.*

NOTE.—These Prices are subject to the fluctuations of the Iron **Market**.

235A   BOULTON & PAUL, MANUFACTURERS,

LIGHT WIRE LATTICE HURDLES,
For Garden Bordering, &c.
No. 5.

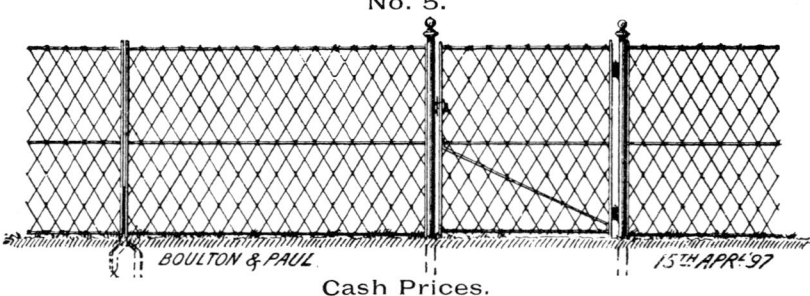

Cash Prices.

6 ft. length—3 ft. high, 3 in. mesh   ...   ...   3/3 per yard.
6 ft. ,,   2 ft. 6 in. high, 3 in. mesh ...   ..   3/-   ,,

Gates and Standards 3 ft. wide, **12/6** each.

No. 1.   No. 2.

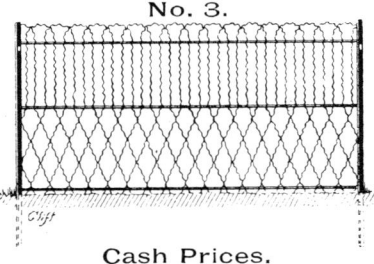   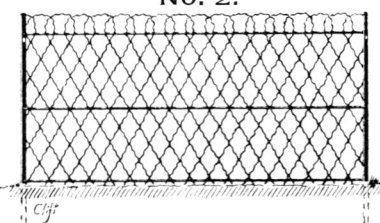

Cash Prices.   Cash Prices.

6 ft. long, 3 ft. high ...   ...   **6/-**   6 ft. long, 3 ft. high **6/-**
6 ft. long, 2 ft. 6 in. high ...   **5/-**   6 ft. long, 2 ft. 6 in. high ...   **5/-**

No. 3.   No. 4.

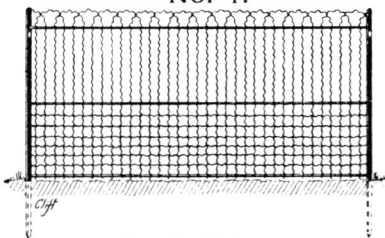

Cash Prices.   Cash Prices.

6 ft. long, 3 ft. high ...   **6** -   6 ft. long, 3 ft. high **6/-**
6 ft. long, 2 ft. 6 in. high ...   **5** -   6 ft. long, 2 ft. 6 in. high ...   **5** -

For other patterns see pages 232, 233, 234.

BOULTON & PAUL, MANUFACTURERS.

IMPERIAL BLACK VARNISH
FOR IRON FENCES, WOOD PALINGS, FARM BUILDINGS, &c.

FOR LANDED PROPRIETORS.

QUALITY GUARANTEED.

THIS Varnish is an excellent substitute for Oil Paint *at one-third the cost*, and should be kept in stock on every Estate and Farm.

It is always in a fit state for use; is applied *cold*, requires no heating, mixing, or other preparation, will dry quickly and leave a bright, hard, Jet-Black surface. Can be laid on by any ordinary labourer.

Cash Prices, delivered to any Railway Station.

In 40-gallon casks, 1/3 per gall. In 18 ditto, 1/6 per gall. Including casks.

Casks not returnable

Steel Scrubs for cleaning Iron Fencing before varnishing, 2/6 each. Tar Brushes, 1/6 each; with long handles, 2/- each.

PAINTS MIXED READY FOR USE.
IN DRUMS.
VERY BEST QUALITY.
Cash Prices.

Bright Green	Light Stone	
Light do.	Middle do.	
Middle do.	Dark do.	
Dark do.	Light Drab	
Bronze do.	Dark do.	
Dark Blue	Salmon	6/6 per gall.
Azure do.	Silver Grey	
Sky do. } 6/- per gall.	Straw	
Chocolate	Dark Slate	
Bright Red	Light do.	
Venetian Red	Buff	
Black	Indian Red	7/- per gall.
Yellow	White	8/- ,,
Umber	Red Lead	8/- ,,
Light Oak	Signal Red	14/- ,,
Dark do.	Any other colours made to order.	

Paint Kettles, half-gallon, 2/- each. Paint Brushes, 1/6 each

Drums charged for and allowed in full when returned. Carriage Paid. 20 and 40 gallon **Casks free**.

Carriage Paid on all Orders above 40/- value to the principal Railway Stations in England and Wales.

MANUFACTURERS AND GALVANIZERS OF

DIAMOND MESH

WIRE NETTING

IRON & WIRE FENCING

FOR THE IMPROVEMENT OF

LANDED PROPERTY, PARK, AND PLEASURE GROUNDS.

BOULTON & PAUL, MANUFACTURERS.

GALVANIZED WIRE NETTING.

Being Manufacturers of Galvanized Wire Netting, we are prepared to execute orders, however large, at the shortest notice.

Cash Prices per Roll of 50 Yards.

Mesh.	Quality.	Gauge.	24 in. wide.	30 in. wide.	36 in. wide.	42 in. wide.	48 in. wide.	60 in. wide.	72 in. wide.
3/8 in. Mesh	Strong	21							
1/2 in. Mesh For Aviaries, Window Guards, &c.	Medium Strong Very Strong	22 20 19							
5/8 in. Mesh For Aviaries, &c.	Medium Strong Very Strong	22 20 19							
3/4 in. Mesh For Aviaries, &c. (Rat Proof).	Light Medium Strong	20 19 18							
1 in. Mesh For Pheasantries, Poultry Enclosures, &c.	Light Medium Strong Very Strong	19 18 17 16							
1 1/4 in. Mesh For Rabbit Warrens. Proof against the smallest Rabbits.	Light Medium Strong Extra Strong	19 18 17 16							
1 1/2 in. Mesh Strongly recommended for Rabbits.	Light Medium Strong Extra Strong	19 18 17 16							
1 5/8 in. Mesh For Ground Game, &c.	Light Medium Strong Extra Strong	19 18 17 16							
2 in. Mesh For Hares, Dogs, Poultry, &c.	Light Medium Strong Extra Strong	19 18 17 16							
2 1/4 in. Mesh For Hares, Dogs, Poultry, &c.	Light Medium Strong Extra Strong	19 18 17 16							
3 in. Mesh For Training Plants, &c.	Very Light Light Medium Strong	19 18 17 16							
4 in. Mesh For Training Plants, &c.	Very Light Light Medium	18 17 16							

SPECIAL PRICE LIST SENT FREE ON APPLICATION.

Orders above 40/- value Carriage Paid to any Railway Goods Station in Great Britain and Principal Ports in Ireland.

GALVANIZED WIRE SHEEP NETTING.

WITH 3-PLY TWISTED SELVAGES.

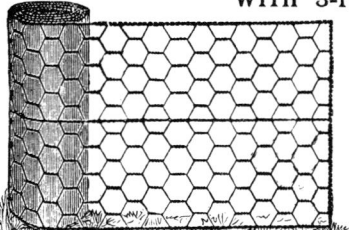

<u>4-inch Mesh</u>, with strong 3-ply Strand woven in the centre.

Cash Price, per Roll of 50 yards,
36 inches wide.

Quality.	Gauge.	Price. s. d.	Length.
Light	16	8 6	Per 50 Yards.
Medium	15	10 4	,,
*Strong	14	12 5	,,
Extra Strong	13	14 9	,,

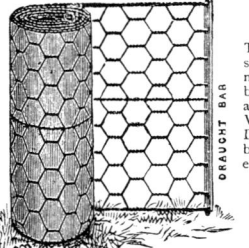

This block represents the same netting as above, but fitted with a Galvanized Wrought-iron Draught Bar, ⅞-in. by 1/10-in., at each end.

Price 1/- per Roll extra.

* Recommended for General Use.

If made in 25 yard Rolls, half above prices.

If without centre Strand, 1/- per Roll of 50 yards less.

Carriage Paid on orders of 40/- value and upwards.

Wider widths may be had at prices in proportion, but the 36-inches wide is found to be sufficient for folding sheep.

WROUGHT-IRON NETTING STAKES.

No. 4.

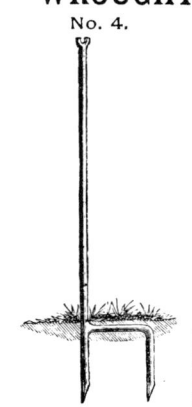

Strong wrought-iron Stakes for supporting Wire Sheep Netting, usually about 6 feet apart.

Price, painted, for 3-ft. Netting, 7/- per doz.

No. 4a.

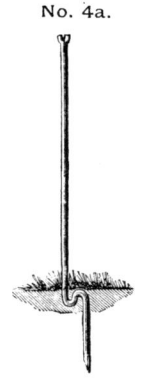

This is a cheaper pattern, which can be recommended.

Price, painted, for 3-ft. Netting, 6/- per doz.

No. 5.

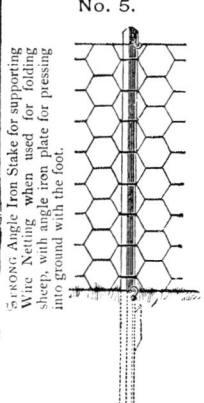

Strong Angle Iron Stake for supporting Wire Netting when used for folding sheep, with angle iron plate for pressing into ground with the foot.

Price, painted, for 3-ft. Netting, 8/6 per doz.

WROUGHT-IRON HOOKS.

For supporting the Netting where wooden stakes are used.

Price 3/6 per Gross, black.

TESTIMONIAL.

From T. J. COOKE, Esq., Flitcham Abbey, King's Lynn.

Gentlemen,—I have much pleasure in stating that I have very great faith indeed in the advantages of your Wire Sheep Netting. I shall be much surprised if it does not prove very superior to the line netting, and much cheaper in the long run. It is quite as easily used as the line netting, and, in my opinion, is even safer against sheep for winter folding, whilst in summer it will be proof against the great enemies of other netting, the hares, which have always prevented me from using ordinary netting at that time of the year.

BOULTON & PAUL, MANUFACTURERS.

No. 53. A NEW RABBIT FENCING FOR ENCLOSING WARRENS, &c.

THE Wire Netting is turned in along the surface of the ground under the lowest wire towards the rabbits. The grass soon grows through the turned-in-wire, and forms a matted surface, underneath which the rabbits will not attempt to scratch. The Wire Netting is turned in at the top of the Fence, just the same as below, but at an obtuse angle, so as to frustrate any attempts to climb up from below. The Barbed Wire is most valuable in any Warrens exposed to Cattle, Sheep, Horses, Dogs, or Foxes.

Fence, 4 ft. high out of ground, with four lines of No. 8 gauge solid annealed drawn wire, and galvanized steel barbed top wire; angle-iron standards, alternately with and without plate under ground, spaced 10 ft. apart; Galvanized Wire Netting, 48 in. wide, 1¼ in. mesh, No. 17 gauge and lacing wire. *Special Estimates on application.*

No. 54. WARREN FENCE FOR WOOD POSTS.

SAME specification as last, but without Standards, and includes, in addition, a Barbed Wire and Staples for fixing to wooden posts outside the Fence as shown in illustration.

If intending purchasers will send a rough plan of ground to be wired in, showing Angles, Gateways, etc., we will give exact estimates (free of cost), to include all material required, delivered at their Railway Station.

HAY RACK FOR WARRENS.   **TESTIMONIAL.**   **TROUGH FOR WARRENS.**

"All the Wire Netting, Barbed Wire, Annealed Wire, Iron Straining Posts, etc., were procured from MESSRS. BOULTON AND PAUL, of Norwich, a well-known firm, which I am glad to take this opportunity of complimenting upon the excellence of all the materials supplied. The order was eventually placed in their hands, after a severe competition with several other firms."—R. J. LLOYD PRICE, ESQ., in *The Field.*

Prices on application.

Prices on application.

BOULTON & PAUL, MANUFACTURERS,

PATENT DROP-FENCE FOR RABBITS, &c.
(PROVISIONAL PROTECTION GRANTED.)

Sketch showing Flap down.

This Fence has many advantages over the string netting used for similar purposes; it requires little or no outlay for repairs. All rabbits can be shot, instead of having to knock many on the head, as in the string netting. No drying of nets is necessary after use. One man can thoroughly and effectually accomplish what many can only do indifferently with string netting. At least 900 yards of the fencing can be dropped at one pull, and consequently by placing the terminus in a central position, and pulling with each hand, about a mile of netting can be dropped in an instant. It is economical, simple, and durable, and, notwithstanding that it can be easily thrown out of gear so as to be useless to poachers, is always in position ready for legitimate use when required.

Sketch showing Flap up.

Cash Prices.

3 ft. 6 in. high, with 1¼ mesh, 17 gauge galvanized wire netting, supported on 1 in. angle-iron standards and stays spaced 12 ft. apart.
£5 0 0 per 100 yards.
Without stays and with standards to drive, **£4 5 0** per 100 yards.
Terminal Winding Pillars to suit, fitted with lever and quadrant, **30/-** each

TESTIMONIAL.

Broomhead Hall, Bolsterstone.

Dear Sirs,—We have been using the Drop-Fence for Rabbits, brought out by your firm, for some time. It has always worked without a hitch and saves an immensity of trouble besides answering its purpose admirably in every way. I have not seen a rabbit get over or under it when dropped, and it is certainly a most deadly contrivance for killing down rabbits. We killed over one thousand here the other day by its use in about four hours, and could have killed more if desired.
(Signed) R. H. REMINGTON-WILSON.

Estimates on application   Estimates for any quantities given.

ROSE LANE WORKS, NORWICH.

DROP HURDLE FOR RABBITS.

This Hurdle is constructed upon the same principal as our Patent Drop Fencing, shown on previous page, but has this advantage, it can be easily erected in an existing Fence, being fixed or removed in a very short space of time. By using the above Hurdle and ordinary $1\frac{1}{4}$ in. mesh Wire Netting with plain wood or iron stakes, a very simple but effective Drop Fence may be obtained, which can be worked with ease on almost any class of ground. Where several Hurdles are used they can all be connected to one working wire, and thus be dropped simultaneously.

A glance at above illustration will explain the working—the pulling wire being carried any distance on wood stakes.

Hurdle 6 ft. long, 3 ft. 6 in. high.

Cash Price 15/- each.

WATSON'S PATENT FENCE,
For Protecting Stacks from Vermin.

Made in hurdles 6 ft. long by 3 ft. high, covered with $\frac{3}{8}$ in. galvanized netting.

Cash Price 1/6 per ft. run.

TESTIMONIAL.
From the
HON. HARBORD HARBORD.

I have pleasure in bearing testimony to your very effectual invention for the prevention of rats or vermin getting into the stacks. I have no hesitation in saying that if properly used, so that the hurdles stand quite clear of the stacks, no rats or vermin can get through or over.

Carriage Paid on all Orders above **40/-** *value to the principal Railway Stations in England and Wales.*

BOULTON & PAUL, MANUFACTURERS,

TOOLS FOR ERECTING WIRE AND WIRE NETTING FENCING.

No. 76. STRONG AUGER.
For Boring Wood Posts.

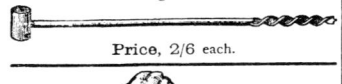

Price, 2/6 each.

No. 77. STEEL SLEDGE HAMMER.
For Driving Standards into the Ground.

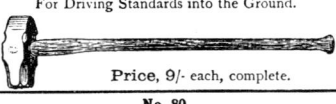

Price, 9/- each, complete.

No. 78. STRONG CUTTING NIPPERS.

For Wire Netting & Solid Wire.
Price, 4/6 per pair.

No. 82. PINCERS AND SPLICER.

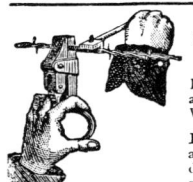

For Splicing Barbed and other Fencing Wires.

Price per set 3/6, at our Works, or delivered with other goods.

No. 80. IMPROVED WIRE STRAINING MACHINE.

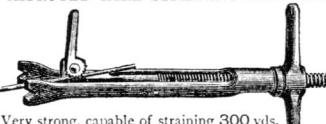

Very strong, capable of straining 300 yds. length of wire. Very portable, weighs only 6½ lbs., and very simple in construction.
Price, 10/- each.

No. 79. STEEL HAND HAMMER.

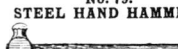

Price, 2/6 each.

No. 81. BAR.
Steeled at both ends.
For fixing Fencing Standards.

Price, 2/6 each.

No. 87. STRAINING SCREWS WITH NUTS FOR WOOD OR IRON POSTS.

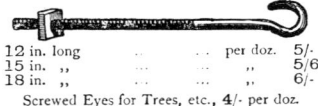

12 in. long	per doz.	5/-
15 in. ,,	,,	5/6
18 in. ,,	,,	6/-

Screwed Eyes for Trees, etc., 4/- per doz.

No. 108. GALVANIZED RAIDISSEURS.

Small size, per doz. 3/- | Extra Large, per doz. 8/6
Large size ,, 5/- | Keys for ditto each 9d.

No. 83. STEEL CHISEL.
To cut Wires.

Price 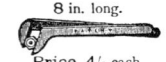 2/- each

No. 84. PATENT SELF-ACTING SPANNER.
8 in. long.

MALLEABLE IRON WEDGES.
No. 88.   **No. 89.**

For Ordinary Standards.
3/- per gross.

For Angle Iron Standards.
3/- per gross.

No. 90. HEATING BARROW.

Price, £2 10 0

Price, 4/- each.

No. 86. WROUGHT-IRON HOOK BOLT.
As used with Angle Iron Standards.

Price, 5/6 per gross, black.

ALL CLASSES OF MATERIAL FOR WIRING WALLS AND GREENHOUSES FOR TRAINING PLANTS, &c.

No. 85. NEW DUPLEX STRAINING SCREW.
For Wood Posts to Strain the Wire both ways.

Price, 1/9 each.

No. 92. WIRE STRAIGHTENING MACHINE.

Price 15/- each.

GALV. WROUGHT-IRON TRAINING HOOKS & EYES.
No. 1.

4 in. long	1/- per doz.	
6 in. ,,	1/3 ,,	
9 in. ,,	1/6 ,,	
12 in. ,,	2/- ,,	

Wrought-iron Plates and Eyes for Straining Wire in Vineries, &c.

Price 3/6 per doz.

Made right and left hand Screws.

MACHINE-MADE STAPLES.
Galvanized.

	per 1000.	cwt.
2 in., No. 6	6/9	19/-
1¾ in., No. 8	4/6	20/-
1½ in., No. 9	3/-	21/-
1¼ in., No. 10	3/6	22/6
1 in., No. 11	2/8	25/6
⅝ in., No. 12	1/10	28/-

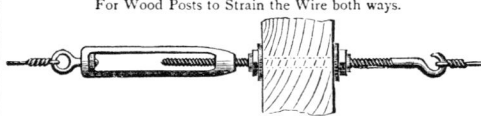

STRAINING BOLT AND HOLDFAST.
Price 4/- per doz.
Keys for turning Nuts, 4d. each.

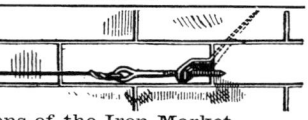

NOTE.—These Prices are subject to the fluctuations of the Iron Market.

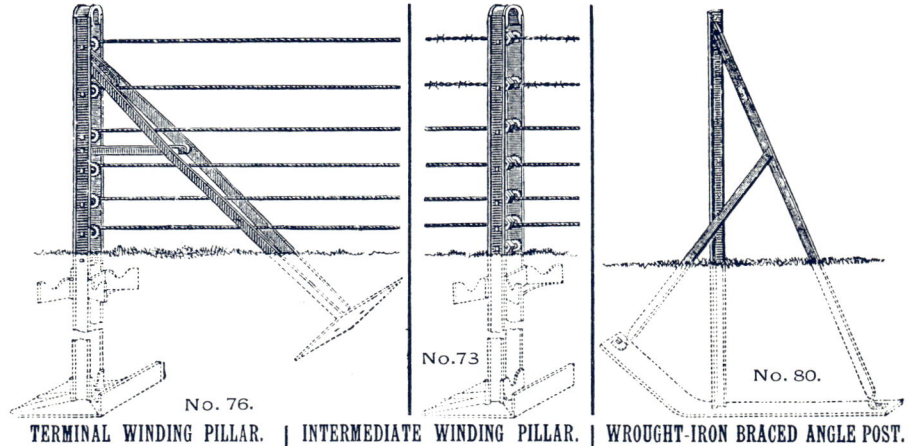

No. 76.	No. 73.	No. 80.
TERMINAL WINDING PILLAR.	INTERMEDIATE WINDING PILLAR.	WROUGHT-IRON BRACED ANGLE POST.
For 3 ft. 6 in. Fence, 5 wires.	For 3 ft. 6 in. Fence, 5 wires.	For 3 ft. 6 in. Fence.
,, 4 ft. 0 in. ,, 6 ,,	,, 4 ft. 0 in. ,, 6 ,,	,, 4 ft. 0 in. ,,
,, 4 ft. 6 in. ,, 7 ,,	,, 4 ft. 6 in. ,, 7 ,,	,, 4 ft. 6 in. ,,

These Pillars take to pieces for convenience in transit. For Prices see accompanying Price List.

WIRE STRAINERS FOR WOOD POSTS.

Cast-iron Brackets with Iron Winders, suitable for straining light Wire Fences.

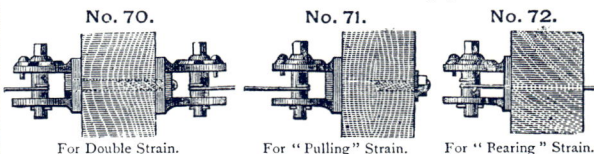

No. 70.	No. 71.	No. 72.
For Double Strain.	For "Pulling" Strain.	For "Bearing" Strain.

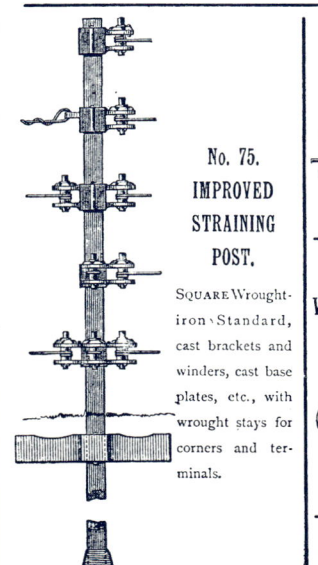

No. 75. IMPROVED STRAINING POST.

SQUARE Wrought-iron Standard, cast brackets and winders, cast base plates, etc., with wrought stays for corners and terminals.

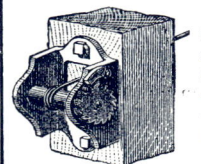

No. 73. BRACKET STRAINERS FOR WOOD POSTS, WITH RATCHET WINDER.

SUITABLE for almost any size wire used for Fencing, and where Trees are converted into Straining Posts, etc.

No. 74. WROUGHT-IRON SIDE STAYS FOR STANDARDS.

With Bolts and Nuts.

For 3 ft. 6 in., 4 ft. and 4 ft. 6 in. Fence.

For Prices see accompanying Price List.

STANDARDS FOR STRAINED WIRE AND WIRE NETTING FENCES.

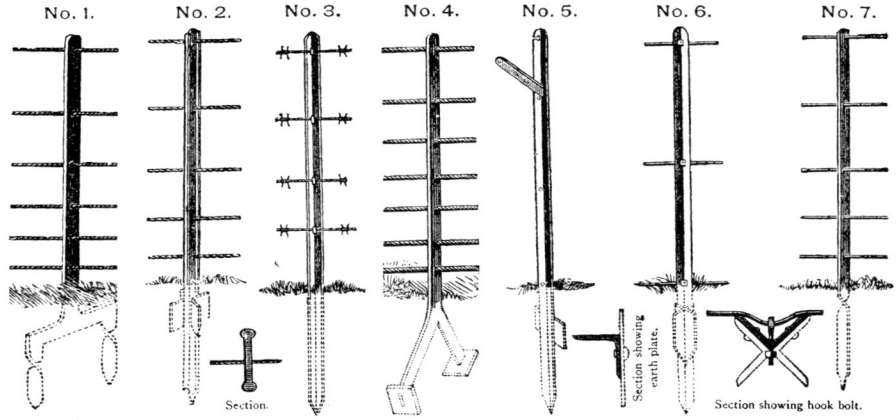

Wire and Standards packed for Shipment.

No. 1. Flat Iron, with double prongs.
No. 2. Bulb Iron, with earth plates.
No. 3. Strong Angle Iron.
No 4. Flat Iron, with anchor feet.
No. 5. Light Angle Iron for Rabbit Fencing.
No. 6. Light Angle Iron for Sheep Fencing.
No. 7. Flat Iron, twisted at base.

For Prices see accompanying Price List.

TESTIMONIAL.

THEODORE R. SAUNDERS, Esq., C.E., M.S.A., Ventnor, Isle of Wight.
I have much pleasure in stating that your workmen have just fixed two and a half miles of Fencing for a new road between Ventnor and Whitwell, and that same has been done in a thoroughly workmanlike and satisfactory manner.

WIRE FOR FENCING.

GALVANIZED STRAND WIRE, 7-PLY.

Gauge.	Yards in One Cwt.
No. 2	About 225
No. 4	,, 300
No. 6	,, 460

BEST SOLID ANNEALED WIRE.

Gauge.	Yards in One Cwt.
No. 2	About 185
No. 4	,, 270
No. 6	,, 340

For Prices see accompanying Price List.

BEST COLD DRAWN ANNEALED WIRE.

THIS wire is drawn from the best quality of iron, and is soft, pliable, and easily fixed by any ordinary labourer. It is much used for light Fences.

GALVANIZED WIRE CABLE, 7-PLY.

THIS is now most extensively used for Fencing, being stronger and more pliable than the solid wire, therefore easier fixed. It is made in very long lengths, from 200 to 1000 yards. It weighs less, and is much stronger than solid wire of the same gauge. Nos. 2 and 4 are used for very strong fences. Nos. 4 and 6 for ordinary cattle fences. The stronger gauge in each case being for the top wire, and the lighter gauge for the lower wires.

ROSE LANE WORKS, NORWICH.

GALVANIZED STEEL BARB FENCING.
FOR FARMERS, GARDENERS, AND COUNTRY GENTLEMEN.

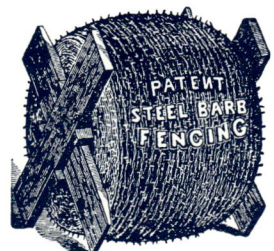

This Fence is now in general use; it possesses an advantage over plain wire, being perfectly impassable by man or beast. It makes a good fence for exposed situations, being a sure defence against trespassers.

By means of this Barb, cattle are prevented from destroying the fence, because, having once come into contact with it, they ever afterwards avoid it, and smaller animals do not dare to pass under or through a fence bristling with sharp barbs or thorns.

The wire composing the strand is made of steel, and is therefore exceedingly strong, much more so than iron wire of a greater thickness. Consisting of two wires, the strand, after once being strained up into its place, always remains straight and firm on the post, through all variations of temperature. The fence is easily seen owing to the projecting barbs. A Barb Fence of three, or at most four lines, is considered equal in point of strength and efficiency to a Seven-line Fence of the ordinary kind. It can be used with either wood or iron posts.

The "Barb Wire" consists of a strand of two wires into which pointed barbs, projecting at right angles, are fixed at intervals as shown in Illustration. The wire neither rusts, stains, decays, nor warps. It is unaffected by fire, wind, or flood, and is a complete barrier against the most unruly stock.

Approximate Lengths and Weights.

	ORDINARY. *i.e.* Barbs 5 in. apart.		THICK-SET. *i.e.* Barbs half the usual distance.	
	Per 100 yds.	Per cwt.	Per 100 yds.	Per cwt.
2-point	20 lbs.	560 yds.	23 lbs.	488 yds.
4-point	21 lbs.	533 yds.	25 lbs.	448 yds.

GALVANIZED STEEL BARB WIRE.

4-pointed.   2-pointed.

Full size of Barbed Wire.

Ordinary set 2-pointed or 4-pointed
100 yards length, 2-pointed ...
280 ,, ,, (½ cwt.)
560 ,, ,, (1 cwt.)
For Prices see accompanying Price List.
1 cwt. Reels free. For smaller quantities, Reels 6d. each extra, not returnable.
2-point Barb Wire always supplied unless otherwise ordered.

BARB FENCING STRETCHER.
For Straining the Fence and Carrying the Reels.

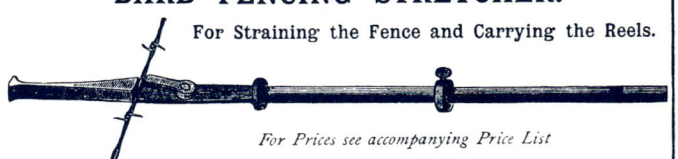

For Prices see accompanying Price List

SWIFT ON WHEELS.
For uncoiling Fencing Wire.

Saves a deal of labour in erecting wire fences. The Swift can be taken off and the reels of Barbed Wire placed on the spindle.

BOULTON & PAUL, MANUFACTURERS.

IMPROVED STRAINED WIRE FENCES.

Winding Pillars are not included in the following estimates, but are charged *extra* (see page 243), because the number required varies greatly according to the situation. For a length of Fencing without gateways, two straining pillars are sufficient for about 300 yards.

SPECIFICATIONS.
Special Estimates on application.

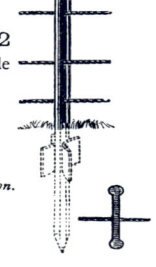

No. 1. For Light Cattle & Sheep. 3 ft. 6 in. high, with 5 lines of wire; two top ones No. 4 gauge, and the lower ones No. 6 gauge. The standards are $1\frac{1}{4}$ in. wide by $\frac{5}{16}$ in. thick, placed 8 ft. apart.

No. 2. For General Purposes. 4 ft. high, with 6 lines of wire; the top one No. 2 gauge, lower ones No. 4 gauge. The standards are $1\frac{1}{4}$ in. wide by $\frac{3}{8}$ in. thick, placed 8 ft. apart.

No. 3. For Horses and Heavy Stock. 4 ft. 6 in. high, with 7 lines of wire, the top one No. 2 gauge, and the lower ones No. 4 gauge. The standards are $1\frac{1}{2}$ in. wide by $\frac{3}{8}$ in. thick, placed 8 ft. apart.

For Prices see accompanying Price List.

If a rough plan of the proposed line of Fence is sent, showing number of yards from point to point, with the Gateways, Angles, &c., we will furnish a special estimate to include the cost o the necessary Pillars and erection.

TESTIMONIALS.

From P. BUCKLEY, ESQ., Engineer, Brundall and Yarmouth Railway.
The Fencing on this line is fixed in a workmanlike manner, it could not be fixed better.

From J. THEOBALD, ESQ., Havering, Bedford.
MR. THEOBALD has much pleasure in saying that the Wire Deer Fencing erected by you has given him great satisfaction.

These Fences can be had with Bulb Iron Standards at $\frac{1}{2}d.$ per yard less.

ROSE LANE WORKS, NORWICH.

WIRE FENCING WITH WOOD POSTS.

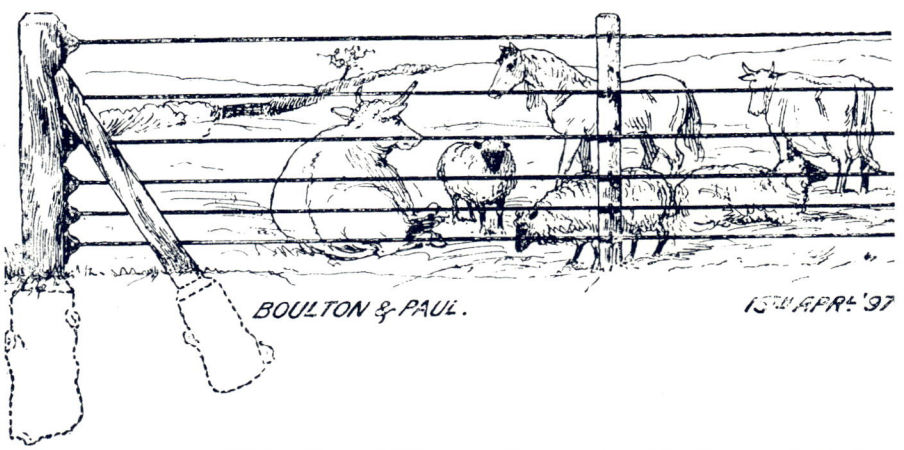

No. 4.
For Sheep.
With 5 lines of Wire, top No. 4, others No. 6 gauge, including Staples, for fixing on purchaser's own wood posts, 3 ft. 6 in. high, spaced about 9 ft. apart.

No. 5.
For General Purposes.
With 6 lines of Wire, all No. 4 gauge, including Staples, for purchaser's posts, 4 ft. high, spaced about 9 ft. apart.

No. 6.
For Horses and Heavy Stock.
With 7 lines of Wires, top No. 2, others No. 4, including Staples for purchaser's wood Posts, 4 ft. 6 in. high, about 9 ft. apart.

For Straining Screws and Winders for tightening Wires, see pages 242 and 243.

No. 81. WROUGHT TUBULAR ANGLE POST.

For Wire Fences.

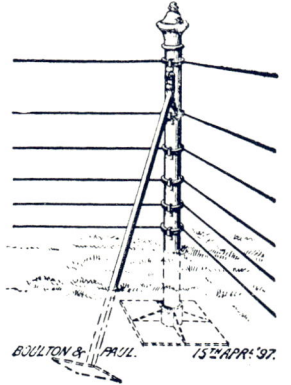

3 ft. 6 in. high	5 wires.
4 ft. 0 in. ,, ...	6 ,,
4 ft. 6 in. ,, ...	7 ,,

STRAINED WIRE FENCES WITH ANGLE-IRON UPRIGHTS.

THESE Fences are more readily fixed than those with anchor feet Standards, no digging being required, and a good foundation is secured by the Standards being driven, without risk of breakage, into the solid subsoil.

The improvement in the appearance of the Fences is very great, and the uprights are made of a superior quality of iron to the ordinary flat bar standards.

As each Standard consists of one solid bar without weld or joint, it cannot be broken at the ground line or bent down by cattle—the V shape being one of the strongest forms in which iron can be rolled.

The system of vertically wedging the wires effectually retains them in a perpendicular position and prevents them ever being moved by the rubbing of cattle.

Section of Standard.

No. 15. For Sheep and Light Cattle.

Special Estimates on application.

3 ft. 6 in. high above ground, 5 lines of wire, all No. 4 Galvanized 7-ply Strand. Angle-iron Standards spaced 7 ft. apart.

No. 16. For General Purposes.

4 ft. high above ground, 6 lines of Galvanized 7-ply Strand Wire, top one No. 2 gauge, lower ones No. 4 gauge. Angle-iron Standards 7 ft. apart.

Section of Standard showing mode of fixing the barbed wire.

THIS Fence is now in general use, and for cheapness is much recommended. With four lines of the barbed wire it is considered equal in efficiency to a seven-line fence of the ordinary kind.

No. 17. For General Purposes.

4 ft. high above ground, 4 lines of four-pointed Barbed Wire Angle-iron Standards, 1¼ in. by 1¼ in. by 7⁄₁₆ in., spaced 10 ft. apart.

For Prices see accompanying Price List.

TESTIMONIAL.

From M. K. DIXON, ESQ., Searles' Estate Office,
Fletching, Sussex.

I am glad to say the Fence has given great satisfaction.

It will save unnecessary correspondence if inquirers will send a Ground Plan with measurements, showing Angles and Gateways if any.

No. 1 STEP-LADDER FOR WIRE FENCING.

For Prices see accompanying Price List.

248   BOULTON & PAUL, MANUFACTURERS.

CONTINUOUS BAR FENCING WITH T IRON UPRIGHTS.

THESE Fences are more readily fixed than those with pronged-feet Standards, no digging being required, and a good foundation is secured by the Standards being driven, without risk of breakage, into the solid subsoil.

The improvement in the appearance of the Fence is very great, and the uprights are made of a superior quality of iron to the ordinary flat bar Standards.

As each Standard consists of one solid bar without weld or joint, it cannot be broken at the ground line or bent down by cattle—the T shape being one of the strongest forms in which iron can be rolled.

The system of vertically wedging the Bars effectually retains them in a perpendicular position, and prevents them ever being moved by the rubbing of cattle.

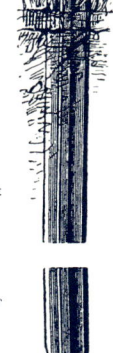

Standards 3 feet apart.

No. 18. For Sheep and Light Cattle.

No. 19. For General Purposes.

Special Estimates on application.

3 ft. 6 in. high, 5 bars, top one $\frac{5}{8}$ in. diameter, and lower ones flat, 1 in. by $\frac{1}{4}$ in.

3 ft. 9 in. high, 6 bars, top one $\frac{3}{4}$ in. diameter, and lower ones flat, 1 in. by $\frac{1}{4}$ in.

All Horizontal Bars of Round Iron   Standards 3 feet apart.

No. 27. For Sheep and Light Cattle.

No. 28. For General Purposes.

3 ft. 6 in. above ground, with 5 bars, top one $\frac{5}{8}$ in., others $\frac{1}{2}$ in. diameter.

3 ft. 9 in. above ground, with 5 bars, top one $\frac{3}{4}$ in., others $\frac{5}{8}$ in. diameter.

For Prices see accompanying Price List.

The above specifications are for Fences in constant demand. Special Estimates for Fences of any other specification, height, or number of bars, will be furnished on application.

Varnishing, painting with oxide of iron paint, and fixing charged extra, according to quantity. Two assistant labourers to be provided by purchaser.

TESTIMONIAL.

From H. A. BARCLAY, ESQ., The Warren, Cromer.

I have much pleasure in certifying that the work you did on my Wire and Iron Fences was done in a FIRST RATE and workmanlike way, and does you great credit.

WROUGHT-IRON CONTINUOUS FLAT-BAR FENCING.

These Fences can be used in any situation; they are as easily fixed in curves or over uneven ground as in straight and level lines; no side stays or extras of any kind are required. The Fences specified in the estimates below are complete in themselves; they can be fixed, taken up, and refixed almost as easily as iron hurdles.

Special Estimates on application.

No. 2. For Sheep and Light Cattle. — 3¼ ft. high above ground, with 5 horizontal bars, the top one round, ⅝ in. diameter, others flat, 1 in. wide by ¼ in. thick. Standards, with welded double prongs, 3 ft. apart; the adjoining uprights are 1¼ in. wide by ⅜ in. thick, and the intermediate uprights 1⅜ in. wide by 1/16 in. thick.

No. 3. For General Purposes. — 3 ft. 9 in. high above ground, with 5 horizontal bars, the top one round, ¾ in. diameter, others flat, 1 in. wide by ¼ in. thick. Standards, with welded double prongs, 3 ft. apart; the adjoining uprights are 1¾ in. wide by ⅜ in. thick, and the intermediate uprights 1½ in. wide by ⅜ in. thick.

No. 4. For Horses and Heavy Stock. — 4 ft. 6 in. high above ground, with 6 horizontal bars, the top on round, ¾ in. diameter, others flat, 1 in. wide by ¼ in. thick. Standards, with welded double prongs, 3 ft. apart; the adjoining uprights are 1¾ in. wide by ⅜ in. thick, and the intermediate uprights 1¼ in. wide by ⅜ in. thick.

Fixing and varnishing charged extra; two assistant labourers to be provided by the purchaser.

For Prices see accompanying Price List.

These Fences can be had with Bulb Iron Standards at 1d. per yard less.

BLACK VARNISH FOR IRON FENCING, &c.

Quality Guaranteed

In 40-gallon and 18-gallon Casks.

For Prices see accompanying Price List, and page 236 of this Catalogue.

This Varnish is an excellent substitute for Oil Paint, *at one-third the cost*, and should be kept in stock on every Estate and Farm. It is always in a fit state for use, is applied *cold*, requiring no heating, mixing, or other preparation, will dry quickly, and leave a bright, hard, Jet-black surface. Can be laid on by any ordinary labourer.

ROSE LANE WORKS, NORWICH. 249A

CONTINUOUS BAR FENCING WITH H IRON UPRIGHTS.

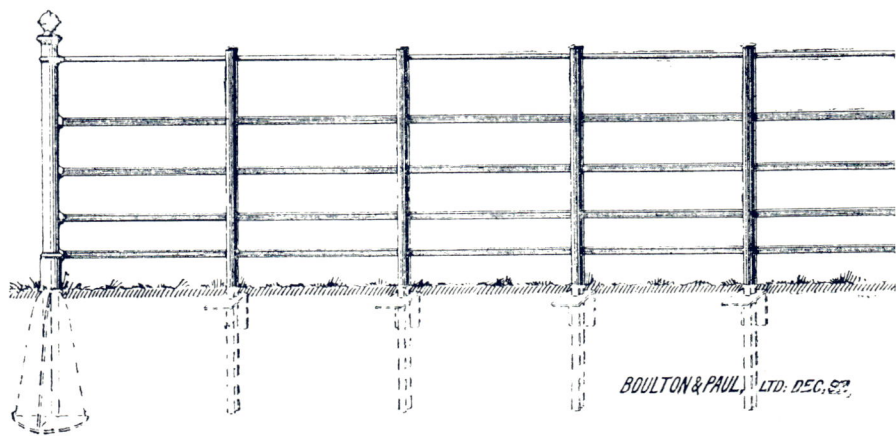

These Fences are a great improvement upon the ordinary Fencing with pronged-feet standards, no digging being required, and a good foundation is secured by the Standards being driven, without the possibility of breakage, into the solid subsoil, and can be easily erected by unskilled workmen.

As will be seen from the engraving, the Standards are in one solid piece of Double Tee or H section iron, no welding being required, the whole being manufactured cold, and the danger of the feet breaking off at the ground line is avoided.

SPECIFICATIONS.

No. 5. For Medium Stock. 3 ft. 6 in. high above ground, with five horizontal bars, top ¾ in. diameter, others flat, 1 in. wide by ¼ in. thick. Standards of H section iron, spaced 3 ft. apart, without earth plates.

No. 6. For General Purposes. 4 ft. high above ground, with five horizontal bars, top ¾ in. diameter, others flat, 1 in. wide by ¼ in. thick. Standards of H section iron, spaced 3 ft. apart, with earth plates.

With six horizontal bars, 2d. per yard extra.

No. 7. For Heavy Stock. 4 ft. 6 in. high above ground, with six horizontal bars, top ¾ in. diameter, others flat, 1 in. wide by ¼ in. thick. Standards of H section iron, spaced 3 ft. apart, with earth plates.

No. 8. Extra Strong. For Heavy Stock. 4 ft. 6 in. high above ground, with six horizontal bars, top ⅞ in. diameter, others flat, 1 in. wide by 5/16 in. thick. Standards of H section iron, spaced 3 ft. apart, with earth plates.

For particulars of Field Gates see page 251.
Fixing and varnishing by our men charged extra.
For Prices see accompanying List.

Half full size section of standard.

WROUGHT-IRON DOUBLE-LEAF GATES.

No. 128.

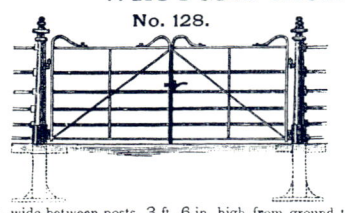

9 ft. wide between posts, 3 ft. 6 in. high from ground to top bar when hung.
Pair Gates only, with hangings for wood posts . £ ,, ,,
Pair Cast-iron Posts with self-fixing bases ,, ,, ,,

No. 129.

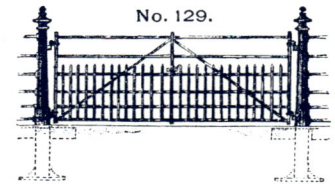

With vertical dog bars, 10 ft. wide between posts, 3 ft. 9 in. high from ground when hung.
Pair Gates only, with hangings for wood posts .. £ ,, ,,
Pair Cast-iron Posts with self-fixing bases ,, ,, ,,

These Gates are used in connection with the Continuous Bar and Wire Fencing described in this Catalogue.

No. 107.
PLAIN WICKET GATE.

This is a Plain Wicket Gate which can be made any required height, and with any number of bars, to match any of the fences described in this Catalogue; or for any other situation, such as entrances to cottage gardens, openings in walls, hedgerows, palings, or other fences.

Gate with hangings and catch, 3 ft. 6 in. high, 3 ft. 6 in. wide ... .. ,,
,, ,, ,, 3 ft. 9 in. ,, 3 ft. 6 in. ,, ... ,,
,, ,, ,, 4 ft. 6 in. ,, 3 ft. 6 in. ,, ... ,,
Standards for above, 3 ft. 6 in. high ... ... per pair ,,
,, ,, 3 ft. 9 in. ,, ... ,,
,, ,, 4 ft. 6 in. ,, ... ,,

No. 30. WROUGHT-IRON FOOTPATH GATE.

This Footpath Gate can be made any required height and with any number of bars; it may be fixed in a wall, hedgerow, paling, or other fence.
Gate with semi-circular guard to suit any fence up to 4 ft. 6 in. high . ,,
Cast-iron pillars with self-fixing bases, ex. ,,

STEP-LADDER OR STILE FOR IRON OR WIRE FENCES.

No. 2.

These are most convenient for crossing where gates have not been provided. They are readily removed from place to place, and take to pieces for packing.
Suitable for any fence up to 4 ft. 6 in. high ,,

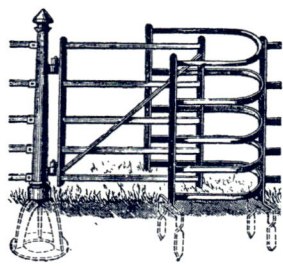

No. 31. WROUGHT-IRON FOOTPATH GATE.

This Footpath Gate can be made any required height, and with any number of bars, so as to match any of the fences described in this Catalogue; or it may be fixed in a wall, hedgerow, paling, or other fence.

Gate with semi-circular hurdle to suit any fence up to 3 ft. 6 in. high ... ,,
Cast-iron pillar with self-fixing base ... ... each ,,
Gate with semi-circular hurdle to suit any fence up to 4 ft. 6 in. high ... ,,
Cast-iron pillar with self-fixing base ... ... each ,,

For Prices see accompanying Price List.

No. 124. WROUGHT-IRON FIELD GATE.

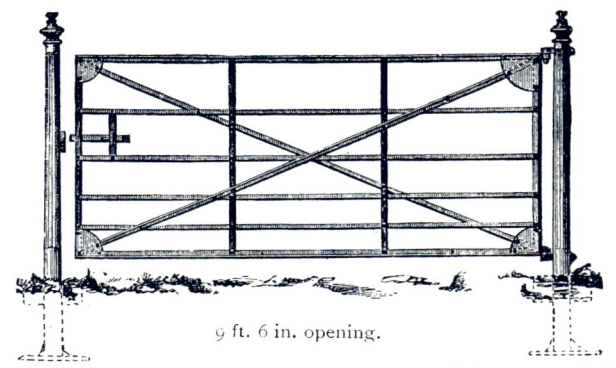

9 ft. 6 in. opening.

This Gate is constructed principally of angle and T iron, with plates in corners, which give great strength without a corresponding increase in weight.

Size—9 ft. wide, 4 ft. high from ground to top bar, top and bottom bar T iron, $1\frac{1}{2}$ in. by $1\frac{1}{2}$ in. by $\frac{1}{4}$ in., heel bar $1\frac{3}{4}$ in. by $\frac{3}{8}$ in., head bar $1\frac{1}{2}$ in. by $\frac{3}{8}$ in., horizontal bars $1\frac{1}{4}$ in. by $\frac{1}{4}$ in. The diagonal bars are angle-iron, 1 in. by 1 in. by $\frac{3}{16}$ in.

No. 125. WROUGHT-IRON FIELD GATE.

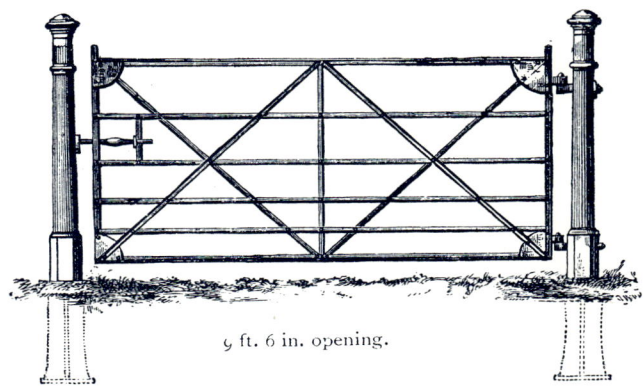

9 ft. 6 in. opening.

This is a strong and substantial Gate, and can be much recommended either as a Field or Entrance Gate. It is constructed upon the same principle as No. 124, but with double diagonal braces.

Size—9 ft. wide, 4 ft. high from ground to top bar, top and bottom bar T iron, $1\frac{1}{2}$ in. by $1\frac{1}{2}$ in. by $\frac{1}{4}$ in., heel bar $1\frac{3}{4}$ in. by $\frac{3}{8}$ in., head bar $1\frac{1}{2}$ in. by $\frac{3}{8}$ in., horizontal bars 1 in. by $\frac{1}{4}$ in., brace bars angle-iron, 1 in. by 1 in. by $\frac{3}{16}$ in.

For Prices see accompanying Price List.

Further Designs of Entrance, Carriage, Park, and Field Gates can be had on application.

BOULTON & PAUL, MANUFACTURERS.

IMPROVED WROUGHT-IRON TREE GUARDS.
Special Estimates on application.

No. 400.
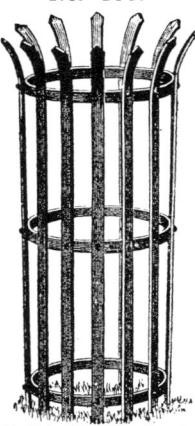

6 ft. high, 2 ft. 0 in. diameter.
6 ft. ,, 2 ft. 3 in. ,,
4 Vertical Bars, 1¼ in. by ¼ in.,
others 1 in. by ¼ in.

No. 405.
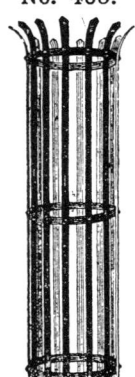

6 ft. high 12 in. diameter,
4 Vertical Bars, 1 in. by ¼ in.,
others ¾ in. by ¼ in.

No. 406.
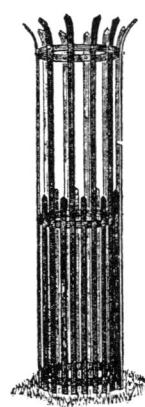

6 ft. high, 12 in. diameter,
4 Vertical Bars, 1 in. by ¼ in.,
others ¾ in. by ¼ in.

No. 407.

6 ft. high, 12 in. diam. at
top, 16 in. diam. at bottom,
4 Vertical Bars, 1 in. by ¼ in.,
others ¾ in. by ¼ in.

All these Guards are supplied with plate feet, but pronged feet can be had if preferred.

TESTIMONIAL.
From THE MOST NOBLE THE MARQUIS OF CHOLMONDELEY,
Houghton Hall.
I have received the Tree Guards all right. I think they are very good and cheap
(*Signed*) J. HARTNELL, Bailiff.

No. 408.
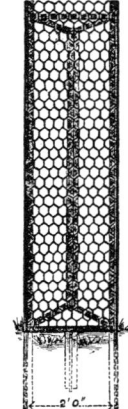

T Iron Uprights covered with
1½-in. strong Galvanized Netting.
5 ft. high, 18 in. diameter.
6 ft. ,, 18 in. ,,

No. 402.
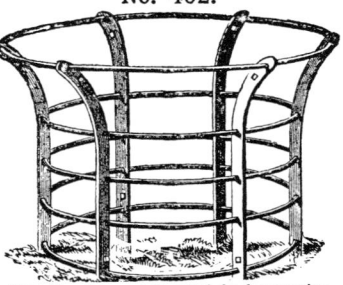

This Guard is specially suited for low-growing
and any bushy trees or for clumps. Very
effective and strong against cattle.

4 ft. high, 4 ft. diameter, 5 rings, top one
⅜ in., others ½ in.
4 ft. 6 in. high, 4 ft. diameter, 5 rings, top
one ⅝ in., others ½ in.

Prices on application.

BARB WIRE TREE GUARD.
6 ft. high, 12 in. square.

No. 409.
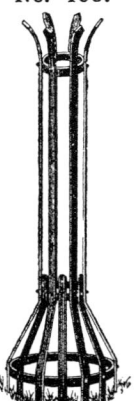

6 ft. high, 6 in. diam. at
top, 18 in. diam. at bottom,
2 Vertical Bars, 1 in. by
¼ in., others ¾ in. by ¼ in.

IMPROVED WROUGHT-IRON FIELD HURDLES.

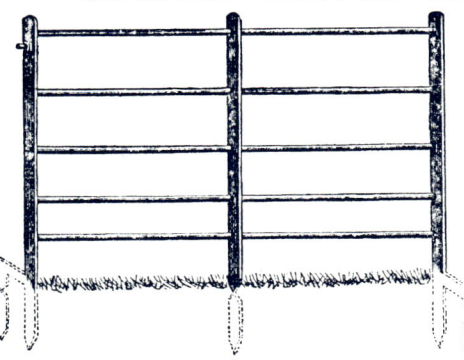

Special Estimates on application.

No. 7a. LIGHT HURDLE FOR GENERAL PURPOSES.

6 ft. long, 3 ft. 6 in. high above ground, with 5 round bars, the top one $\frac{3}{4}$ in. diameter, and the lower ones $\frac{5}{8}$ in. diameter, 3 uprights $1\frac{1}{4}$ in. wide by $\frac{3}{16}$ in. thick, with one bolt and nut, painted with oxide of iron paint.

No. 8a. STRONG HURDLE FOR HORSES & HEAVY CATTLE.

6 ft. long, 3 ft 9 in. high above ground, with 5 round bars, all $\frac{3}{4}$ in diameter, and 3 uprights $1\frac{1}{2}$ in. wide by $\frac{3}{4}$ in. thick, with one bolt and nut, painted with oxide of iron paint.

No. 10. LIGHT CATTLE HURDLE.

6 ft. long, 3 ft. 6 in. high above ground, with 5 bars, top one $\frac{3}{4}$ in. round, others flat, 1 in. wide by $\frac{1}{4}$ in. thick, end uprights $1\frac{1}{4}$ in. wide by $\frac{1}{4}$ in. thick, middle upright $1\frac{3}{8}$ in. wide by $\frac{3}{16}$ in. thick, with one bolt and nut, painted with oxide of iron paint.

No. 11. STRONG OX HURDLE.

6 ft. long, 4 ft. high above ground, with 5 bars, top one $\frac{3}{4}$ in. round, others flat, 1 in. wide by $\frac{1}{4}$ in. thick, end uprights $1\frac{1}{2}$ in. wide by $\frac{3}{16}$ in. thick, middle upright $1\frac{1}{2}$ in. by $\frac{3}{4}$ in., with one bolt and nut, painted with oxide of iron paint.

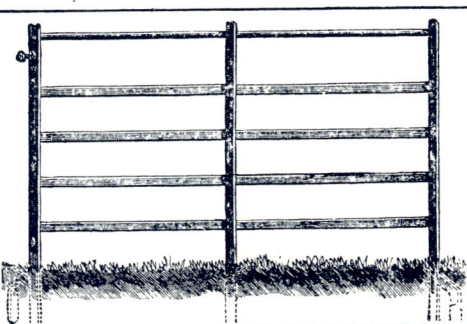

No. 28. LIGHT CATTLE HURDLE.
(New Pattern.)

THESE Hurdles can be driven without risk of breakage into the solid subsoil, a good foundation is secured and no digging required.
6 ft. long, 3 ft. 6 in. high, $\frac{3}{4}$ in. round top bar, four $\frac{1}{2}$-in. round lower bars, angle-iron ends.

No. 29. STRONG CATTLE HURDLE.

6 ft. long, 3 ft. 9 in. high, $\frac{3}{4}$ round top bar, 4 flat bars, angle-iron ends, painted with oxide of iron paint.

FOR GATES TO MATCH THESE HURDLES, SEE PAGE 250

For Prices see accompanying Price List.

No. 14. SUPERIOR WROUGHT-IRON SHEEP-FOLD HURDLES

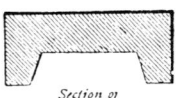

Section of Channel Steel Bottom Bar.

Specification of the cheapest and best Hurdle in the Market.

THE top and ends being made of T shaped steel bar in one piece, give great strength to the hurdle ; the rounded corners are a great improvement on those hitherto made, being much less dangerous than the sharp corners, as well as adding to the strength. Each Hurdle is 12 ft. long, 3 ft. 6 in. high, and weighs nearly 1½ cwt. ; they readily take to pieces, and are easily packed to travel long distances by rail or road. T shaped steel top and ends in one length, 1¼ in. wide by ¼ in. thick ; four middle bars 1 in. wide by ¼ in. thick ; bottom bar of channel section steel, 1¼ in. wide by ¼ in. thick, which adds greatly to the strength ; wheels 10 in diameter ; with chains and all necessary bolts and nuts ; painted with oxide of iron paint.

For Prices see accompanying Price List.

Sample Hurdle sent Carriage Free.

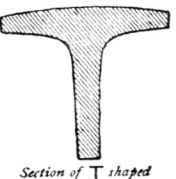

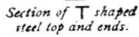

Section of T shaped steel top and ends.

No. 126.

STRONG WROT.-IRON UNCLIMBABLE GATE.

Suited to situations exposed to trespassers.

THE vertical bars are flat, pointed at top, and riveted upon a framework of angle-iron, sufficiently close to prevent the insertion of the foot.

For 9 ft. 6 in. opening, 4 ft. 6 in. high,

With Hangings for wood or stone posts,

For Prices of Cast-iron Posts see page 267.

No. 12c. WROUGHT-IRON HURDLES FOR FOLDING SHEEP AND GENERAL PURPOSES.

THESE Hurdles are specially recommended to Farmers, as they are made extra strong, so as not to be injured by constant removal, and will be found much better adapted for Folding Sheep than Hurdles with light uprights and projecting feet. They are connected by a link attached to the top bar, which is preferable to bolts and nuts where Hurdles are frequently moved.

Specification—6 ft. long, 3 ft. high, with Angle-iron Ends, ⅝-in. round top bar, 3 flat bars.

For Prices see accompanying Price List.

Carriage Paid on Ten Hurdles. Special Estimates on application.

No. 16. WROUGHT-IRON LATTICE HURDLE.

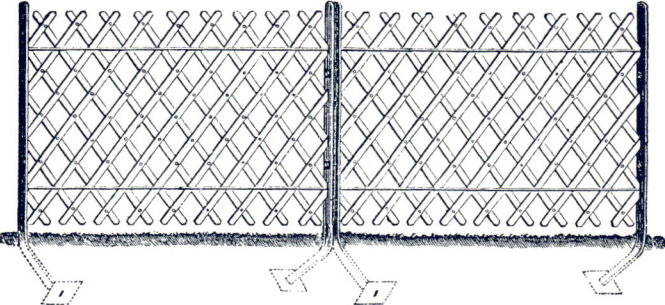

A SUITABLE Fence for enclosing gardens, excluding dogs, poultry, &c. Made in Hurdles 5 ft. long, 4 ft. high above ground, with anchor feet, the end uprights and bottom bar are made of angle-iron. These Hurdles are made with but one horizontal bar riveted to bottom of lattice bars, and not as shown in illustration. The other bars which form the diamond pattern are flat, $\frac{3}{4}$ in. wide by $\frac{3}{16}$ in. thick, riveted upon each other and into the end uprights. It makes a very strong and ornamental Fence. Painted with oxide of iron paint, and with all necessary bolts and nuts.

For Prices see accompanying Price List.

No. 17. EXTRA STRONG WROUGHT-IRON HURDLE.

WROUGHT-IRON Fence made in form of Hurdles, suitable for gardens. 4 ft. high above ground, with anchor feet. The vertical bars are of round iron, placed about 6 in. apart at top and 2¼ in. at bottom, and pointed at the top. The end uprights and horizontal bars are 1½ in. wide by $\frac{3}{16}$ in. thick. Very strong, and difficult to climb. Painted with oxide of iron paint, and with all necessary bolts and nuts.

Made with ½-in. or ⅝ in. round vertical bars, also with 1-in. by ¼-in. flat bars placed flat-ways towards you.

Hand Gates and wrought Standards to match.

For Prices see accompanying Price List.

TESTIMONIAL.

From MR. JAMES CLARK, Gardener, Distington Hall.

We received the Hurdles all right, and in good condition. They are just what we wanted, and they give satisfaction.

Special Estimates on application.

BOULTON & PAUL, MANUFACTURERS.

WROUGHT-IRON UNCLIMBABLE GATES AND RAILINGS.
No. 91.

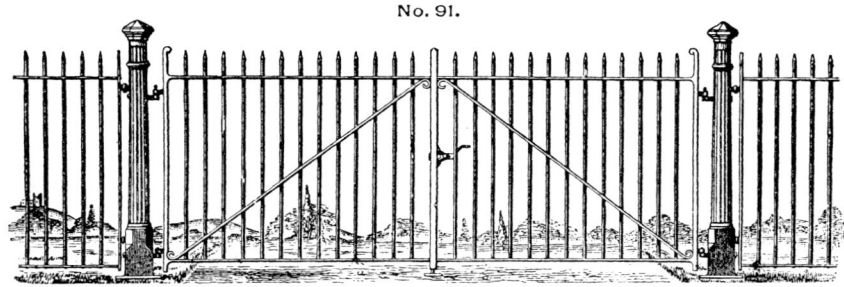

These Gates are cheap and substantial, for use with the Unclimbable Hurdles, for public situations. The bars are made with sharp-pointed tops, where trespassers may be expected, but can be made with rounded points if desired. They are hung to handsome fluted cast-iron posts, and are for 9 ft. opening, by 4 ft. high from ground.

Special Estimates on application.

Pair of Gates with ½-in. round vertical bars spaced 3½ in. apart, rails and braces 1¼ in. by ⅜ in. £ ,, ,, ,,
Pair of Gates with ⅝-in. round vertical bars spaced 3⅝ in. apart, rails and braces 1½ in. by ⅜ in. ,, ,, ,,
Cast-iron Fluted Posts, No. 11b, with self-fixing iron bases (see page 260) ... ... ,, ,, ,,

No. 15. WROUGHT-IRON UNCLIMBABLE HURDLES.

NOTE.—The Prongs for these Hurdles are sent loose to prevent breakage in transit.

This is an excellent Fence for a roadside, for enclosing churchyards, schools, &c., as it is most difficult to climb. The vertical bars are of round iron placed 3¼ in. apart, and are pointed at top, end uprights and horizontal bars are 1½ in. by ⅜ in. Painted with oxide of iron, and all necessary bolts and balls, ready for fixing.

Hurdles 6 ft. long, 4 ft. high above ground, with ½-in. vertical bars.   Hurdles 6 ft. long, 4 ft. high, with ⅝-in. vertical bars.
Hurdles 6 ft. long, 5 ft. high, with ⅝-in. vertical bars.

Extra strong Fence, with ⅝-in. square bars placed angle-ways.   Hand Gates to match, with wrought-iron standards.

For Prices see accompanying Price List.

TESTIMONIAL.
From FREDERICK COLMAN, Esq., Kew.

Our Churchyard Fence being now completed, I feel it only my duty to express our entire satisfaction with the work, and also with the able manner in which your fixer has carried it out. It is due to so steady and competent a workman that I should say this much.

WROUGHT-IRON UNCLIMBABLE PARK FENCING.

Special Estimates on Application.

No. 26.   No. 27.   No. 26.

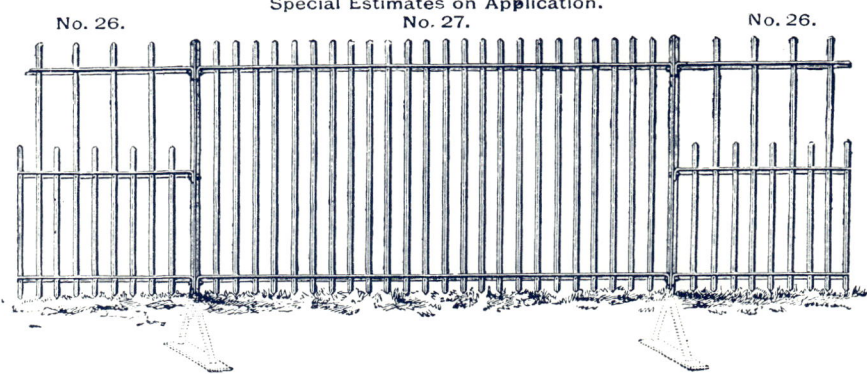

This Fencing is suited for very exposed situations, and will resist the most expert climber. The vertical bars are flat towards you, pointed top and bottom, and riveted to angle-iron horizontal bars. Made in panels 9 ft. long, 4 ft. high above ground; the ends of the horizontal bars being firmly bolted to T iron standards, made with self-fixing anchor bases.

For Prices see accompanying Price List.

AUTOMATIC UNCLIMBABLE RAILING.

SIDE VIEW   FRONT VIEW.

This Fencing can be arranged to fix to any fall of ground.

For Prices see accompanying Price List.

HAND GATE FOR UNCLIMBABLE FENCING.

No. 127.

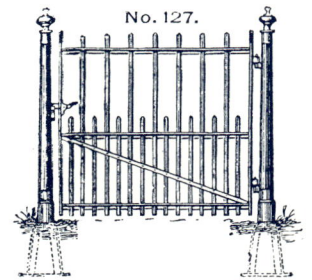

The vertical bars are flat, pointed at the top, and riveted upon a framework of angle-iron.

4 ft. high, for 3 ft. 6 in. opening, each Cast-iron posts, with self-fixing iron bases ... ... per pair

TESTIMONIALS.

From THE RIGHT HON. THE EARL OF EFFINGHAM, Tusmore Park.
The Fencing arrived safely and to his Lordship's entire satisfaction.

Portmore Lodge, St. Filliers.
Miss Mackintosh is very much pleased with the Gates and Palisading, just put up by Messrs. Boulton and Paul, and also with their prompt attention to her telegram. Everything is most complete.

BOULTON & PAUL. MANUFACTURERS.

ORNAMENTAL GARDEN AND PARK HURDLES.
WITH PRONGED FEET FOR FIXING INTO THE GROUND.
Special Estimates on application.

No. 19.

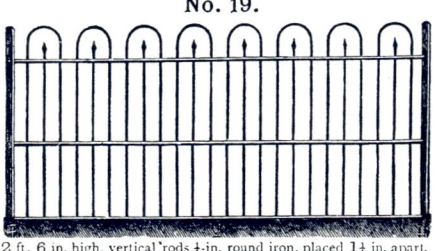

2 ft. 6 in. high, vertical rods ½-in. round iron, placed 1¼ in. apart.
3 ft. high, with ⅝-in. round vertical bars, 1½ in. apart.
Hand Gate to match, with wrought-iron standard.

No. 20.

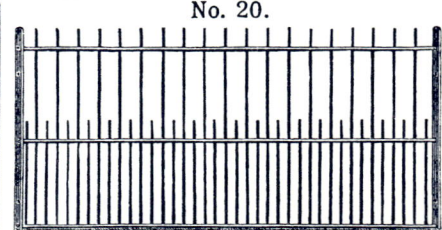

2 ft. 6 in. high, vertical rods ½-in. round iron, placed 1¼ in. apart.
3 ft. high, with ⅝-in. round vertical rods placed 2 in. apart.
Hand Gate to match, with wrought-iron standard.

No. 22.

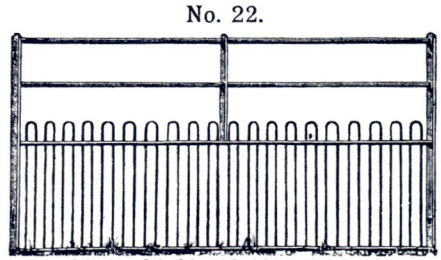

6 ft. long, 3 ft. 6 in. high, ¼-in. vertical wires.
Hand Gate to match, with wrought-iron standard.

No. 25.

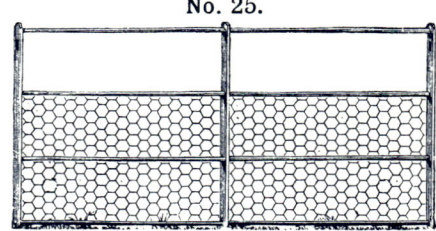

6 ft. long, 3 ft. high, covered to second bar with 1½ in. netting.
Hand Gate to match, with wrought-iron standard.

No. 18.

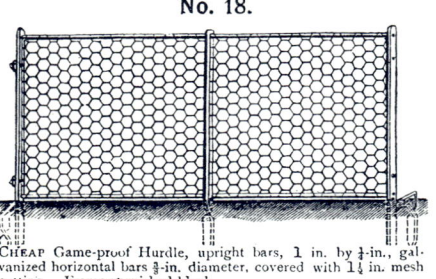

CHEAP Game-proof Hurdle, upright bars, 1 in. by ¼-in., galvanized horizontal bars ⅜-in. diameter, covered with 1¼ in. mesh netting. Frame varnished black.
2 ft. 6 in. high.
3 ft. high, with 3 horizontal bars.

No. 8.

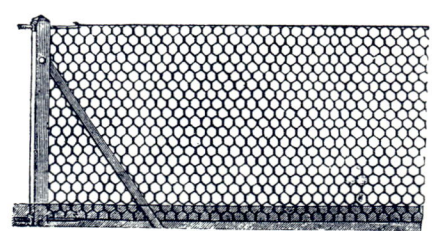

CHEAP Rabbit and Game-proof Fence, 2 ft. 6 in. high, for surrounding flower gardens, pleasure grounds, etc., including stout galvanized wire top and bottom, galvanized netting, 1½ in. mesh, No. 18 gauge, with standards two yards apart.

For Prices see accompanying Price List.

WROUGHT-IRON PALISADING FOR COTTAGES & GARDENS.

Special Estimates on application.

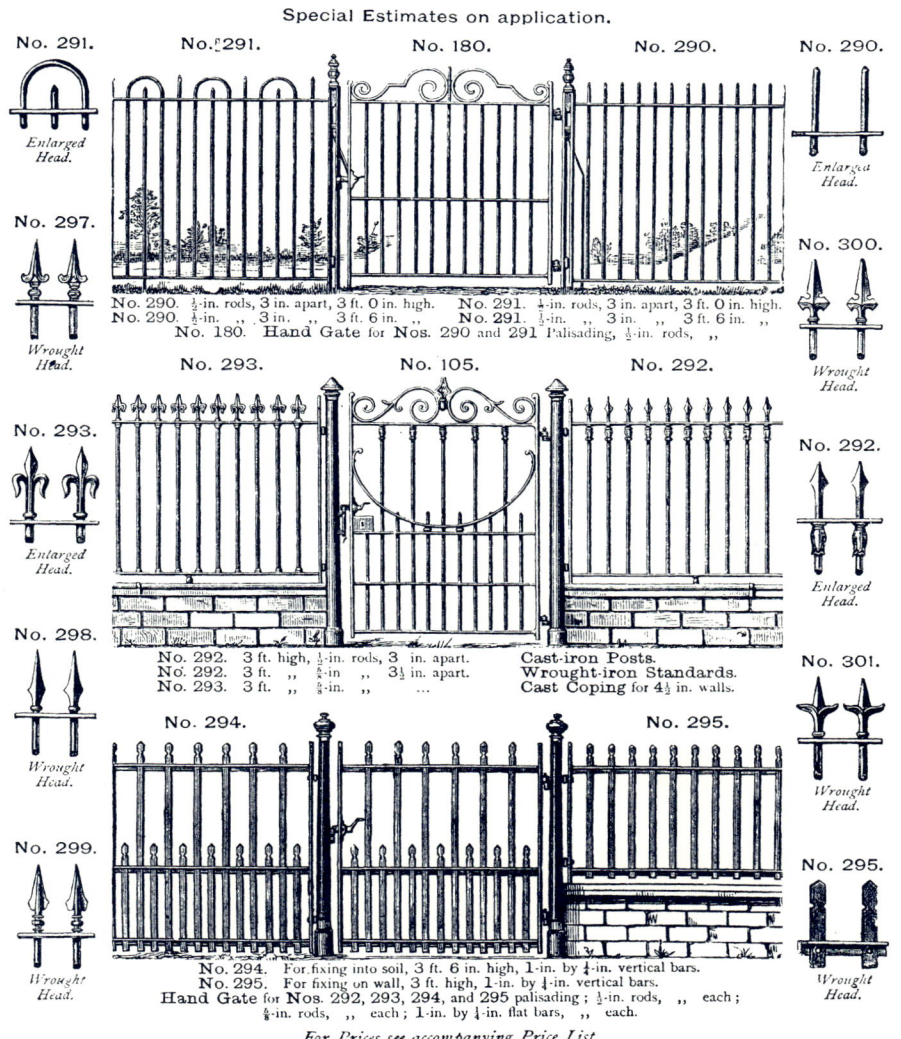

No. 290. ½-in. rods, 3 in. apart, 3 ft. 0 in. high.   No. 291. ½-in. rods, 3 in. apart, 3 ft. 0 in. high.
No. 290. ⅝-in. ,, 3 in. ,, 3 ft. 6 in. ,,   No. 291. ⅝-in. ,, 3 in. ,, 3 ft. 6 in. ,,
No. 180. Hand Gate for Nos. 290 and 291 Palisading, ½-in. rods, ,,

No. 292. 3 ft. high, ½-in. rods, 3 in. apart.   Cast-iron Posts.
No. 292. 3 ft. ,, ⅝-in ,, 3½ in. apart.   Wrought-iron Standards.
No. 293. 3 ft. ,, ⅝-in. ,, ...   Cast Coping for 4½ in. walls.

No. 294. For fixing into soil, 3 ft. 6 in. high, 1-in. by ¼-in. vertical bars.
No. 295. For fixing on wall, 3 ft. high, 1-in. by ¼-in. vertical bars.
Hand Gate for Nos. 292, 293, 294, and 295 palisading ; ½-in. rods, ,, each ;
⅝-in. rods, ,, each ; 1-in. by ¼-in. flat bars, ,, each.

For Prices see accompanying Price List.

ORNAMENTAL WROUGHT-IRON HAND & WICKET GATES.
SUITABLE FOR GARDENS AND VILLA RESIDENCES.

These Gates are of good useful design and are well made; they are constructed with round upright bars, the frames of flat bar-iron; made with latch and hangings for wood, stone, or iron posts.

No. 11b.   No. 104.   No. 94.   No. 11b.

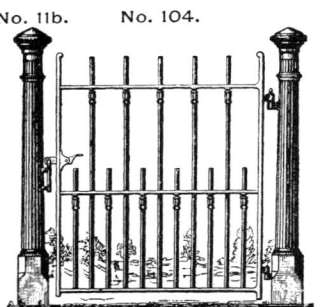

CAST-IRON BASES.

For Iron Pillars. No Wood or Stone Blocks required.

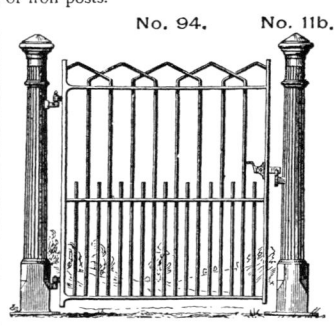

Gate for 3 ft. 6 in. opening, 4 ft. high from ground when hung. With heels and shell ornaments.

With ½-in. round vertical bars ... £ ,, ,, ,,
With ⅝-in. round vertical bars ... ,, ,, ,,
If with lock and key ... extra ,, ,, ,,

Gate for 3 ft. 6 in. opening, 4 ft. high from ground when hung.

With ½-in. round vertical bars ... £ ,, ,, ,,
With ⅝-in. round vertical bars ... ,, ,, ,,
If with lock and key ... extra ,, ,, ,,

Cast-iron Fluted Posts, No. 11b, with self-fixing bases as illustrated, £ ,, ,, ,, per pair.

No. 1.   No. 93.   No. 92.   No. 12.

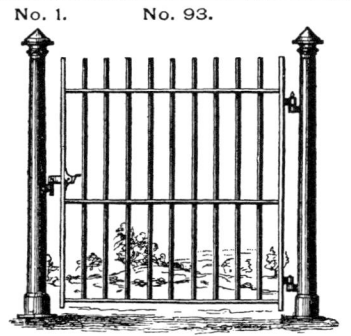

For 3 ft. 6 in. opening, 4 ft. high

With ¾-in. round vertical bars... .. £ ,, ,, ,,
With ½-in. round vertical bars... ,, ,, ,,
With ⅝-in. round vertical bars .. ,, ,, ,,
Cast posts, No. 1, for wood or stone blocks per pair ,, ,, ,,
Cast posts, No. 1, with self-fixing iron bases per pair from ,, ,, ,,

This Gate is suited for situations where architectural effect is desired; it is constructed entirely of wrought-iron, and is very light and strong.

5 ft. 6 in. high from ground, 4 ft. opening.
Cast-iron posts, No. 12, for leading to stone.

Prices on application.

For Prices see accompanying Price List.

ROSE LANE WORKS, NORWICH.

LIGHT ORNAMENTAL GATES.

Constructed with wrought-iron frames filled in with wirework.   Sizes, 3 ft. 6 in. high by 3 ft. openings between posts.

No. 2b. CAST POSTS.

Suitable for all above Gates, having self-fixing bases   per pair.

For Prices see accompanying Price List.

No. 115. ORNAMENTAL WROUGHT-IRON ENTRANCE GATE.

PREPARED FOR WOOD, STONE, OR IRON POSTS.

No. 7.   No. 7.

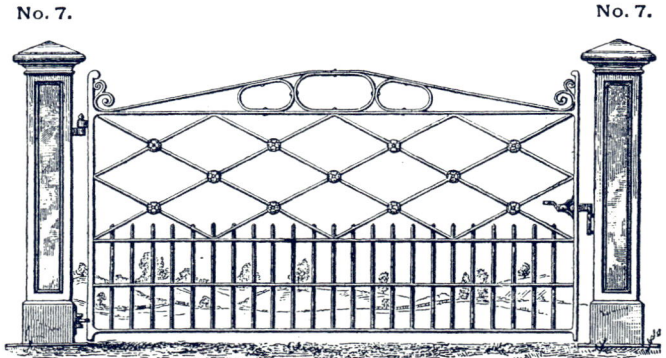

Gate for 10 ft. opening, 5 ft. 6 in. high from ground at centre when hung ... ... £ ,, ,, ,,
No. 7 pattern massive cast-iron posts, 12 in. square at base, for fixing to stone blocks ... per pair ,, ,, ,,
No. 7 pattern massive cast-iron posts, 12 in. square, with iron self-fixing bases ... ... ,, ,, ,, ,,

No. 108. ORNAMENTAL WROUGHT-IRON ENTRANCE GATES.

No. 8.   No. 8.

Gate for 10 ft. opening, 5 ft. 6 in. high from ground at centre when hung, long bars ¾-in., dog bars ⅝ in. ... £ ,, ,, ,,
Double-leaf Gates made to this pattern; with ⅝-in. long vertical bars, and ½-in. dog bars ... ... ,, ,, ,,
Double-leaf Gates made to this pattern, with ¾-in. long vertical bars, and ⅝-in. dog bars ... ... ,, ,, ,,
No. 8 pattern ornamental cast-iron posts, 12 in. square at base, with bolts for stone blocks ... per pair ,, ,, ,,
No. 8 pattern ornamental cast-iron posts, 12 in. square, with self-fixing iron bases, no blocks required ,, ,, ,, ,,

For Prices see accompanying Price List.

ENTRANCE GATES.
No. 2.
REGISTERED DESIGN.

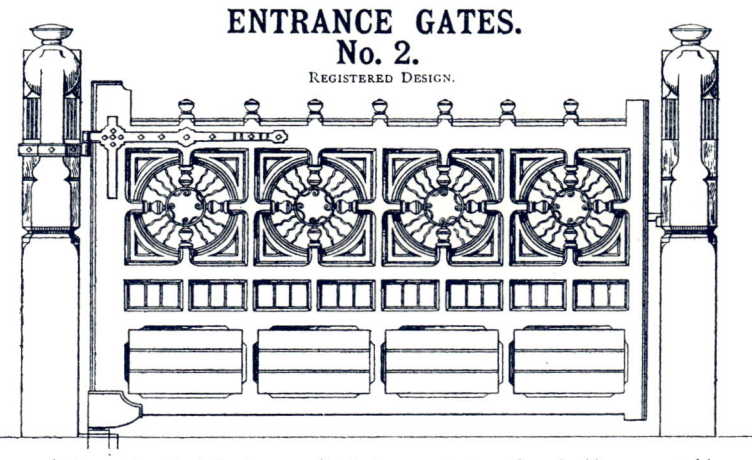

CONSTRUCTED of best selected pitch pine, varnished three coats, strengthened with ornamental ironwork, and fitted with strong hinges and latch. Sizes—9 ft. 6 in. between posts, 5 ft. 6 in. high when fixed.
Ornamental, 9-in. cut, shaped, and turned Oak Posts.

For Prices see annexed List.

Packing Cases and Mats are charged at cost price extra, which will be allowed in full if returned to Works Carriage Paid.

No. 3.
REGISTERED DESIGN.

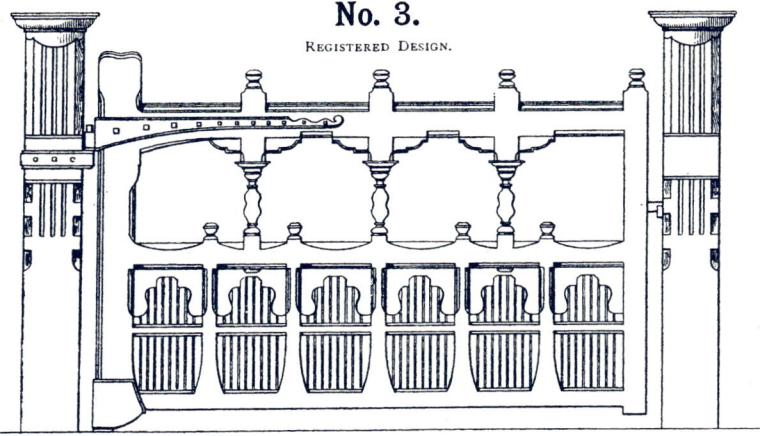

Special Designs prepared to suit any situation.

WICKET GATES.

No. 1. WICKET GATE.
REGISTERED DESIGN.

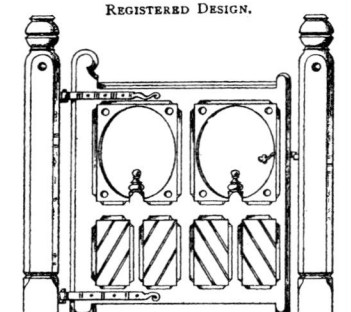

No. 2. WICKET GATE.
REGISTERED DESIGN.

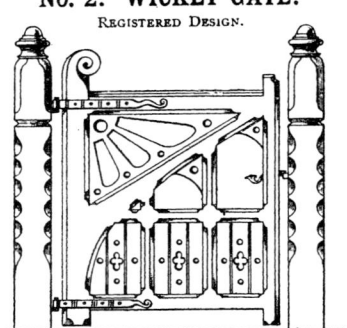

CONSTRUCTED of selected pitch pine, varnished three coats, ironwork bronzed, fitted with ornamental hinges and latch. Size—3 ft. 6 in. between posts, 3 ft. 9 in. high when fixed.

For Prices see annexed List.

No. 3. WICKET GATE.
REGISTERED DESIGN.

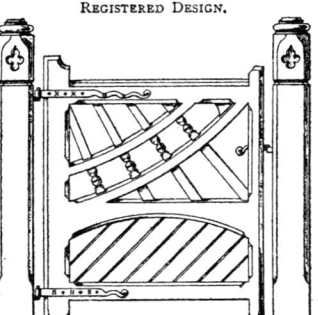

No. 4. WICKET GATE.
REGISTERED DESIGN.

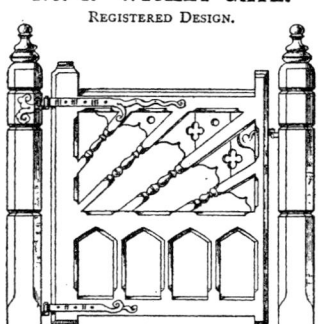

CONSTRUCTED of selected pitch pine, varnished three coats, fitted with ornamental hinges and latch, ironwork bronzed. Size—3 ft. 6 in. between posts, 3 ft. 9 in. high when fixed.

For Prices see annexed List.

Packing Cases and Mats are charged at cost price extra, which will be allowed in full if returned to Works Carriage Paid.

BOULTON & PAUL, MANUFACTURERS.

WICKET GATES.

No. 5. WICKET GATE.
REGISTERED DESIGN.

No. 6. WICKET GATE.
REGISTERED DESIGN.

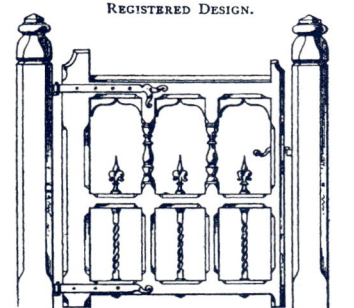
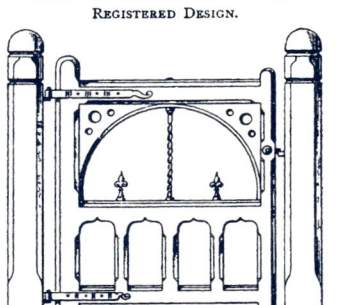

CONSTRUCTED of selected pitch pine, varnished three coats, fitted with ornamental hinges and latch, and strengthened with ironwork, bronzed.

Size—3 ft. 6 in. between posts, 3 ft. 9 in. high when fixed.

For Prices see annexed List.

No. 1. ENTRANCE GATE.
REGISTERED DESIGN.

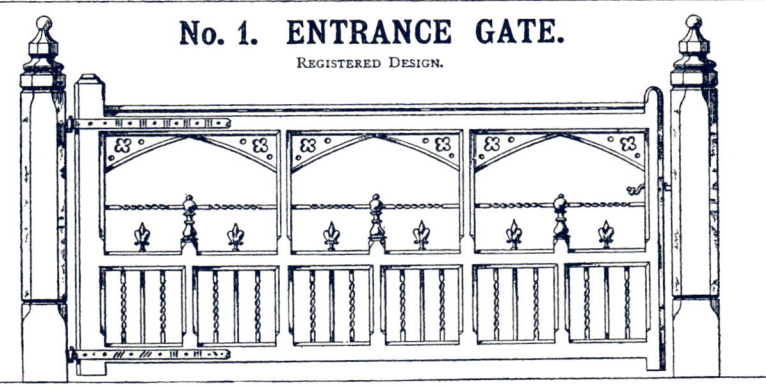

CONSTRUCTED of best selected pitch pine, varnished three coats, strengthened with ornamental ironwork, and fitted with strong hinges and latch.

Size—9 ft. 6 in. between posts, 4 ft. 9 in. high when fixed.

For Prices see annexed List.

Packing Cases and Mats are charged at cost price extra, which will be allowed in full if returned to Works Carriage Paid.

BOULTON & PAUL, MANUFACTURERS,

No. 6. PALED FENCING.

PALES 3 in. by 1 in., rails 10 ft. long, 3 in. by 2 in., posts 8 ft. long, 4 in. by 5 in., mortised and painted.

Red deal, rough sawn, with pointed tops. 3 ft. 6 in. high, 4 ft. high, 5 ft. high.

Planed and painted, or varnished.

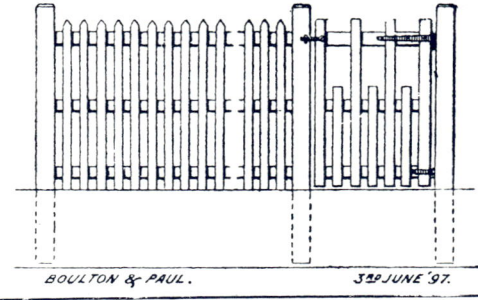

No. 8. ENTRANCE OR STABLE YARD DOORS.

MADE of red deal, planed and painted one coat. Made to any size at **1/-** per square foot.

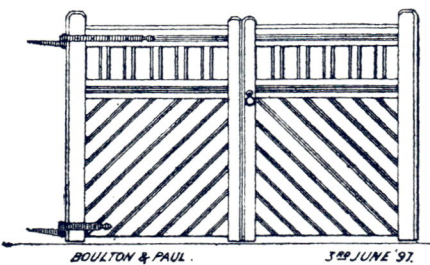

No. 2. FIELD GATE.

10 ft. wide by 4 ft. high, good sound red deal, planed and painted.

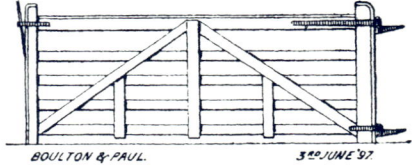

For Prices see accompanying List.

WOOD TRELLIS FENCING AND CLOSE BOARDED WOOD FENCING.

Estimates given upon receipt of particulars of requirements.

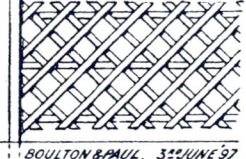
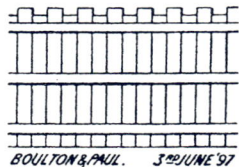

No. 118. WROUGHT-IRON ENTRANCE OR CARRIAGE GATE.

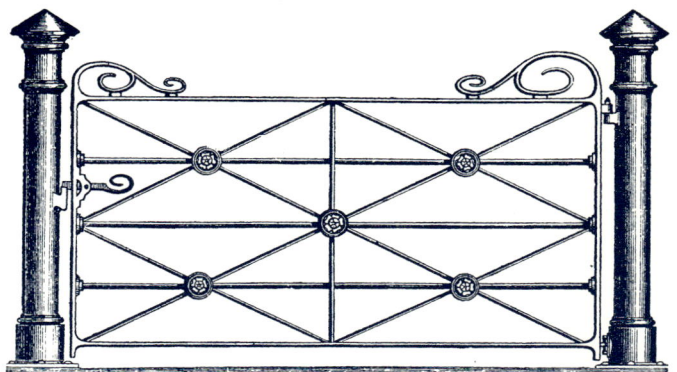

This Gate is remarkably neat in appearance, constructed entirely of wrought-iron, with massive cast-iron pillars. It swings both ways, and shuts itself.

Gate only, for 10 ft. opening, 4 ft. 6 in. high in centre from ground when hung, mounted for iron piers £ ,, ,, ,,
Mountings for wood or stone posts ... ... extra ,, ,, ,,
No. 4 pattern round cast-iron pillars, 6 in. diameter, prepared to fix to either wood or stone ... ... ,, ,, ,,
No. 4 pattern round cast-iron pillars, 6 in. diameter, with self-fixing iron bases .. .. ,, ,, ,,

No. 119. WROUGHT-IRON ENTRANCE OR CARRIAGE GATE.

This Gate is in constant request. It is very neat in appearance, constructed entirely of wrought-iron, and made with cross stays, not with bow stay as illustrated. It swings both ways, and shuts itself; is proof against dogs, etc.

Gate only, for 10 ft. opening, 4 ft. 6 in. high in centre from ground when hung, mounted for iron piers £ ,, ,, ,,
Mountings for wood or stone posts ... ... extra ,, ,, ,,
No. 3 pattern hexagon cast-iron pillars, 7 in. in diameter, prepared to fix to either wood or stone .. ,, ,, ,,
No. 3 pattern hexagon cast-iron pillars, 7 in. in diameter, with self-fixing iron bases ... ,, ,, ,,

Further Designs of Entrance, Carriage, and Park Gates can be had on application.
For Prices see accompanying Price List.

BOULTON & PAUL, MANUFACTURERS.

No. 109. WROUGHT-IRON ENTRANCE GATES.

No. 4   No. 4

Cast-iron Gate Stop.

With self-fixing base.
Price 6/- each.

Cast-iron Gate Stop.

For letting into stone.
Price 5/6 each.

Prepared for Wood, Stone, or Iron Posts.

Double-leaf for 10 ft. opening, 5 ft. 6 in. high from ground when hung, ⅝-in. long bars, ½-in. dog bars £ ,, ,, ,,
Double-leaf for 10 ft. opening, 5 ft. 6 in. high from ground when hung, ¾-in. long bars, ⅝-in. dog bars ,, ,, ,,
No. 4 pattern massive cast posts, for bolting to wood or stone blocks ... ... ,, ,, ,,
No. 4 pattern massive cast posts, with self-fixing iron bases, requiring no blocks ... ... ,, ,, ,,
        Extra, brass-warded lock and sham lock, with Gothic furniture, ,,

No. 110. WROUGHT-IRON ENTRANCE GATES.

No. 3.   No. 3.

Iron Catches.

For keeping gates open with self-fixing base
Price 5/- each

Iron Catches.

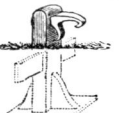

For keeping gates open with self-fixing base.
Price 5/- each.

Prepared for Wood, Stone, or Iron Posts.

Double-leaf for 10 ft. opening, 5 ft. 6 in. high from ground when hung, ⅝-in. bars ... ... £ ,, ,, ,,
Double-leaf for 10 ft. opening, 5 ft. 6 in. high from ground when hung, ¾-in. bars ... ... ,, ,, ,,
        Extra, brass-warded lock and sham lock, with Gothic furniture, ,,
No. 3 pattern hexagon cast-iron posts, 5¼-in. diameter, for bolting to wood or stone blocks per pair ,, ,, ,,
No. 3 pattern hexagon cast-iron posts, 7-in. diameter, for bolting to wood or stone blocks ,, ,, ,, ,,
No. 3 pattern hexagon cast-iron posts, 10¼-in. diameter, for bolting to wood or stone blocks ,, ,, ,, ,,
        If with self-fixing iron bases, extra, per pair, ,, ,, ,,

Further Designs of Entrance, Carriage, and Park Gates can be had on application.

For Prices see accompanying Price List.

No. 121. WROUGHT-IRON ENTRANCE GATE.

No. 2. No. 2.

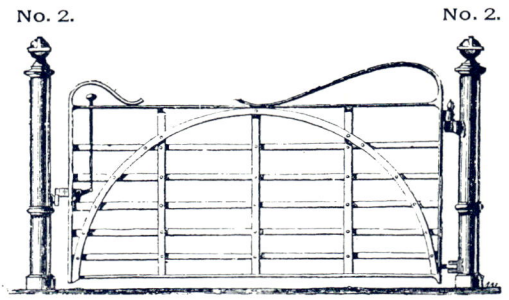

For **9** ft. **6** in. opening, **4** ft. high to top rail above ground when hung, **6** bars   each   £ ,,   ,,   ,,
For **9** ft. **6** in. opening, **4** ft. **6** in. high above ground, **7** bars   ...   ...   ,,   ,,   ,,   ,,
Cast-iron posts for ditto, No. **2**, for bolting to wood or stone blocks   ... per pair   ,,   ,,   ,,
If with self-fixing iron bases   ...   .   ,,   ,,   ,,   ,,

No. 123. WROUGHT-IRON ENTRANCE GATE.

No. 3. No. 3.

For **9** ft. **6** in. opening, **4** ft. **6** in. high from ground to top rail when hung,
    7 bars (as illustration)   ...   .   each   £ ,,   ,,   ,,
For **9** ft. **6** in. opening, **5** ft. high from ground to top rail when hung, **8** bars   ,,   ,,   ,,   ,,
Hexagon cast-iron posts, No. **3**, **5**¼ in. diameter, for wood or stone blocks   per pair   ,,   ,,   ,,
If with self-fixing iron bases   ...   .   ,,   ,,   ,,   ,,

For Prices see Price List.

No. 113. STRONG WROUGHT-IRON DOUBLE-LEAF ENTRANCE GATES.

Gates, 10 ft. to 11 ft. wide, 6 ft. high in centre from ground when hung, ¾-in. vertical bars, with hangings for stone or brick piers, including stop for sliding bolt ... ... £ ,, ,,
Gates, 10 ft. to 11 ft. wide, 6 ft. high in centre from ground when hung, 1-in. vertical bars, with hangings for stone or brick piers, including stop for sliding bolt ... ... ,, ,, ,,
Fitted with brass-warded lock and sham lock, with Gothic furniture.

No. 95. DOUBLE-LEAF ENTRANCE GATES.

No. 11   No. 11.

Gates for 9 ft. opening, 5 ft. high from ground to top of scroll at centre when hung, with ½-in. vertical bars... £ ,, ,,
Gates for 9 ft. opening, 5 ft. high from ground to top of scroll at centre when hung, with ⅝-in. vertical bars... ,, ,, ,,
Gates for 9 ft. opening, 5 ft. high from ground to top of scroll at centre when hung, with ¾-in. vertical bars... ,, ,, ,,
Strong cast-iron fluted posts for bolting to stone blocks, No. 11, 5½-in. diameter ... ... per pair ,, ,, ,,
Strong cast-iron fluted posts for bolting to stone blocks, No. 11b, 4¼-in. diameter ... ... ,, ,, ,,
Self-fixing iron bases for bolting to posts, no stone required, for No. 11 ,, for No. 11b. ,,
Extra, brass-warded lock and sham lock, with Gothic furniture, ,,

For Prices see accompanying Price List.

No. 114. EXTRA STRONG WROUGHT-IRON DOUBLE-LEAF ENTRANCE GATES.

Gates for 9 ft. or 10 ft. opening, 8 ft. high from ground to top of heel ornament, ¾-in. long bars, ⅝-in. short bars, with hangings for stone or brick piers, including stop for sliding bolt ... ... £ ,, ,, ,,
Gates for 9 ft. or 10 ft. opening, 8 ft. high from ground to top of heel ornament, 1-in. long bars, ¾-in. short bars, with hangings for stone or brick piers, including stop for sliding bolt ... ... ,, ,, ,,

Fitted with brass-warded lock and sham lock, with Gothic furniture.

No. 112. WROUGHT-IRON ENTRANCE GATES.

No. 4.   No. 4.

Gates for 10 ft. opening, 5 ft. 6 in. high from ground when hung, ⅝-in. long bars, ½-in. dog bars per pair £ ,, ,, ,,
Gates for 10 ft. opening, 5 ft. 6 in. high from ground when hung, ¾-in. long bars, ⅝-in. dog bars ,, ,, ,, ,,
Massive cast posts, No. 4, for bolting to stone or wood blocks ... ... ,, ,, ,, ,,
Massive cast posts, No. 4, with self-fixing iron bases, bolted to posts, no blocks required ... ,, ,, ,, ,,

Extra, brass-warded lock and sham lock, with Gothic furniture.

For Prices see accompanying Price List.

No. 117. DOUBLE PARK GATES.

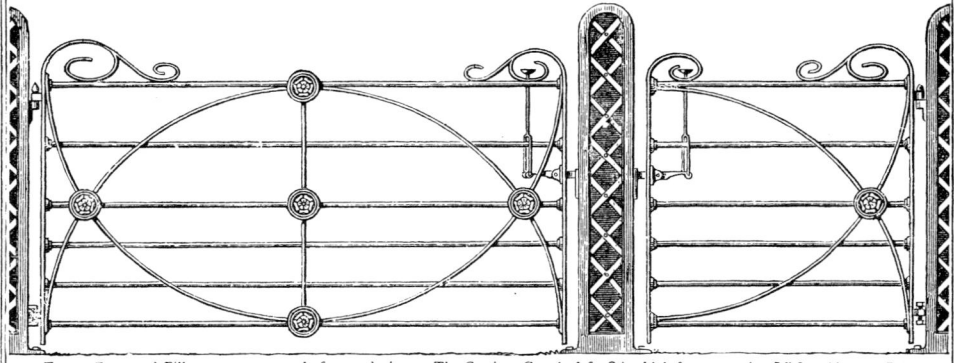

THESE Gates and Pillars are constructed of wrought-iron. The Carriage Gate is 4 ft. 6 in. high for an opening 10 ft. wide, and Side Gate 4 ft. 6 in. high, 4 ft. wide. Top and bottom bars 1-in. round, intermediate bars ¾-in. round, head bar 1¾ in. wide by ½ in. thick, heel bar 1¾ in. wide by ⅞ in. thick, cross braces and centre upright 1¾ in. wide by ⅜ in. thick, with ornamental scrolls and bosses as shown. Pillars are 10 in. square and 5 ft. high.

Carriage Gate for 10 ft. opening, 4 ft. 6 in. high from ground in centre when hung ... £ ,, ,, ,,
Small Gate for 4 ft. opening, 4 ft. 6 in. high from ground in centre when hung ... ,, ,, ,,
Pair of Pillars, 10 in. square, prepared to fix to either wood or stone blocks ... ,, ,, ,,
Ditto, with self-fixing iron bases ... ... ,, ,, ,,

No. 116. DOUBLE PARK GATES.

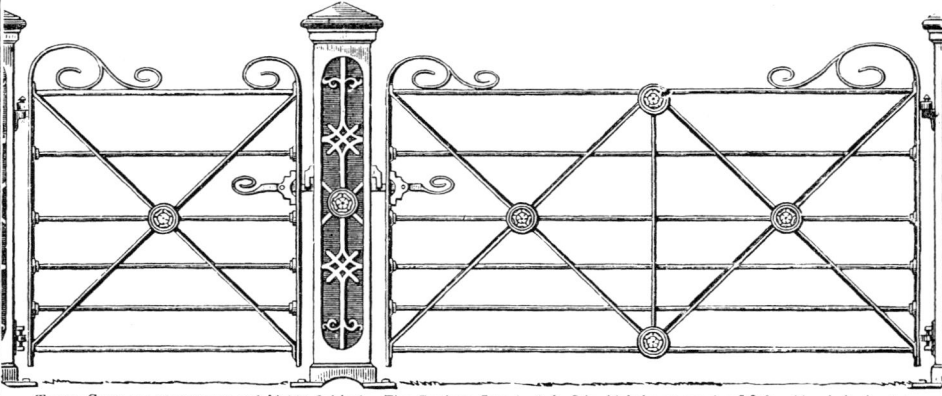

THESE Gates are very strong and highly finished. The Carriage Gate is 4 ft. 6 in. high for an opening 10 ft. wide, six horizontal bars, the top and bottom bars 1-in. round, four middle bars ¾-in. round, bottom bar three inches from ground, heel bar 1¾ in. wide by ⅞ in. thick, head bar 1¾ in. wide by ½ in. thick, cross braces and centre uprights 1¾ in. wide by ⅜ in. thick, with ornamental cast-iron bosses and roses. Pillars are 10 in. square, suitable for fixing to either wood or stone blocks.

Carriage Gate for 10 ft. opening, 4 ft. 6 in. high from ground in centre when hung ... £ ,, ,, ,,
Small Gate for 4 ft. opening, 4 ft. 6 in. high from ground in centre when hung ... ,, ,, ,,
Pair of cast-iron pillars, fitted for either wood or stone bases ... ... ,, ,, ,,

For Prices see accompanying Price List.

BOULTON & PAUL, MANUFACTURERS.

STRONG WROUGHT-IRON ORNAMENTAL RAILING.
SUITABLE FOR STREETS, SQUARES, PARKS, OR ENCLOSING GRAVES, &c.
No. 301. Unclimbable Pattern.

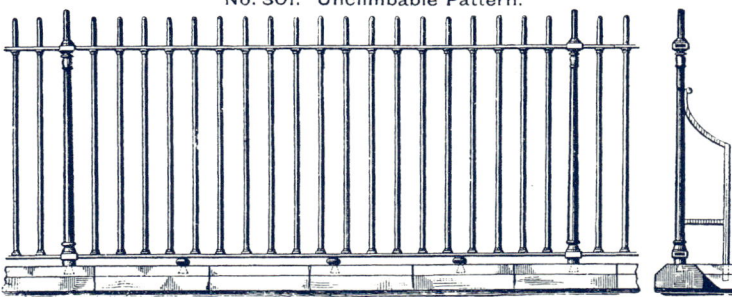

Made in Panels 8 ft. long, for leading into cast-iron standards fitted with stay, the whole fixing on stone coping.
3 ft. high above coping, round bars ⅞-in. diameter, 5 in. from centre to centre, prepared for stone coping ... per yard £ .. .. ..
4 ft. high above coping, round bars ⅞-in. diameter, 5 in. from centre to centre, prepared for stone coping ... ,, ,, ...
If with square upright bars set angleways, per yard extra.
Extra, cast-iron main standards with stay 8 ft. apart, either height, ,, each.

No. 302 Pattern.

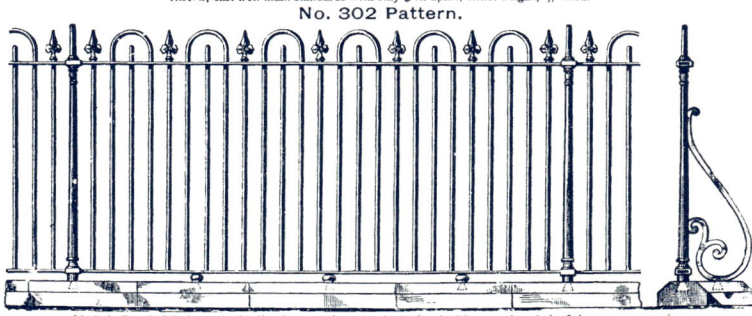

Made in Panels 8 ft. long, for leading into cast-iron standards fitted with stay, the whole fixing on stone coping.
3 ft. high, ⅞-in. bars spaced 4 in. from centre to centre, prepared for stone coping ... ... per yard £ .. .. ..
4 ft. high, ⅞-in. bars spaced 4 in. from centre to centre, prepared for stone coping ... ,, ,, ...
Extra, cast-iron main standards with stay 8 ft. apart, either height, ,, each.

No. 303 Pattern.

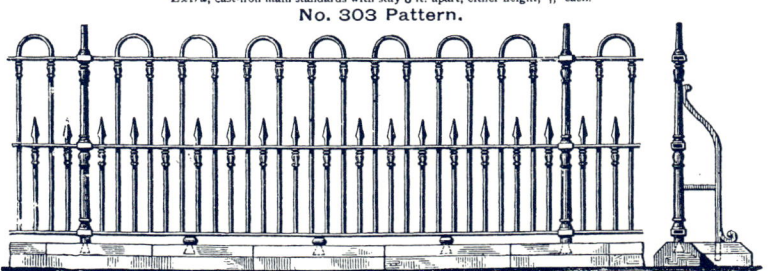

Made in Panels 8 ft. long, for leading into cast-iron standards fitted with stay, the whole fixing on stone coping.
3 ft. high, ⅝-in. bars, lower bars placed 3 in. from centre to centre, top bars 6 in. from centre to centre ... ... per yard £ .. .. ..
3 ft. 6 in. high, ⅝-in. bars, lower bars placed 3 in. from centre to centre, top bars 6 in. from centre to centre ... ,, ,, ...
Extra, cast-iron main standards, 8 ft. apart, with stay, either height, ,, each.
Strong cast-iron Coping, 9 in. wide, for either of the above Railings, ,, per yard.

Special Estimates given. *For Prices see accompanying Price List.*

STRONG WROUGHT-IRON RAILING.
SUITABLE FOR PUBLIC PLACES.

No. 123.   As Erected at St Andrew's Church, Norwich.   No. 123.

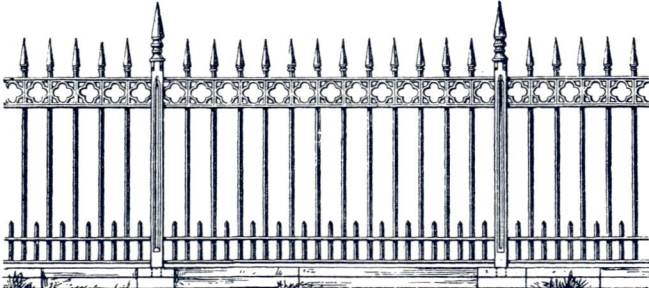

No. 124.   As Erected at Coltishall Schools.   No. 124.

No. 125.   As Erected at Southend Water Works.   No. 125.

Special Estimates given.

NEW ORNAMENTAL WROUGHT-IRON RAILINGS
FOR VILLA RESIDENCES.

Pattern No. 160.

Standing 2 ft. high above coping, in about 6 ft. lengths, with Cast-iron Junction Standards.

Pattern No. 161.

2 ft. high above coping, in 6 ft. lengths, with Cast-iron Junction Standards.

Strong Cast-iron Coping for 9 in. walls.

Gates and Posts to match above patterns.

Prices on application.

BOULTON & PAUL, MANUFACTURERS,

ORNAMENTAL WROUGHT-IRON RAILINGS.
FOR VILLA RESIDENCES.

Pattern No. 162.

2 ft. high, with ½-in. square bars, points finished as No. 164, in 6 ft. lengths, with Cast-iron Junction Standards.

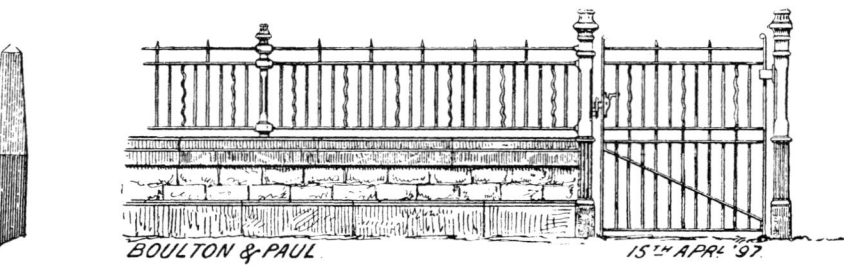

Pattern No. 163.

2 ft. high above coping, ½-in. square, and ½-in. by ¼-in. bars.

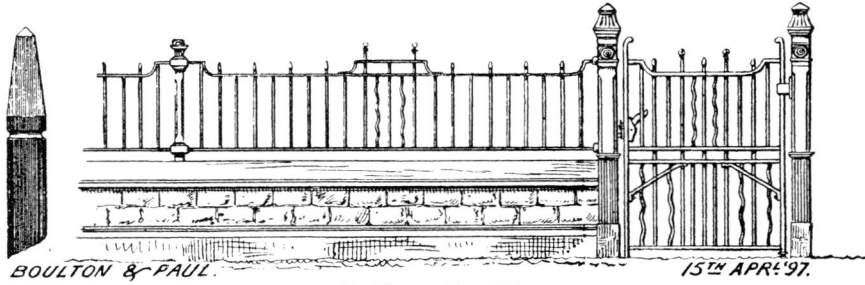

Pattern No. 164.

2 ft. high above coping, with ½-in. square bars, made in about 8 ft. lengths, with Cast-iron Junction Standards. Strong Cast-iron coping suitable for any of the above Railings, 9 in. wide. Gates and Iron Posts to match any of the above patterns.

Prices on Application.

ORNAMENTAL WROUGHT-IRON RAILING.

FOR VILLA RESIDENCES.

Pattern No. 165.

2 ft. high, with ½-in. square bars, points finished as No. 163, made in 6 ft. lengths, with Cast-iron Junction Standards.

Pattern No. 166.

2 ft. high, ½-in. square bars, made in 6 ft. lengths, with Cast-iron Junctions. Strong Cast-iron Coping suitable for any of the above Railings, 9 in. wide. Gates and Iron Posts to match any of the above patterns.

Prices on application.

ORNAMENTAL WROUGHT-IRON RAILING.

Pattern No. 167.

15 in. high, ½-in. square upright bars, ½-in. by ¼-in. rings.

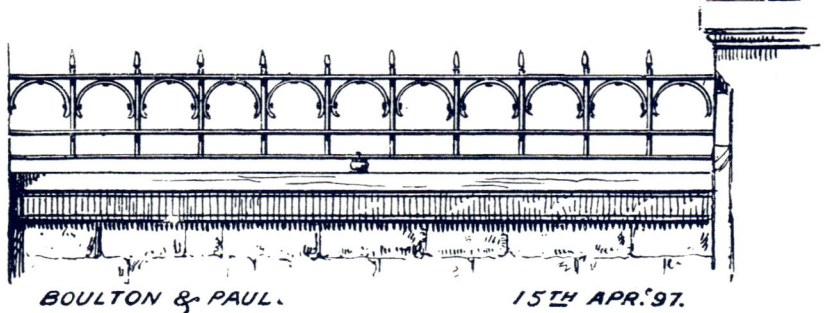

Pattern No. 168.

15 in. high, ½-in. square, upright bars, ½-in. by ¼-in. bows.

Prices on application.

ORNAMENTAL WROUGHT-IRON RAILING.

Pattern No. 169.

Made in about 8 ft. lengths, with one ornamental panel in centre. Height, 4 ft. above stone plinth, vertical bars $\frac{5}{8}$-in. square, spaced about 5 in. centres.
Cast-iron Junction Standard with stays for connecting lengths.

Ornamental Wrought-iron Gate.

Pattern No. 170.

Made in 8 ft. lengths, with ornamental panel in centre. Height, 4 ft. above stone plinth, vertical bars $\frac{5}{8}$-in. and $\frac{1}{2}$-in. square.
Cast-iron Junction Standard with stays for connecting lengths.

4 ft. high, to suit opening 3 ft. 6 in. wide.

Pattern No. 173.

Pattern No. 171.

2 ft. 3 in. high above coping, $\frac{1}{2}$-in. square and $\frac{1}{2}$-in. by $\frac{1}{4}$-in. bars; in 6 ft. lengths, with Cast-iron Junction Standards.

Pattern No. 172.

2 ft. 3 in. high above coping, $\frac{1}{2}$-in. square verticals, $\frac{3}{8}$-in. by $\frac{1}{4}$-in. rings; made in 8 ft. lengths, with Cast-iron Junction Standards.

Strong Cast-iron Coping suitable for any of the above Railings, 9 in. wide.
Gates and Iron Posts to match any of the above patterns.

Prices on application.

BOULTON & PAUL, MANUFACTURERS,

ORNAMENTAL WROUGHT-IRON RAILING.

For leading in Stone Curb, or with Feet for Fixing in Ground.

3 ft. 6 in. high.

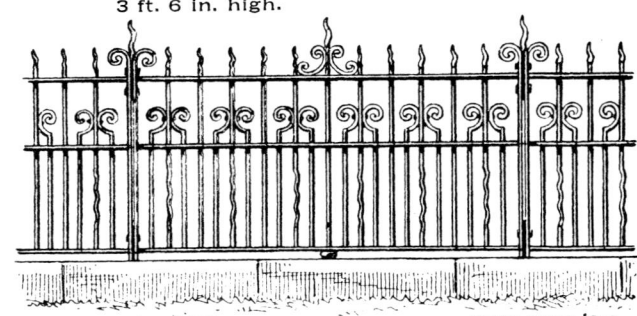

Pattern No. 174.

Made in 6 ft. lengths, vertical bars of $\frac{3}{4}$-in. by $\frac{1}{4}$-in., spaced 3 in. centres at bottom.

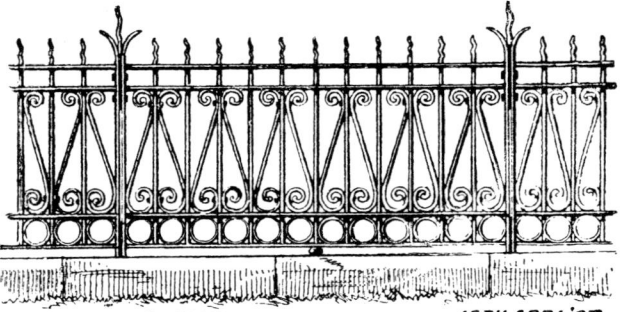

Pattern No. 175.

Made in 6 ft. lengths, verticals of $\frac{3}{4}$-in. by $\frac{1}{4}$-in., spaced 6 in. centres.

Pattern No. 176.

Made in 6 ft. lengths, vertical bars of $\frac{1}{2}$-in. square spaced about 5 in. centres.

Strong Cast-iron Coping suitable for any of the above Railings, 9 in. wide. Gates and Iron Posts to match any of the above patterns.

Prices on application.

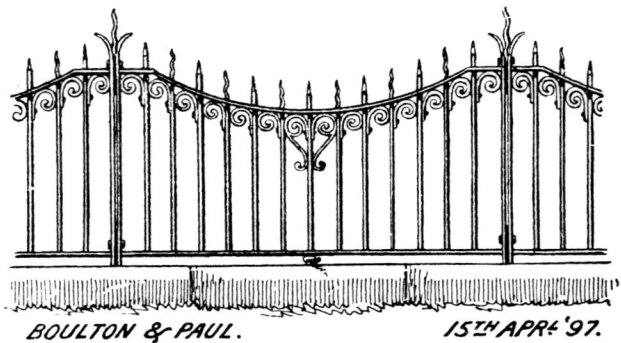

ORNAMENTAL WROUGHT-IRON RAILING.

For leading in Stone Curb, or with Feet for Fixing in Ground.
3 ft. 6 in. high.

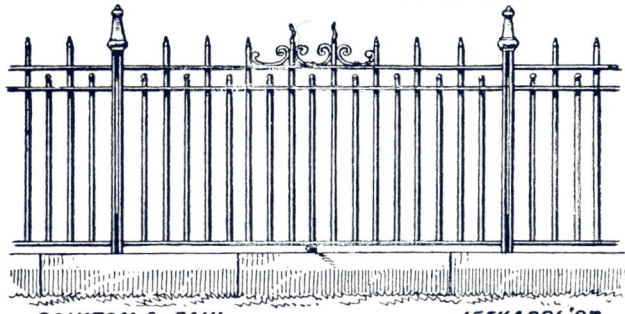

Pattern No. 177.

Made in 6 ft. lengths, vertical bars of $\frac{1}{2}$-in. square spaced 4 in. centres, with ornamental cast head for bolting between.

Pattern No. 178.

Made in 6 ft. lengths, with $\frac{1}{2}$-in. square vertical bars, spaced $4\frac{1}{2}$-in. centres.

Pattern No. 179.

Made in 6 ft. lengths, vertical bars of $\frac{1}{2}$-in. square, spaced 5 in. centres.

Strong Cast-iron Coping suitable for any of the above Railings, 9 in. wide.

Gates & Iron Posts to match any of the above patterns.

Prices on application.

BOULTON & PAUL, MANUFACTURERS.

ORNAMENTAL IRON RAILING FOR DWARF WALLS.
No. 119. Suitable for Gothic Mansions and Villa Residences.

Is chiefly made of wrought-iron, and is very strong. 1 ft. 5 in. high to top of horizontal bar from stone coping, in. 8 ft. lengths.

No. 120. Suitable for Villa Residences.

The standards cast-iron, with wrought-iron horizontal bars. 2 ft. 9 in. high to top of standard from coping, prepared ready for fixing.

No. 121. As Erected at St. Giles' Church, Norwich.

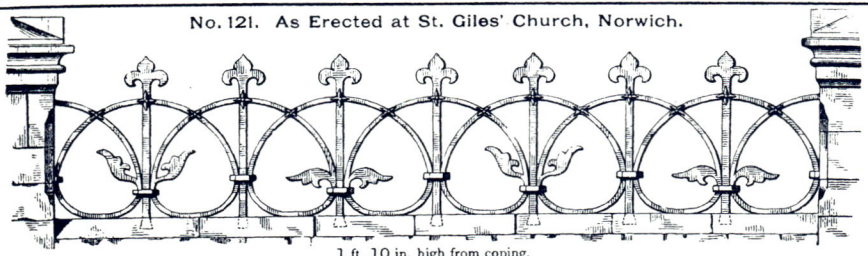

1 ft. 10 in. high from coping.

WROUGHT-IRON RAILING WITH MALLEABLE-IRON ORNAMENTS.
No. 122. As Erected at St. Stephen's Church, Norwich.

Strong Cast-iron coping suitable for any of the above Railings, 9 in. wide, ,, per yard. Gates and Iron Posts to match any of the above patterns. Prices on application.

Special Estimates given.

IRON RAILINGS FOR TOMBS, MONUMENTS, &c.
TO STAND ON STONE CURBING, PROPERLY FITTED AND PAINTED, READY FOR FIXING.

No. 319.

Standards 2 ft. apart.
Cast-iron Standards, 10½ in. high.
Wrought-iron Tubular Rail, 7 in. from coping.
Cast-iron Cable Rail.

No. 320.
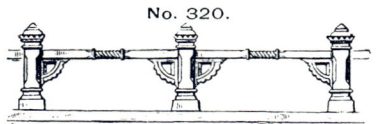

Standards 18 in. apart.
Cast-iron Standards, 11 in. high.
Cast-iron Rail, with Brackets, 7 in. from coping.

No. 315.
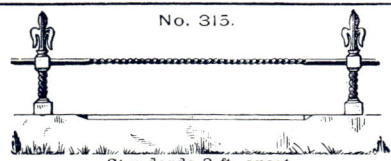

Standards 3 ft. apart.
Cast-iron Standards, 15 in. high.
Cast-iron Cable Rail, 8 in. from coping.
Wrought-iron Twisted Rail, 8 in. from coping.

No. 323.

Standards 3 ft. apart.
Cast-iron Standards, 16 in. high.
Cast-iron Rail, 12 in. from coping.

IRON HEADS FOR STONE POSTS.
WITH FASTENINGS FOR CHAINS.
No. 318.

No. 317.

Prices on application.

BOUNDARY RAILING.
FOR CARRIAGE DRIVES, PUBLIC ROADS, OR ENCLOSURES.

3 ft. 6 in. high, with 1½-in. wrought-iron tubular bar

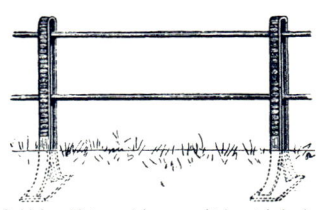

3 ft. high, with two ⅞-in. wrought-iron tubular bars.

Further Designs of Tomb Railing, and Railing for Churches, etc., can be had on application.
For Prices see accompanying Price List.

BOULTON & PAUL, MANUFACTURERS.

GALVANIZED WROUGHT-IRON CATTLE FEEDING TROUGHS, &c.

Wrought-iron Calf Rack.

MADE entirely of wrought-iron, with galvanized iron trough for feeding calves and young stock.
Cash Price, 8 ft. long, £3 17 6

No. 35. Galvanized Iron Sheep Rack.

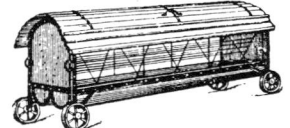

This Rack is made entirely of wrought-iron, frame of angle iron, roof of corrugated iron, which throws the water off the ends, the trough of galvanized iron. By an improved arrangement in the rack the sheep cannot waste the hay.
Cash Price, to feed 20 Sheep, £3 17 6

Galvanized Double Sheep Rack.
For 24 Sheep.

CAN be used as a hay rack or as feeding troughs.
8 ft. long, without wheels £2 0 0
If with two wheels ... 2 4 0

No. 38. Covered Calf Trough.

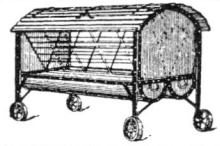

Cash Price, 8 ft. 6 in. long, £5 0 0

No. 39. Calf Trough.

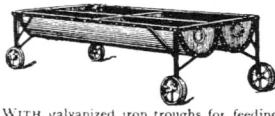

WITH galvanized iron troughs for feeding calves, cows, horses, and colts.
Cash Price, 8 ft. 6 in. long, £2 2 6

Wrought-iron Cattle Drinking Trough.

20 in. wide, 30 in. deep, 6 ft. long.
Cash Price £2 0 0

No. 32. Lamb or Sheep Trough.

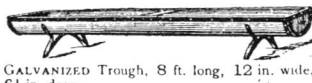

GALVANIZED Trough, 8 ft. long, 12 in. wide, 6½ in. deep.
Cash Price 15/6 each.

No. 31. Wrought-iron Sheep Trough.

8 ft. long, 14 in. wide, 8½ in. deep.
Cash Price 17/6

Galvanized Wrot.-iron Pig Troughs.

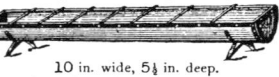

8 in. wide, 5 in. deep.
2 ft.   3 ft.   4 ft.   6 ft. long.
3/6   4/6   5/6   6/6 each.

Wrought-iron Sheep Trough.
(Galvanized.)

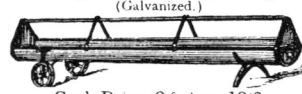

Cash Price, 8 ft. long, 18/6

Wrot.-iron Calf Crib.

Cash Price,
8 ft. 6 in. long . £1 18 6
If made water-tight 2 0 0

Galvanized Wrought-iron Pig Troughs.

10 in. wide, 5½ in. deep.
3 ft.   4 ft.   5 ft.   6 ft.   8 ft. long.
5/6   6/6   7/6   8/6   10/6 each.

No. 24. Covered Sheep Trough.

Cash Price, 8 ft. 6 in. long, £1 13 0

Curved Iron Shelter for Sheep.
Complete with Wood Sills and Spikes.

Circular Cattle Crib.
5 ft. diameter.

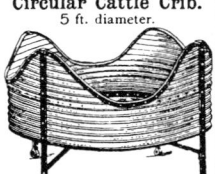

THE advantage of this shape is that it can be easily rolled about without injury. It is very strong and durable.
Cash Price £3 3 0

Portable Shelter for Sheep and Lambs.

MADE with wrought-iron hurdles mounted on wheels, and covered with galvanized corrugated sheet-iron. 12 ft. long, 4 ft. wide.
Price, Carriage Paid, £6 0 0

20 ft. long, 10 ft. wide, 5 ft. high   £6 10 0
20 ft.   ,,   15 ft.   ,,   5 ft.   ,,   8 0 0
Special Estimates for larger sizes on application.

Carriage Paid on all Orders above 40/- value to the principal Railway Stations in England and Wales.

NOTE.—The above goods are made of Galvanized Sheet-iron with wrought-iron framework painted.

No. 300. CORRUGATED STEEL CISTERNS.

Galvanized after manufactured.

These Cisterns are designed to meet the demand that exists for something cheap, and yet at the same time serviceable.

Cash Prices.

Galls.	Outside Dimensions.					£	s.	d.
	Length.		Width.		Depth.			
	ft.	in.	ft.	in.	ft.	in.		
about 20	1	8	1	3	1	6	0	10	6
25	1	10	1	2	2	2	0	11	6
30	2	0	1	4	2	2	0	12	6
50	2	4	1	8	2	2	0	17	0
60	2	6	2	0	2	2	0	19	6
80	2	10	2	4	2	2	1	3	6
100	3	2	2	8	2	2	1	8	6

No. 302. GALVANIZED AND PAINTED CATTLE TROUGHS.

With strong safe edge binding round top.

Cash Prices.

Length.		Width.		Depth.		Ordinary Strong Quality.					
						Galvanized No. 14 Gauge.			Painted. No. 14 Gauge.		
ft.	in.	ft.	in.	ft.	in.	£	s.	d.	£	s.	d.
3	0	1	6	1	4	1	14	6	1	4	6
4	0	1	6	1	4	2	2	6	1	10	6
5	0	1	6	1	4	2	10	0	1	16	6
6	0	1	6	1	4	2	18	0	2	2	6
7	0	1	6	1	4	3	6	0	2	8	6
8	0	1	6	1	4	3	14	0	2	14	6

If mounted on four wheels and fitted with a ring at one end, 20/- each extra.

No. 301. WROUGHT-IRON WATER TANKS OR CISTERNS

Galvanized after made.

Cash Prices.

Galls.	Outside Dimensions.						No. 14 Gauge.
	Length		Width.		Depth.		
about 20	ft.	in.	ft.	in.	ft.	in.	
20	1	8	1	3	1	6	16/-
25	2	0	1	8	1	3	17/6
30	2	0	2	0	1	3	19/-
40	2	0	2	0	1	7	22/6
50	2	7	2	0	1	7	25/6
80	3	0	2	2	2	0	34/6
100	3	0	2	6	2	2	39/9
150	3	7	2	10	2	5	52/-
200	4	0	3	0	2	8	64/-
250	5	0	3	0	2	8	73/-
300	6	0	3	0	2	8	82/6

Other sizes can be supplied to order.

Loose Covers for Cisterns, to fit on angle-iron Frame, 1d. per gallon extra.

No. 303. GALVANIZED CORRUGATED IRON CORN BINS

Round Square, or Rectangular shape

For Cash Prices see page 320.

Carriage Paid on all Orders above **40/-** value to the principal Railway Stations in England.

BOULTON & PAUL, MANUFACTURERS.

THE LATEST IMPROVED CHAFF-CUTTING MACHINES.

No. 2a.

No. 2a. HAND-POWER CHAFF CUTTER.

This Machine has a rising mouth, in which the pressure is brought to bear upon the feed by means of a lever and weight, and not by the uncertain and cheap method of springs. Cuts two lengths of chaff.
Width of mouth 8¼ in., rising from 2¼ to 3¼ in.
Cash Price £2 4 0

No. 3.

No. 3. HAND-POWER CHAFF CUTTER.
WITH LARGE FLY-WHEEL.

This Machine is fitted with a large fly-wheel to suit the requirements of those who prefer a heavy wheel. It cuts two lengths of chaff without change wheels.
Width of mouth 9 in., rising from 2¼ in. to 3 in.
Cash Price, Carriage Paid, £3 11 6

CORN GRINDING AND CRUSHING MACHINES.

GRINDING MILLS.

The **Pioneer** may be driven by a Two or Three Horse Gear or Small Engine, the **Farmer** by an Engine of Two or Three Horse-power, and when driven at about 500 revolutions per minute it will grind from 10 to 12 bushels per hour, the **Pioneer** proportionately less.

If grinding coarse Meal the above quantities are largely increased.

Cash Prices.
The **Pioneer**, with Pulley £6 6 0
The **Farmer**,   ,,   8 0 0

MAIZE AND OAT MILLS.

These Mills will grind Maize sufficiently fine for all feeding purposes.

Cash Prices.
M M 0. For Hand-power, wooden hopper ... £1 17 6
M M 1. For Hand-power, iron hopper and feed regulator 2 6 0
Large Fly-wheel, much recommended extra 0 5 0
M M 2. For Hand, Horse or Steam power, large Fly-wheel included ... 2 16 6
Pulleys extra according to size.

STOCK FEEDING MACHINERY.

THE "ADVANCE" MACHINES.

Contain all the latest improvements, have large hoppers, and will be found strong and substantial, very fast cutting, and amongst the most efficient in the trade.

Cash Prices.
No. 0. The "Advance" Root Pulper £2 14 6
No. 1. The "Advance" Root Pulper 3 3 0
No. 0. The "Advance" Root Grater 2 18 6
No. 1. The "Advance" Root Grater 3 10 0
(*With special steel discs, fine or coarse.*)
No. 0. The "Advance" Root Slicer £2 13 0
No. 1. The "Advance" Root Slicer 3 3 0
Slicers made to cut finger pieces ... ... extra 3 10 6
Ornamental Covers included with all the above.

THE "TINY" AND "GEM" MACHINES.

For small occupiers. Rigid in work, large hoppers. Special covers for shooting the roots into a basket.

Cash Prices.
The "Tiny" Root Pulper £2 0 0
The "Tiny" Root Grater 2 4 0
(*With special steel discs, fine or coarse.*)
The "Gem" Root Pulper £2 4 0
The "Gem" Root Grater 2 8 6
(*With special steel discs, fine or coarse.*)
The "Gem" Root Slicer £1 16 6
Ornamental Covers included with all the above, except the Slicer.

These Machines are not manufactured by ourselves, but are sent direct from Makers' Works. Carriage Paid on all Orders above 40/- value to the principal Railway Stations in England and Wales.

Kennel, Poultry Yard.

Pigeon Loft, Pheasantry, and Aviary.

MANAGEMENT OF DOGS.

REGISTERED COPYRIGHT.

The Kennel.—A dog must be housed well—sleeping quarters dry, airy, yet free from draught, ample light, and as cheerful as possible. So much intelligence and animation has the dog, that it is too bad to locate him in an out-of-the-way corner. He loves to see and be seen, and takes the liveliest interest in everything going on around him. So when you decide on the site for your "kennel," remember our friend's claim to a prominent position.

Aspect.—Let it be a genial spot, facing south, if possible, with good drainage and water supply, and so near your own dwelling that, being frequently under your inspection, you may insist upon scrupulous cleanliness.

Feeding.—No hard and fast principle can be here laid down as to the proper food for dogs, varying as it does with the breed, size, and weight of each of them, and in the exercise or work which may be expected of them. We will, however, remark that pet dogs especially, as a rule, are cruelly overfed, under a sense of mistaken kindness. It should be remembered that the process of digestion in the dog is a very slow one; two feeds in twenty-four hours is the utmost that a dog in health can benefit by. Lap dogs ought not to have solid meat at any time—broth, bread, milk, and biscuits. Large dogs, taking good exercise, can do with a small proportion of flesh, but never a meal of meat alone. Dog biscuits are perhaps the most convenient form of food as a staple, containing as they do a judicious proportion of animal food, blended with meal. Care should be taken that they are stored in a very dry place, as when once musty they are hurtful, and dogs dislike them very much in that state.

Treatment.—All authorities agree in recommending an occasional change in diet, and in giving cabbage, boiled and mixed in with other food now and then. Water, pure and fresh, should always be within reach of dogs, and feeding and drinking vessels scalded and cleaned daily. Food to be given regularly. The daily exercise and scamper by no means omit, nor a daily grooming with hound gloves, a brush, or wisp of straw.

Washing.—When, by reason of dirt or vermin, a dog requires washing (never wash one on a full stomach) there are several soaps recommended to destroy parasites and beautify the coat; and if, in addition to this, a final rinse in a decoction of Quassia Chips be resorted to, no vermin will remain in a coat so treated.

Disease.—Dogs have almost as many recognised forms of disease as the human race. The more frequent complaints are:—Distemper, Catarrh, Cough, Mange, Inflammation, &c. Our space at command will not allow us to enter fully upon the diagnosis of these ailments.

Remedies.—We will, however, give a recipe for a *simple purgative*, to be given on the first appearance of the dog being amiss. Syrup of Buckthorn, three parts; Syrup of Squills, one part; Castor Oil, two parts. Dose—a teaspoonful for Lap Dog; two for a larger Dog; up to half a wineglassful for a Retriever. Shake up well before pouring out, as the oil will float. *To be given fasting.* Also a good pill for coughs, colds, lung affections, will be found in powdered Liquorice, 24 grains; powdered Ipecacuanha, 6 grains; powdered Opium, 6 grains; powdered Rhubarb, 12 grains; powdered Gum Ammoniacum, 24 grains; powdered Gum Acacia, 24 grains; mix and make into twenty pills. One for a 16 to 30-lb. dog; half-one for a Lap Dog. And a good tonic pill is—Quinine, 12 grains; Sulphate of Iron, 18 grains; Extract of Gentian, 24 grains; powdered Ginger, 18 grains; make twelve pills—Half-one for a small dog; one for a larger; one and a half for a large dog.

Never delay to seek veterinary skill where a dog is exhibiting urgent symptoms, or in case of injury from accident.

Advice given gratis for arranging Kennels, and gentlemen waited upon in any part of the Kingdom.

"In the Canine Department, MESSRS. BOULTON & PAUL of Norwich may claim the credit of being the Dog's true friends, for they have certainly brought to perfection every appliance connected with the Kennel."—*Australian Gazette.*

CAUTION.—Beware of inferior imitations. All articles bear our name as a guarantee of good workmanship

No. 125.
WOOD KENNELS FOR HOUNDS.

HOUSES constructed of red deal framing, covered outside with stained weather-boarding, lined corrugated iron roof, whitened inside, wrought-iron kennel railing.

Estimates and Specifications free on application.

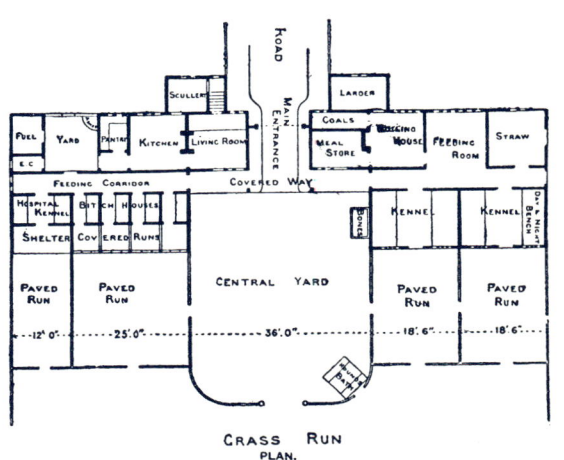

THIS plan can be modified to suit any situation, or special designs will be prepared free of charge.

SITES surveyed and Gentlemen waited upon in any part of the country.

TESTIMONIAL.
From G. O. MEARES, ESQ., Southampton.
The Kennels arrived safely, and I have had them put up, and they give great satisfaction.

BOULTON & PAUL, MANUFACTURERS.

No. 126. RANGE OF KENNELS FOR HOUNDS.

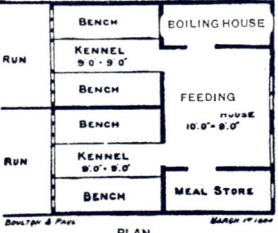

These Kennels are strongly constructed. They are made in sections, and can be readily taken to pieces for removal, and re-erected without injury to the materials.

Approximate Price £47 0 0

TESTIMONIAL.

From J. W. P. PAGE, Esq., Stockton-on-Tees.

The Kennel arrived safely and is now fixed; we are very much pleased with it, and I consider the price most reasonable.

No. 127. RANGE OF KENNELS FOR BITCHES.

Constructed on the principle recommended in the Hunting Series of the Badminton Library.

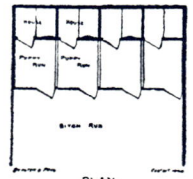

Special Estimates on application.

TESTIMONIAL.

From Mrs. C. F. C. CLARKE, Westwood.

My husband was so pleased with the two Kennels we bought of you for the Pointers, that we want some more.

CAUTION.—Beware of inferior imitations.   All articles bear our name as a guarantee of good workmanship.

ROSE LANE WORKS, NORWICH.

No. 128.
HOUND KENNELS.

As supplied to M. VAUGHAN DAVIES, ESQ., Tan-y-Bwlch.

WALLS covered outside with rustic joint weather-boarding. Kennel Houses lined inside with match-boarding. Corrugated iron roof, with matchboard and felt lining. Day and Night Benches. Corrugated iron fencing with wood posts, upper part kennel railing.

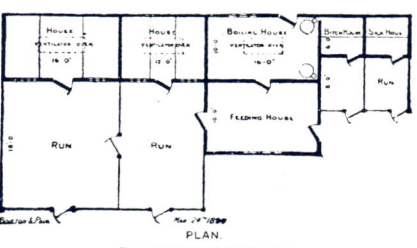

PLAN.
REGISTERED COPYRIGHT.

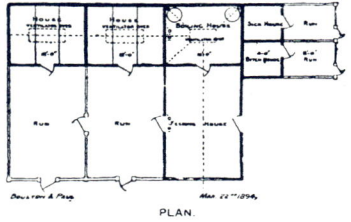

PLAN.
REGISTERED COPYRIGHT.

THIS is a smaller arrangement of the above, and is constructed in the same way.

Estimates and Special Plans on application.

TESTIMONIAL

From Miss P. HANCKENHAGEN, Bersbergen.

The Kennel I had from you has now been in use for a few weeks, and I unhesitatingly pronounce it to be the best Kennel I have seen; it is much admired, and my dogs have never been in better health.

CAUTION.—Beware of inferior imitations. All articles bear our name as a guarantee of good workmanship.

BOULTON & PAUL, MANUFACTURERS.

No. 129.
BLOCK OF BUILDINGS FOR A HUNTING ESTABLISHMENT.

Comprising Huntsman's Cottage, Stables, Kennel and Yards, &c.

REGISTERED DESIGN, No. 5945.

TESTIMONIALS.

From MR. S. J. BISHOP, Keeper to J. Freme, Esq., Perrtreoselas.

I had a gentleman to look at my Kennels yesterday, who lives close to me. I assured him with confidence that they are the best kennels I ever saw, they are healthy, harbour no vermin, no trouble to keep clean, and very little expense putting up. The Benches are the best I ever saw in all my experience.

From H. G. TATHAM, ESQ., Chesfield.

The Dog Kennels that you have erected for me are everything that could be desired. I have not a fault to find, and hope shortly to extend them.

REGISTERED COPYRIGHT.

Plan showing inside arrangements.

WE shall be pleased to furnish estimates for the above Buildings, or portion of same, or for other size Buildings on application.

TESTIMONIALS.

From R. WARREN VERNON, ESQ., Shawford.

I am of course very much pleased with all the things you have made me, the Kennels and Fowls' House, but I cannot help writing to say how very pleased I am with the Dog Boxes. Several people have told me they could not make any such thing at the price.

From EDWARD OLIVER, ESQ., Holloway.

I must compliment you upon the very efficient manner in which you execute the orders entrusted to you, everything arrived in perfect order, beautifully packed. I shall have great pleasure in recommending you to my friends.

Carriage Paid to the principal Railway Stations in England and Wales.

CAUTION.—Beware of inferior imitations.   All articles bear our name as a guarantee of good workmanship.

No. 132. COMPLETE HOUND KENNELS.

REGISTERED COPYRIGHT.

ELEVATION.

THESE Buildings are warm and well ventilated, and comprise all the accommodation needed for a pack of hounds.

This plan can be enlarged or modified to suit requirements.

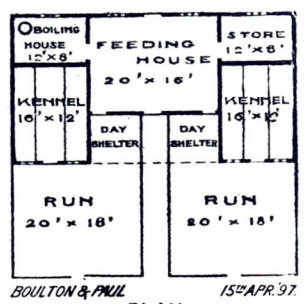

PLAN.

TESTIMONIAL.

From H. B. BUSH, ESQ., Hon. Sec., Aldershot Division, Foot Beagles.

I should like to take this opportunity of informing you of the general and complete satisfaction of the Officers of the Aldershot Division with the new Hound Kennels, and also to thank you for the promptness with which you have carried out our orders.

Approximate Price £110.

See also plans on page 281C.

CAUTION.—Beware of inferior imitations. All articles bear our name as a guarantee of good workmanship.

BOULTON & PAUL, MANUFACTURERS,

No. 133. RANGE OF KENNELS AND YARDS.

REGISTERED COPYRIGHT.

These Kennels are made with strong deal framing, covered on outside with rustic-jointed weather boarding. The roof is of corrugated iron, lined inside with felt and matchboarding. They are strongly and well made, warm and well ventilated. Each house is 6 ft. wide, 5 ft. deep, 6 ft. high to eaves, and 7 ft. 6 in. high at ridge. Yards 6 ft. long, 6 ft. wide, and 6 ft. high, and fitted with single folding benches. Each yard is provided with gate and dog-trough. The whole are well painted and prepared ready for easy erection by purchaser. The yards can be extended if required.

Cash Prices, Carriage Paid.

Two Houses and Yards complete	... ...	£15 10 0
Three Houses and Yards ,,	... ...	22 15 0
Six Houses and Yards ,,	... ...	42 0 0

Open Air Benches for Runs, 7/6 each extra.
Wood Batten Floors for Yards, 20/- each extra.

Cement concrete makes the best flooring, and it is advisable to place the whole building and yards upon it.

PLANS OF HOUND KENNELS.

No. 134.

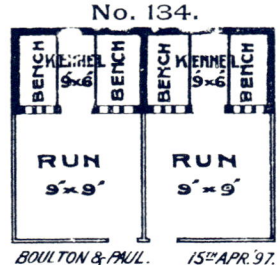

The elevation of these Buildings will be similar in construction to No. 126, page 280.

No. 135.

No. 136.

Approximate Prices, Carriage Paid, see Hound Kennel List.

CAUTION.—Beware of inferior imitations. All articles bear our name as a guarantee of good workmanship.

No. 137. BLOCK OF KENNELS AND YARDS.

REGISTERED COPYRIGHT.

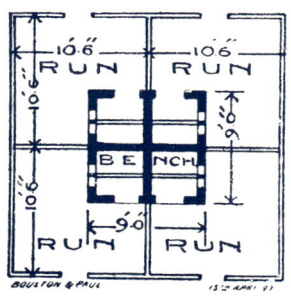

THE above illustration shows a block of four Kennels and Yards, suitable for sporting dogs. The Kennels form a square in the centre of the runs, as shown on plan. The Building is strongly constructed, and is specially suitable for placing in the park or paddock, being of a picturesque appearance. Each kennel is comfortable and well ventilated, and fitted with single folding bench.

Prices on application.

TESTIMONIAL.
From G. F. GOOCH, ESQ., Stamford.
The Kennels are now erected, and everything was complete and satisfactory.

CAUTION.—Beware of inferior imitations. All articles bear our name as a guarantee of good workmanship.

BOULTON & PAUL, MANUFACTURERS,

No. 138. KENNELS AND RUNS.

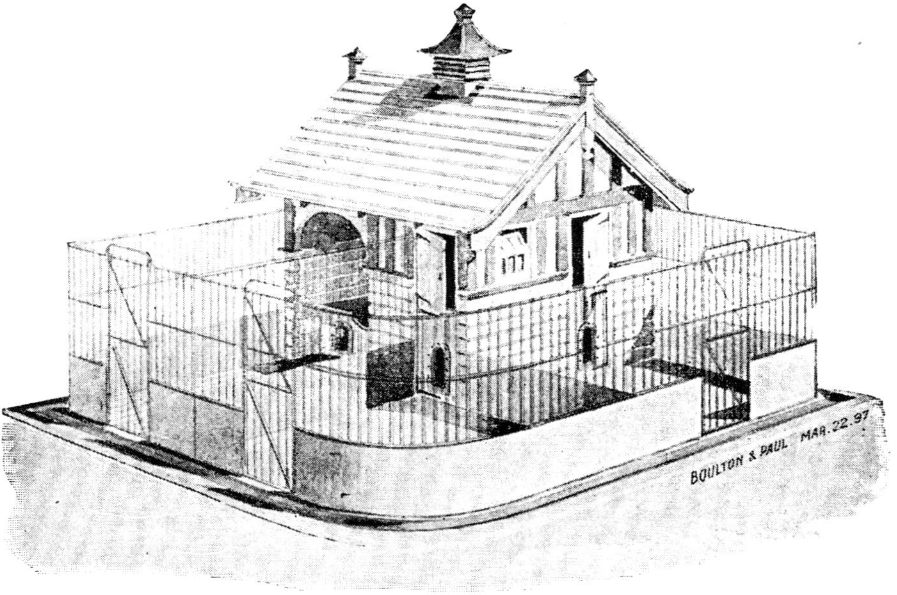

REGISTERED COPYRIGHT.

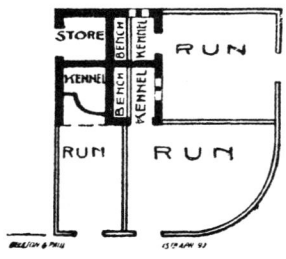

THESE Kennels are of artistic construction, and would make a handsome addition to a gentleman's stables or pleasure grounds. The whole construction is arranged to take to pieces easily, thereby rendering it a tenant's fixture. The design may be altered, if desired, to suit the taste of the purchaser. The Kennels are very comfortable and well ventilated, and provided with single folding benches.

Prices on application.

TESTIMONIAL.

From A. ROBINSON, ESQ., Northants.

The Kennel arrived safe last week, and I am very pleased. It is very handsome now it is put up; it puts all others in the shade.

CAUTION.—Beware of inferior imitations. All articles bear our name as a guarantee of good workmanship.

No. 124. LEAN-TO HOUSE FOR BEARS, KANGAROOS, &c.
Can also be adapted to accommodate Dogs.

REGISTERED COPYRIGHT.

THE Fencing is carried over to completely enclose the Run, a portion of which is glazed above the railing, as shown. The houses are constructed with a view to warmth and ventilation, and can be fitted to accommodate any animal.

Special Prices on application.

TESTIMONIAL.
From J. R. BLANEY, Esq., Pensiloe.
The Dog's House you supplied me with last year I am very pleased with, and all who have seen it say it is a marvel of cheapness.

NEW No. 96a. PORTABLE DOUBLE KENNELS,
Without Back, for Standing against a Wall.

REGISTERED COPYRIGHT.

THIS arrangement is a tenant's fixture, and consists of two Kennels, each 4 ft. by 4 ft., by 6 ft. 3 in. high at eaves, 7 ft. 6 in. at back, with matchboarded sides, and galvanized iron roof. Ventilator, door with strong hinges, and outside bench to each Kennel. Painted three coats, and the inside whitened. The Yards are each 6 ft. by 4 ft., with galvanized corrugated iron sheets, 2 ft. high at bottom, and fitted with feeding troughs.

Cash Price, £9 10 0. Carriage Paid.

TESTIMONIAL.
From J. L. PATINSON, Esq., Longton.
The Kennel has arrived. We have had it fixed, and are much pleased with it.

Carriage Paid to the principal Railway Stations in England and Wales.

CAUTION.—Beware of inferior imitations. All articles bear our name as a guarantee of good workmanship.

BOULTON & PAUL, MANUFACTURERS.

No. 120. RANGE OF ORNAMENTAL DOG KENNELS AND COVERED RUNS.

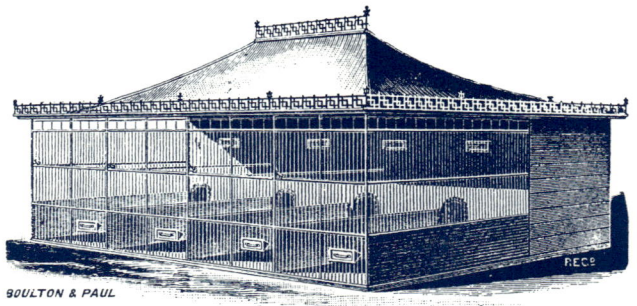

REGISTERED DESIGN, No. 209,450.

FITTED with doors front and back, sliding ventilators to admit light and air to each Kennel. Matchboard and felt ceiling at eaves level, ensuring a comfortable temperature at all times of the year.

Cash Price, Carriage Paid.

Four Kennels and Runs (as illustrated), each house 6 ft. by 5 ft., and each run 8 ft. by 5 ft. ... £52 10 0

TESTIMONIAL.
From K. PINCHBECK, ESQ., Northumberland Avenue.
I have received the Dog Kennels, and am more than pleased with them—they are simply perfection.

No. 121. RANGE OF DOG KENNELS & COVERED RUNS.

REGISTERED DESIGN, No. 209,451.

THIS HOUSE IS constructed similarly to No. 120, but is of a cheaper design. It is also fitted with matchboard and felt ceiling at eaves level.

Cash Price, Carriage Paid.

Four Kennels and Runs (as illustrated), each house 6 ft. by 5 ft., and each run 8 ft. by 5 ft. ... £39 15 0
Two Kennels and Runs ,, ,, ,, ... 22 0 0
One Kennel and Run ,, ,, ,, ... 12 0 0

Prices of larger sizes on application.

TESTIMONIAL.
From H. L. HORSFALL, ESQ., Glanrafon.
The Range of Kennels, No. 121, are now finished, they are very nice and entirely to my satisfaction.

Carriage Paid to the principal Railway Stations in England and Wales.

CAUTION.—Beware of inferior imitations. All articles bear our name as a guarantee of good **workmanship**.

ROSE LANE WORKS, NORWICH.

No. 88. RANGE OF DOG HOUSES AND YARDS.

FOR TERRIERS. **FOR SPORTING DOGS.**

REGISTERED COPYRIGHT.

Section showing interior arrangements.

THIS is a plan whereby a Range of Kennels can be carried out to any scale required. It will be seen that the dogs are kept quite apart and distinct, and that the building arrangements can be enlarged or diminished to suit the situation. A passage, 3 ft. 6 in., runs from end to end of the structure, and dividing it entirely. The most important object attained is to secure a plentiful supply of fresh air without draughts, and a ready access to each kennel. Fitted with improved day and night benches, A and B, described on page 296. In sections, ready for easy erection by purchaser.

Beyond the building are the open Yards, with gate to each, and fitted with revolving troughs, so that the dogs can be fed without entering the yards.

Cash Prices, Carriage Paid.

	For Large Dogs.		For Terriers.
Four Kennels, each 5 ft. by 4 ft., and Yards 6 ft. by 4 ft.	£30 0 0	... 4 ft. by 4 ft., and Yards, 6 ft. by 4 ft.	£28 0 0
Six Kennels ,, ,, ,, ,,	45 0 0	,, ,, ,,	39 0 0
Eight Kennels ,, ,, ,, ,,	55 0 0	,, ,, ,,	51 0 0

We have erected Kennels on this principle for W. WRIGHT, ESQ., Sussex; HARDING COX, ESQ., Missenden Abbey; R. WARREN VERNON, ESQ., Shawford; and others.

TESTIMONIALS.

From SYDNEY H. SLOCOCK, F.R.C.V.S., Hounslow.

I enclose Cheque for Dog Cages with which I am very pleased. I find they answer my requirements in all ways, and are a wonderful saving of room.

From J. F. SIMPSON, ESQ., M.R.C.V.S., J.P., Maidenhead, Berks.

GENTLEMEN,—I am much pleased with the two Dog Hospital Wards (consisting of thirteen distinct kennels) which you have recently supplied me with; they have a light and neat appearance, which all my clients admire, and there is an appearance of comfort about them which I confess I have never seen in any other Dog Infirmary.

Testimonials received from all parts of the Kingdom.

REGISTERED COPYRIGHT.
Outside View.

Kennels have been erected by us for most of the largest Breeders and Exhibitors of the present day.

Carriage Paid to the principal Railway Stations in England and Wales.

CAUTION.—Beware of inferior imitations. All articles bear our name as a guarantee of good workmanship

BOULTON & PAUL, MANUFACTURERS.

RANGE OF KENNELS AND YARDS.

"Messrs. BOULTON & PAUL of Norwich have recently introduced a most complete range of Dog Kennels and Yards This firm have provided dog and poultry keepers with many new things, all of which are not only novelties but useful accessories; **and this range of Dog Kennels must be classed as one of the most valuable of their productions.**"—*Live Stock Journal.*

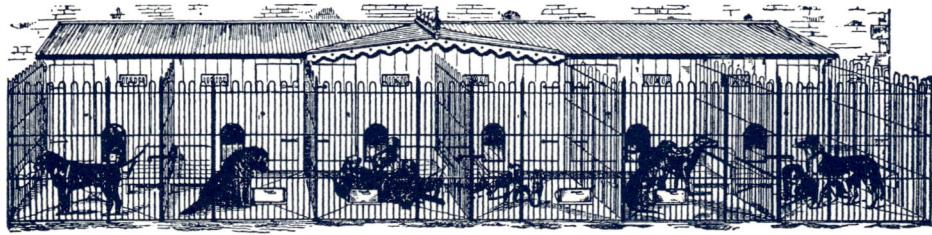

REGISTERED COPYRIGHT.

THE House is made of wood to place against a wall in such a manner that any handy workman can erect it. The roof is of **galvanized corrugated iron**, with iron gutters and leading down-pipe. A sliding ventilator is provided to admit air and light. The inside is fitted with a loose floor and wooden bench of improved construction, which can be removed for cleaning purposes. The railing is of wrought-iron, with ¾-in. vertical bars, 2 in. apart, fitted with gate and padlock; also improved reversible troughs (as illustrated) for feeding from the outside. Cement is a good flooring for the yards, and is easily laid. Each house is 6 ft. wide, 5 ft. deep, 6 ft. high at eaves, 7 ft. 3 in. high at back. Yards 6 ft. long, 6 ft. wide, 6 ft. high. Gates 2 ft. opening. Well painted and prepared **ready for fixing**. Yards can be made larger if required.

Reduced Cash Prices.

No. 82.	One House and Yard, complete	...	...	...	...	...	£7 10 0
No. 83.	Two Houses and Yards, complete	...	...	...	...	...	12 15 0
No. 84.	Three Houses and Yards, complete	...	...	...	...	...	18 18 0
No. 85.	Six Houses and Yards, complete (as illustrated)	...	...	...	...	35 0 0

Open-Air Benches, 7/6 each extra. Wood Batten Floors for Yards, 20/- each extra (see page 296). **Boarded Backs** with doors behind (if preferred), 25/- per Kennel extra. Galvanized Corrugated Sheet Iron at bottom of Fencing (as shown page 295), 10/- per Kennel extra.

TESTIMONIAL.

From H. F. LAWFORD, ESQ., Kinellan.

Mr. Lawford is much pleased with the Kennels. The dogs are in perfect comfort, and the Kennels are rather an ornament than the reverse. Messrs. Boulton and Paul have carried out exactly what was required.

REGISTERED COPYRIGHT.

Section showing inside arrangements.

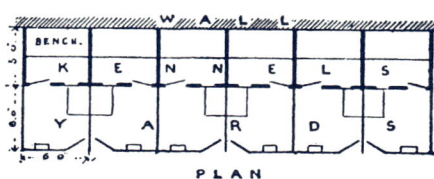

REGISTERED COPYRIGHT.

TESTIMONIAL.

From REV. GEORGE MILLS, North Barrow Rectory, Bath.

The Kennel arrived to-day, and has been put up, together with the Run; and you must allow me to add **my testimony** to those of others as to the completeness of the whole thing. It stands an ornament to my lawn, and I can only reiterate the many flattering statements contained in your catalogue respecting your workmanship.

Carriage Paid to the principal Railway Stations in England and Wales.

CAUTION.—Beware of inferior imitations. All articles bear our name as a guarantee of good **workmanship**.

No. 102.
RANGE OF KENNELS AND YARDS.

REGISTERED DESIGN, NO. 102.

For Placing against a Wall.

TESTIMONIALS.

From R. J. BARBER, ESQ., Belmont.

I am very pleased with the Kennels, and my dogs seem to approve of the extra comfort to an ordinary kennel.

From J. A. BROUGHTON, ESQ., Bradford.

I am delighted with the Kennel and consider it extraordinarily cheap. I shall be glad to recommend you to any one requiring a similar article.

Cash Prices, Carriage Paid.

	With Yards 6 ft. long.	With Yards 12 ft. long.
One Kennel 6 ft. by 4 ft. 6 in., 6 ft. at eaves, 7 ft. to top	£7 10 0	£9 0 0
Two Kennels 12 ft. by 4 ft. 6 in., 6 ft. at eaves, 7 ft. to top	14 0 0	16 10 0
Three Kennels 18 ft. by 4 ft. 6 in., 6 ft. at eaves, 7 ft. to top	20 0 0	23 0 0

Fitted with eaves-gutters and down-pipes.

No. 106.
NEW DOUBLE KENNELS AND COVERED RUNS.

RUSTIC joint weather-board walls, corrugated iron roof with gutters to eaves, matchboard ceiling, batten floor, and benches to houses.

REGISTERED COPYRIGHT.

TESTIMONIAL.

From ROBERT PARR, ESQ., Over Silton, Northallerton.

We received Dog Kennel quite safe, and we are highly pleased with it.

Cash Prices, Carriage Paid.

For St. Bernards, each house 6 ft. by 5 ft. and each run 8 ft. by 5 ft. ...	...	£22 0 0
For Colleys and Retrievers, each house 5 ft. by 4 ft. and each run 7 ft. by 4 ft. ...	...	18 0 0

Carriage Paid to the principal Railway Stations in England and Wales.

CAUTION.—Beware of inferior imitations. All articles bear our name as a guarantee of good workmanship.

BOULTON & PAUL, MANUFACTURERS.

No. 100. LEAN-TO COMPOSITE KENNEL
FOR PLACING AGAINST A WALL.

TESTIMONIAL.

From
W. L. MACKAY, Esq.,
Dublin.

I send you a line to say that the Kennel arrived, and was put together without any difficulty. I am very pleased with it.

TESTIMONIAL.

PUTNEY HILL.

MISS A. HAY begs to thank Messrs. Boulton and Paul for the Kennel sent. Everything fits beautifully, and she is much pleased with it.

THIS Kennel is constructed upon the same principle as Nos. 80, 81, and 89, but is made to place against a wall, which is an advantage where space is an object. Well painted and prepared ready for fixing. Fitted with revolving trough for feeding outside, and with eaves-gutters and down pipes.

Reduced Cash Prices, Carriage Paid.

House and Yard for Terriers, 8 ft. long, 3 ft. 6 in. wide, 4 ft. high at back	£5 0 0
House and Yard for Retrievers and Spaniels, 9 ft. 6 in. long, 4 ft. wide, 5 ft. high at back	5 10 0
House and Yard for Mastiffs and St. Bernards, 12 ft. long, 5 ft. wide, 5 ft. 6 in. high at back	6 10 0

Galvanized Sheet Iron around Runs, 18 in. high, 10/-, 8/6, and 7/- extra respectively.
Wood Batten Floors, 20/-, 15/-, and 10/- extra respectively.

No. 103. COMPOSITE KENNELS.

TESTIMONIAL.

From J. CARMICHAEL, Esq., Kingstown.

I am glad to tell you that the Kennel and Run arrived all right, and I am much pleased with it.

TESTIMONIAL.

From
C. GARTSIDE,
Esq., Glen Garth.

I purchased from you in 1883 some Dog Houses, and I take this opportunity of saying that they are still in good condition, and have given me every satisfaction.

Cash Prices, Carriage Paid.

	With Yards 6 ft. long.	With Yards 12 ft. long.
One Kennel 8 ft. long, 3 ft. 4 in. wide, 3 ft. 4 in. high	£7 15 0	£9 10 0
Two Kennels 16 ft. long, 3 ft. 4 in. wide, 3 ft. 4 in. high	14 0 0	17 10 0

Carriage Paid to the principal Railway Stations in England and Wales.

CAUTION.—Beware of inferior imitations. All articles bear our name as a guarantee of good workmanship.

ROSE LANE WORKS, NORWICH. 286A

No. 139. LEAN-TO DOUBLE KENNEL WITH COVERED RUNS.

For placing against a wall.

STRONGLY constructed, of superior make and finish, comfortable and well ventilated, each Kennel provided with single folding bench. Each House 5 ft. by 4 ft., 5 ft. high in front, and 7 ft. high at back; each run 5 ft. square.

Cash Prices, Carr. Pd.
Single House and Run, £7 10 0
Double House and Run, £15 0 0

REGISTERED COPYRIGHT.

No. 140. LEAN-TO KENNEL WITH COVERED RUN.

For placing against a wall.

12 ft. long by 5 ft. wide, 5 ft. high in front, 6 ft. 6 in. high at back, fitted with single folding bench.

Cash Price, Carr. Paid.
£7 10 0

For smaller Kennel see No. 100, page 286.

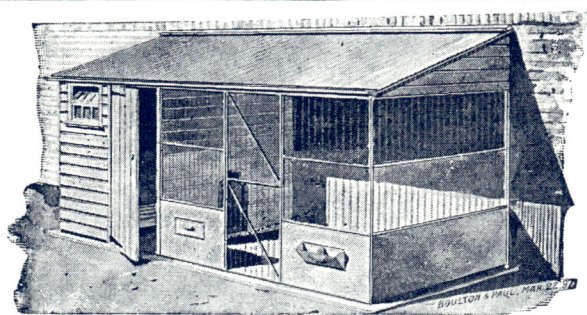

REGISTERED COPYRIGHT.

No. 141. LEAN-TO DOUBLE KENNEL AND RUNS.

THESE Kennels have a passage behind, and are similar in construction to No. 104 Kennels, on page 287, but are larger and of stronger construction, and in addition have specially constructed ventilators over passage as shown. Each Kennel is 5 ft. by 5 ft., and has a yard 6 ft. by 5 ft., and fitted with single folding bench.

Cash Prices, Carr. Paid.
Two Kennels & Yards ... £14
Three Kennels & Yards ... 20
Four Kennels & Yards ... 25

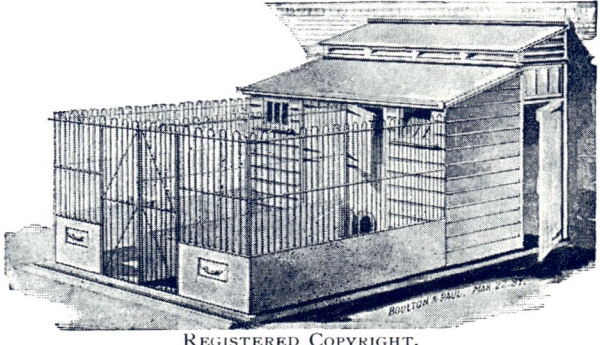

REGISTERED COPYRIGHT.

CAUTION.—Beware of inferior imitations. All articles bear our name as a guarantee of good workmanship.

No. 104. RANGE OF LEAN-TO KENNELS.

TESTIMONIAL.

From
J. NEWSTEAD, Esq.,
Clifton.

I am very pleased with the Kennels. They are the most convenient and comfortable Kennels I have seen; and I consider them very cheap.

TESTIMONIAL.

From
D. W. FOX, Esq.,
Birkdale.

The Kennels are as nice and complete as they possibly can be in every respect. The dogs are very contented with their new quarters.

REGISTERED COPYRIGHT.

WITH passage 2 ft. 6 in. wide at the back, to fix against a wall. Height at eaves, 6 ft.; at back, 7 ft. 6 in. Each Kennel is 4 ft. by 4 ft., and has a yard 6 ft. by 4 ft., a wood batten floor in Kennel, revolving feeding trough in yard, also fitted with eaves-gutters and down-pipes.

Cash Prices, Carriage Paid.

One Kennel and Yard	£6 10 0
Two Kennels and Yards	12 0 0
Three Kennels and Yards	17 10 0

Wood Batten Floors for Yards, 12/- each extra. Boarded Backs, 15/- per Kennel extra. Open Air Benches extra, see plan.

TESTIMONIAL.

From J. C. GILMORE, Esq., Bristol.

The Dog's House and Enclosure have been now up some little time, and I am very pleased with them. Every one who has seen them agrees with me that they have never seen a better thing of the kind.

No. 105. RANGE OF LEAN-TO KENNELS.

TESTIMONIAL.

From
C. C. CLARKE, Esq.,
Bungay.

The Kennel arrived safely, and I am very much pleased with it. It took very little time to put together and is most effective (besides being well made) and handsome in appearance.

TESTIMONIAL.

From C. H. C. HARRISON, Esq., Holme-on-Sevale.

I received the Dog Kennel yesterday. We have put it together, and I must say it is the most perfect Kennel I have ever seen.

REGISTERED COPYRIGHT.

To fix against a wall. Height at eaves, 6 ft.; at back 7 ft. Each Kennel is 4 ft. by 4 ft., and has a yard 6 ft. by 4 ft. a wood batten floor in the Kennel, revolving feeding trough in yard, also fitted with iron eaves-gutters and down-pipes.

Cash Prices, Carriage Paid.

One Kennel and Yard	£5 0 0	Three Kennels and Yards	£13 10 0
Two Kennels and Yards	9 0 0	Four Kennels and Yards	16 15 0

Batten Floor for Yards, 12/- each extra. Boarded Backs with Doors behind, 15/- per Kennel extra.

Carriage Paid to the principal Railway Stations in England and Wales.

CAUTION.—Beware of inferior imitations. All articles bear our name as a guarantee of good workmanship

BOULTON & PAUL, MANUFACTURERS.

COMPOSITE KENNEL.—THE CHAIN DISPENSED WITH.

THE House or Shelter is boarded at the back, so need not be placed against a wall; roof covered with corrugated iron. The bed is recessed under shelter of roof, the top forming a shelf for dog to lie upon, which is readily taken off for getting at the inside. Cement makes a good flooring for the yards. Each Kennel is provided with two revolving troughs for feeding from the outside, and is highly finished.

Reduced Cash Prices, Carriage Paid.

No. 80.	Double House and Yards for Mastiffs and St. Bernards (as illustrated above), 12 ft. long, 10 ft. wide, 5 ft. 6 in. high	£15 0 0
No. 81.	Single House and Yard, 12 ft. long, 5 ft. wide	7 15 0
	Single House and Yard for Retrievers and Spaniels (as shewn below), 9 ft. 6 in. long, 4 ft. wide, 5 ft. high	6 0 0
No. 89.	Single House and Yard for Terriers (as below), 8 ft. long, 3 ft. 6 in. wide, 4 ft. high	4 10 0

Wood Batten Floors for the Yards, 20/-, 15/-, and 10/- each respectively, extra.
Corrugated Sheets for covering runs of No. 80 Kennel, 16/- per Kennel extra.
Galvanized Crimped Sheet Iron around runs of No. 80 Kennel, 18 in. high, as a protection from draught, 10/- each Kennel, extra.

TESTIMONIAL.

From REV. J. CUMMING MACDONA,
Cheadle Rectory, Cheshire.

I am extremely pleased with the Kennel you sent me. It is more perfect than any Kennel I have yet seen; it is comfortable, commodious, convenient, and cheap at the price.

REGISTERED COPYRIGHT.
View of Second size Kennel, No. 81.
Price £6 0 0

REGISTERED COPYRIGHT.
Plan of Single Kennel.

REGISTERED COPYRIGHT.
View of Terrier Kennel, No. 89.
Price £4 10 0

Carriage Paid to the principal Railway Stations in England and Wales.

CAUTION.—Beware of inferior imitations. All articles bear our name as a guarantee of good workmanship.

ROSE LANE WORKS, NORWICH

No. 122. RANGE OF IMPROVED KENNELS, WITH RUNS, FOR TERRIERS.

REGISTERED DESIGN, No. 207,092.
Front View.

THE doors are arranged to open at the back for cleaning the houses, which are fitted with sliding floor. The runs are wired in at the top.

Cash Price, Carriage Paid.

Four Kennels and Runs, as shown above, each Kennel 4 ft. by 2 ft. 9 in.
and each Run 6 ft. by 4 ft. £14 10 0

TESTIMONIAL.

From PAUL BEVAN, ESQ., London.

The Kennel duly arrived, and is a most ingenious contrivance; and I am very well pleased with its appearance now that it is erected. It is the best Kennel I ever saw.

No. 122. IMPROVED DOUBLE KENNEL, WITH RUNS, FOR TERRIERS.

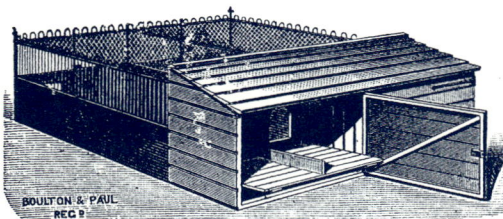

REGISTERED DESIGN, No. 207,092.
Back View.

THE doors are arranged to open at the back for cleaning the runs, and are wired in at the top.

Cash Price, Carriage Paid.

Two Kennels and Runs, as shown above, each Kennel
4 ft. by 2 ft. 9 in., and each run 6 ft. by 4 ft. ... £7 10 0

TESTIMONIAL.

From DR. FLETCHER, Derby.

I am very much pleased with the Kennel and Run, and consider it to be the neatest one of its kind that I have ever seen.

TESTIMONIALS.

From JOHN A. C. RAYNER, ESQ., Ambleside.

The Kennel has arrived, and is very satisfactory, will you please make me another exactly like it, and when ready I will give you the address where to send it.

From J. M. HERBERT, Esq., Wick.

The Kennel Railings have come safely to hand, and I am very pleased with them.

From J. G. PHILPOTT, ESQ., Brenchley.

Some few months back you forwarded me two of your Dog Kennels; I like them so much that I should be glad if you would send one more of the same size and two more rather smaller, for terriers.

From CAPTAIN P. COKE, Leeds.

I think the Kennels are perfect in their way.

Carriage Paid to the principal Railway Stations in England and Wales.

CAUTION.—Beware of inferior imitations. All articles bear our name as a guarantee of good workmanship

ROSE LANE WORKS, NORWICH.   289A

No. 142. INTERIOR OF WOODEN BUILDING FOR HOUSING TERRIERS.

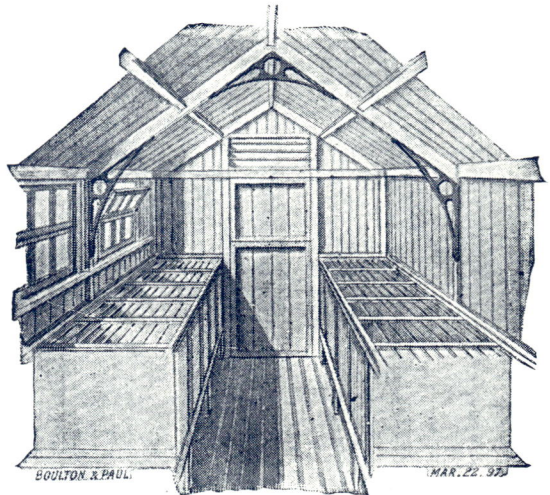

REGISTERED COPYRIGHT.

THE above shows the interior of a wooden building specially adapted or housing terriers. The pens are fixed on a raised bench, with separate door to each. Outer yards can also be provided if required. We shall be glad to give special estimates for a building of this kind upon receipt of particulars of requirements.

TESTIMONIALS.

From J. E. NEWKINS, Esq., St. Ives.

On my return this evening, I found you had promptly executed my order for Kennel and Travelling Box, for which I beg to thank you. I am very pleased with both, and consider they are splendid value combined with latest improvements.

From J. G. BAYLEY, Esq., Glyndhurst.

I purchased from you a short time ago a Terrier House which is the most perfect Kennel and Run I ever saw.

CAUTION —Beware of inferior imitations. All articles bear our name as a guarantee of good workmanship.

BOULTON & PAUL, MANUFACTURERS.

No. 93. NEW REGISTERED DOG KENNEL.

NOTE.—These Kennels cannot be compared with the cheap class of kennels advertised.

BED ALWAYS DRY.
REGISTERED, NO. 30,550.

THESE Kennels are made of best red deal, sides varnished, roof painted maroon, thoroughly seasoned and highly finished. All parts are accessible for cleansing and disinfecting. This is a great feature in the construction, the entrance being at the side, with a *folding* inside partition, affords a warm and dry bed in all weathers. They take to pieces, and pack flat for travelling. The Bench slides underneath the Kennel when not in use, and is an important addition to the Kennel, and adds greatly to the comfort of the dog.

Reduced Cash Prices, including Registered Sliding Bench.

	Length. ft. in.	Width. ft. in.	Height. ft. in.	
For Terriers...	2 6	1 4	2 5	£1 1 0
For Colleys, Spaniels, or Retrievers	3 6	2 3	3 4	1 12 6
For St. Bernards or Mastiffs	4 6	2 8	4 6	2 15 0

Chains and Collars not supplied. Feeding Pan as shown attached to Kennel, 3/- each. If on Castors, 7/6 extra.

TERRIER SIZE.

TESTIMONIALS.

From J. D. LUMSDEN, ESQ., Huntingtowerfield.

I have great pleasure in stating that the five Terrier Kennels give me great satisfaction. I have been able to keep my dogs in much better health ever since I got your Kennels.

From A. G. GOLD, ESQ., London.

Your ingenious Dog Kennel has given complete satisfaction, it was so easily put together; and the front bench is much appreciated by the dog.

From EDW. OLIVER, ESQ., Haverstock Hill.

The Dog Kennel received safely. It is a very good one, and a great improvement on the old style. The Registered Bench is a capital idea, and a great boon to the dogs.

COLLEY SIZE.

REGISTERED COPYRIGHT.

Plan showing open Run, made of our No. 86 Pattern Kennel Railing as shown on page 295, attached to a Portable Kennel.

Size—6 ft. square, 6 ft. high, with gate, feeding trough, and day bench.

Prices Complete, Carriage Paid.

For Terriers	...	...	£5 0 0
,, Retrievers	...	...	5 10 0
,, St. Bernards	...	...	6 10 0

MASTIFF SIZE.

Carriage Paid on all Orders above 40/- *value to the principal Railway Stations in England and Wales.*

CAUTION.—Beware of inferior imitations. All articles bear our name as a guarantee of good workmanship.

ROSE LANE WORKS, NORWICH.   200A

No. 143. THE KEEPER'S KENNEL.

TESTIMONIAL.
From
A. MACKINTOSH, Esq.
Cardiff.
The three Kennels are very good and give great satisfaction.

TESTIMONIAL.
From
Mr. THOS. PUNTON,
Belford.
The Dog Kennels arrived here quite safely and in good condition. They please me very much.

THIS is a strong serviceable Kennel adapted for keepers' use and rough wear, it is made entirely of wood. Size, 3 ft. 6 in. long, 2 ft. 3 in. wide, 3 ft. 4 in. high. The walls are covered with our staining preparation, and the roof painted red.

Cash Price 25/- each.

BENCHES FOR DOG SHOWS OR KENNELS.

Estimates and full particulars on application.

TESTIMONIAL.
From S. E. HURNDALL, Esq., Gosforth.
I am very pleased with the dog railings which are now up and are most compact.

CAUTION.—Beware of inferior imitations. All articles bear our name as a guarantee of good workmanship.

ROSE LANE WORKS, NORWICH.

No. 123.
COMBINED KENNEL AND RUN.

REGISTERED COPYRIGHT.

THIS is a very convenient arrangement, and comprises a small Run with sleeping accommodation. It is specially adapted for Terriers and pet dogs. It also makes an excellent Puppy Kennel, and also Cat Kennel.

Cash Prices, Carriage Paid.
Size, 5 ft. long, by 3 ft. 6 in. wide ... £3 10 0
  ,,   4 ft.   ,,   2 ft. 6 in. ,,   ... 2 10 0
If on Castors, 7/6 extra.

TESTIMONIAL.
From F. A. KELTON, ESQ., Belmont House.
The Kennel came safely to hand last Thursday. It was beautifully packed. The illustration does not do justice to the article. It is thoroughly well got up and finished—a really handsome ornament and strong to boot.

KENNEL WITH RUN.

REGISTERED COPYRIGHT.

OUTER Run for attaching to our No. 93 Kennels, with wood batten floor and feeding pan.

Prices of Runs only.
Terrier size. 5 ft. long ... ... £1 10 0 extra.
Retriever size, 6 ft. long   ... 2 0 0   ,,
Mastiff size, 6 ft. long   ...   2 10 0   ,,
For Prices of Kennels, see page 290.

TESTIMONIAL.
From CAPTAIN F. R. MOTT, Rock Ferry.
I received the Kennel, with which I am very much pleased, and I shall take great pleasure in showing it to my friends, as well as recommend it.

REGISTERED DOG KENNEL.

REGISTERED, NO. 30,551.

THIS Kennel is specially suited for Terriers, but can be made for larger dogs.
The top of bed forms a day shelf for the dog to lie upon. It is hinged for cleaning out the inside of the Kennel.

Cash Prices.
	Length.	Height.	
For Terriers ...	2 ft. 5 in.	2 ft. 8 in.	£1 5 0
For Retrievers...	4 ft. 0 in.	3 ft. 6 in.	2 0 0

THE KEEPER'S DOG KENNEL.

FOR use during the breeding season. Can be shifted about as desired. The Kennel can be detached, and the barrow used for removing Coops or other Kennels.

Cash Price 45/-

TESTIMONIAL.
From E. J. BROON, ESQ., Hoddon Castle.
I have received the Kennel and Barrow and am surprised at their cheapness, they are so substantial.

No. 33. HOUSE FOR ST. BERNARDS OR MASTIFFS.

Price £2 15 0

Carriage Paid on all Orders above 40/- *value to the principal Railway Stations in England and Wales*

CAUTION.—Beware of inferior imitations.   All articles bear our name as a guarantee of good workmanship.

ROSE LANE WORKS, NORWICH. 292A

No. 94.
KENNEL AND TRAVELLING BOX COMBINED.

This is a Kennel arranged as a Travelling Box, with ventilator in roof.

Cash Prices.

Terrier size ...	£1	1	0
Retriever size	1	12	6
St. Bernard size	2	15	0

No. 97. CAT'S HOUSE WITH RUN.

Constructed of red deal framing. House walls covered with tongued and grooved matchboarding, weatherboarded roof, doors in house and run for cleaning purposes.

Cash Prices.

Double House (as illustrated), 7 ft. 6 in. long by 5 ft. wide.	£4	10	0
Single House ...	2	10	0

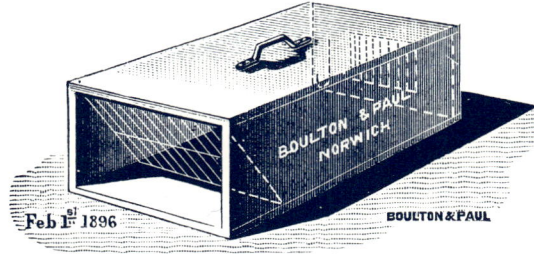

TRAP FOR CATCHING UP FERRETS.

A Gamekeeper writes: "I have caught up over twenty ferrets in your box during the past month."

Price 5/- each.

CAUTION.—Beware of inferior imitations. All articles bear our name as a guarantee of good workmanship.

BOULTON & PAUL, MANUFACTURERS.

TRAVELLING BOX FOR DOGS.
No. 130. New Article.

REGISTERED COPYRIGHT.

A PROJECTING bar is fixed in front of door to prevent Luggage being placed close to bars.

Cash Prices.
No. 1. 22 in. long, 20 in. wide, 24 in. high £0 14 6
No. 2. 36 in. long, 20 in. wide, 27 in. high 1 5 0
No. 3. 42 in. long, 30 in. wide, 33 in. high 1 15 0

TESTIMONIAL.
From J. R. WINDLE, ESQ., Gordle.

The new pattern Box is, in my opinion, everything that could be desired, and the improvements you have introduced are at once manifest, and will I have no doubt, be greatly appreciated by exhibitors.

EXHIBITION TRAVELLING BOX AND KENNEL COMBINED.

TESTIMONIAL.
From ALFRED H. TWINING, ESQ., Salcombe.

SIRS,—I received the Dog-Box safely last night, and am much pleased with it. I shall have much pleasure in recommending it to friends.

NOTE. The No. 94a Box is only fitted as a Kennel to order at an extra charge.

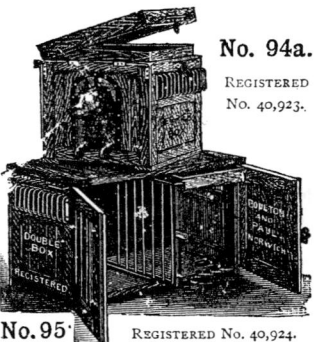

No. 94a.
REGISTERED No. 40,923.

No. 95. REGISTERED No. 40,924.

Cash Prices.
No. 94a. For Terriers. Single Box, 20 in. high, 22 in. long, 20 in. wide £1 10 0
No. 95. For Spaniels, Colleys or Retrievers, or with partition for two Terriers, 36 in. long, 20 in. wide, 21 in. high 2 2 0
For Bloodhound or Mastiff, 3 ft. 6 in. long, 2 ft. 9 in. high, 2 ft. 6 in. wide... ... 3 3 0

No. 63. DOUBLE FERRET KENNEL WITH RUNS.

NO FOOT ROT.
NO DISEASE.

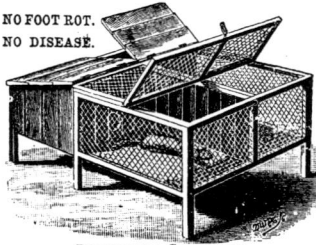

REGISTERED COPYRIGHT.

From T. L. RUDSTON READ, ESQ., Allerton Hall.

I have received the Ferret House with which I am well pleased I consider it a most ingenious invention.

THIS is a well constructed Kennel for Ferrets, every arrangement being made for cleanliness. The floor of the runs slope to the front, so they can be frequently rinsed down.

Cash Prices.
4 ft. long, 3 ft. wide, 2 ft. high ... £1 15 0
Larger size 2 5 0

IMPROVED GALVANIZED IRON FERRET FEEDING PAN.

REGISTERED COPYRIGHT.

Cash Price.
Size 6 in. by 4 in., and 2 in. deep, 1/6 each.

No. 64. FERRET HUTCH.
As described in the Book on *Ferrets and Ferreting*, published by L. Upcott Gill, 170 Strand, W.C.

MADE of best red deal, painted three coats; divided into two equal compartments.

From J. W. WILKES, ESQ., Bayswater, W.

I am very pleased with the Ferret Kennel you sent to my address in Wales.

Cash Price.
3 ft. long, 18 in. high, 18 in. deep, £1 1 0

PORTABLE FERRET BOXES OF THE LATEST DESIGN.

Cash Prices.
Single Box, 5/- each.   Double Box, 10/- each.
Rabbiting Spades 5/6 each.   Ratting Spades, 4/6 each.
Ferret Lines, 1/6 each.

Carriage Paid on all Orders above 40/- value to the principal Railway Stations in England and Wales.

CAUTION.—Beware of inferior imitations.   All articles bear our name as a guarantee of good workmanship

ROSE LANE WORKS, NORWICH 293

No. 60c. NEW IMPROVED PUPPY HOUSE WITH RUN.

REGISTERED COPYRIGHT.

TESTIMONIAL.
From
CHARLES COOK, Esq.,
Windygates.
Some years ago I bought a small Terrier Kennel from you which gave and continues to give me every satisfaction.

TESTIMONIAL.
From
ARCHIE B. BOYD, Esq.,
Aylsham.
Within the last few years I have had two Dog Kennels from you, and have been highly satisfied with them.

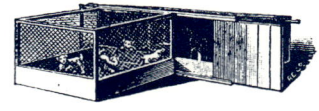

THIS House has been specially designed to combine cheapness with strength and durability. It is fitted with sliding doors, and is 6 ft. long, 3 ft. wide, 2 ft. 6 in. high.

Cash Price (as illustrated above), **£1 10 0** each.

Two, Carriage Paid.

Outer Run to No. 60c Puppy House (as illustrated), **15/-** each, extra.

No. 62. PUPPY HOUSE.

REGISTERED COPYRIGHT.

THIS makes a capital Kennel for young Puppies. Being portable, it can be shifted to fresh ground as often as required, every facility being provided for keeping scrupulously clean. Reversible trough is fitted for feeding from the outside.

Cash Price, Carriage Paid.

6 ft. long, 2 ft. 6 in. wide, 2 ft. 6 in. high .. **£2 5 0**

See Puppy Fencing.

TESTIMONIALS.

From C. W. MANSELL, Esq., Victoria Barracks.
The Puppy House arrived all right, and gives every satisfaction.

From REV. H. B. BUSH, Farnborough.
I am very much pleased with the Kennel and Run, and wish another of a similar kind forwarded at once.

No. 101. KENNEL RAILING FOR PUPPIES.

REGISTERED COPYRIGHT.

THIS Fencing will supply a want long felt in many Kennels, as most breeders who have not sufficient accommodation at hand have found to their cost. By the use of these hurdles, temporary pens can be formed in a few minutes, and with the use of No. 62 Puppy House a litter of pups can be comfortably housed and attended to without the least trouble. The ground may also be changed at will.

Hurdles 6 ft. long, 3 ft. high ; vertical bars $\frac{1}{4}$-in. diameter.

Cash Price 5/6 each.

Gates and Standards to match, **10/-**.

Carriage Paid on all Orders above **40/-** *value to the principal Railway Stations in England and Wales.*

CAUTION.—Beware of inferior imitations. All articles bear our name as a guarantee of good workmanship.

BOULTON & PAUL, MANUFACTURERS,

ORNAMENTAL WIRE LATTICE HURDLES,

FOR ENCLOSING GARDENS, PLEASURE GROUNDS, BORDERS, POULTRY YARDS, KENNELS, &c.

No. 1 HURDLE.

Cash Prices.

6 ft. long, 3 ft. high 6/-
6 ft. long, 2 ft. 6 in. high ... 5/-

No. 2 HURDLE.

Cash Prices.

6 ft. long, 3 ft. high 6/-
6 ft. long, 2 ft. 6 in. high ... 5/-

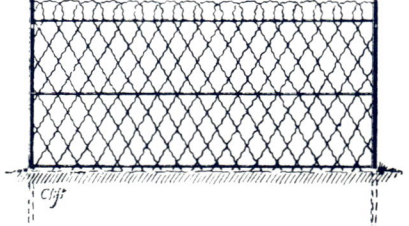

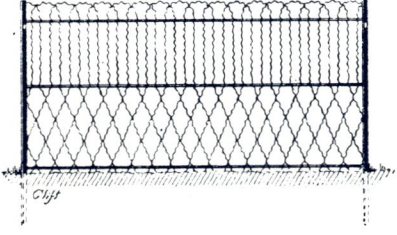

No. 3 HURDLE.

Cash Prices.

6 ft. long, 3 ft. high 6/-
6 ft. long, 2 ft. 6 in. high ... 5/-

No. 4 HURDLE.

Cash Prices.

6 ft. long, 3 ft. high 6/-
6 ft. long, 2 ft. 6 in. high ... 5/-

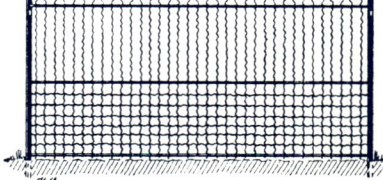

Carriage Paid on all Orders above **40/-** *value to the principal Railway Stations in England and Wales.*

CAUTION.—Beware of inferior imitations. All articles bear our name as a guarantee of good workmanship.

ROSE LANE WORKS, NORWICH. 294A

ORNAMENTAL WROT.-IRON KENNEL RAILING.

Cash Prices.

Nos. 93 and 94, with ⅜ in. vertical bars, and ¼ in. dog bars 2 ft. high at bottom.
5 ft. high, 9/- per yard.   6 ft. high, 10/- per yard.
Gates to match, 2 ft. wide, 22/6 each.

Cash Prices.

No. 95, with ⅜ in. vertical bars, for placing on a wall or fixing into the ground.
4 ft. high, 6/6 per yard.   5 ft. high, 7/6 per yard.   6 ft. high, 8/6 per yard.

No. 96, with ⅜ in. vertical bars and corrugated sheet at bottom 2 ft. high.
4 ft. high, 7/6 per yard.   5 ft. high, 8/6 per yard.   6 ft. high, 9/6 per yard.

Gates to match, 2 ft. wide—4 ft. high, 18/6 ; 5 ft. high, 20/6 ; 6 ft. high, 22/6.

Carriage Paid on all Orders above **40/-** value to the principal Railway Stations in England and Wales.

CAUTION.—Beware of inferior imitations. All articles bear our name as a guarantee of good workmanship.

BOULTON & PAUL, MANUFACTURERS.

No. 87. WROUGHT-IRON KENNEL RAILING.

REGISTERED COPYRIGHT.

No. 2 Pattern, to fix on Stone or Brick Wall.

Reduced Cash Prices.

With ⅜-in. vertical bars, spaced **2** in. apart .. **4** ft. high, **6/6** .. **5** ft. high, **7/6** .. **6** ft. high, **8/6** per yard.
With ½-in. vertical bars, spaced **2½** in. apart .. **4** ft. high, **7/-** .. **5** ft. high, **8/6** .. **6** ft. high, **10/6** ,,
With ⅝-in. vertical bars, spaced **3** in. apart .. **4** ft. high, **10/6** .. **5** ft. high, **12/6** .. **6** ft. high, **15/6** ,,
Single Gate, **2** ft. wide (as shown above) with padlock, **4** ft. high, **18/6**; **5** ft. high, **20/6**;
6 ft. high, **22/6** each.
Cast-iron Coping for **4½**-in. wall, **3/-** per yard; ditto for **9**-in. wall, **6/6** per yard.

TESTIMONIAL.

From MAJOR-GEN. MACDONALD, Dunalastair, Perthshire.

I have great pleasure in saying that your Dog Kennel Fencing is most admirable, and much approved of by the shooting tenants of Lochgary and myself. The price was very reasonable. The materials seem durable, and cleanliness and absence of Dog Ticks are gained by the system.

No. 90. ENCLOSURE FOR DOGS.

Cash Prices.

6 ft. long, 6 ft. wide, 5 ft. high	..	...	...	£2 12 6
6 ft. ,, 6 ft. ,, 6 ft. ,,	..	..	..	3 2 6
8 ft. ,, 3 ft. ,, 5 ft. ,,	..	...	...	3 4 6
8 ft. ,, 8 ft. ,, 6 ft. ,,	..	...	...	3 17 0

Including Gate.
Revolving Trough, fitted to railing, 5/- extra.

TESTIMONIAL.

From E. H. HILL, ESQ., Broadway Court, Worcester.

The Dog Kennel Fencing you supplied me with is most satisfactory, and forms a capital yard for the dogs.

Carriage Paid on all Orders above **40/-** *value to the principal Railway Stations in England and Wales.*

CAUTION.—Beware of inferior imitations.   All articles bear our name as a guarantee of good workmanship.

No. 86. WROUGHT-IRON KENNEL RAILING.

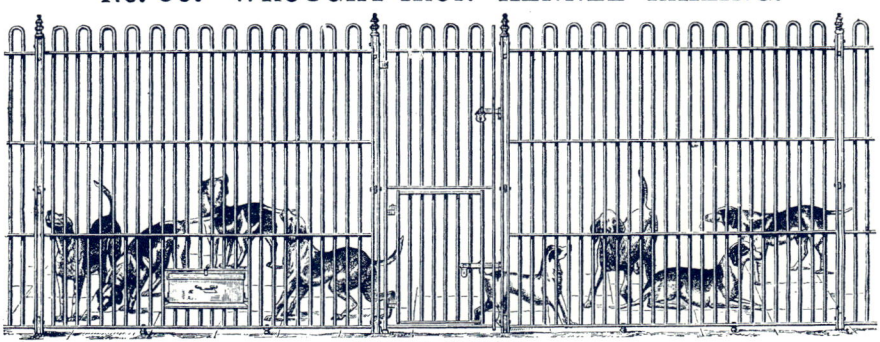

REGISTERED COPYRIGHT.
No. 1 Pattern, to fix into the Ground.

These Railings are constructed in the simplest manner, so that any handy workman can put them together. They are made in 6 ft. lengths, and jointed together with bolts and nuts; feet are provided for fixing into the ground. Can be readily removed. Made to suit any situation without extra charge.

Reduced Cash Prices.

With 7/16-in. vertical bars, spaced 2 in. apart .. 4 ft. high, 5/3   5 ft., 5/9   6 ft., 6/6 per yard.
With 3-in. vertical bars, spaced 2 in. apart .. 4 ft. high, 6/6   5 ft., 7/6   6 ft., 8/6   7 ft., 9/9 per yard.
With ¼-in. vertical bars, spaced 2½ in. apart .. 4 ft. high, 7/-   5 ft., 8/6   6 ft., 10/6   7 ft., 11/6   ,,
Single Gates, 2 ft. wide (as shown on page 294) and padlock, 5 ft. high, 20/6; 6 ft. high, 22/6;
7 ft. high, 25/6 each.
If double Gates as shown above, 5/9 extra. Reversible Troughs for feeding from outside fitted to Railing, 8/6 each.

TESTIMONIAL.

From E. H. HILL, Esq., Broadway Court.
The Dog Kennel Fencing you supplied me with is most satisfactory, and forms a capital yard for the dogs.

No. 86a. WROUGHT-IRON KENNEL RAILING.
With Corrugated Iron Bottoms.

REGISTERED COPYRIGHT.

The lower portion of the fencing is covered with Galvanized Corrugated Sheet Iron specially prepared for the purpose, which shelters dogs from draughts, and in partition fences prevents fighting. It is also less expensive than brickwork.

Cash Price, 1/6 per yard on above prices.

TESTIMONIALS.

From A. J. CARVER, Esq., Beckenham.
I am very well satisfied with the Dog Kennel Fencing you sent me; it forms an excellent run for my St. Bernards. The Fencing is rather an ornament to my garden than otherwise.

From G. HERBERT LOFTUS, Esq., Co. Clare.
The Kennel Fencing which I had from you some months ago has in every way given me satisfaction, and I now send you a further order, for which I enclose plan.

Carriage Paid on all Orders above 40/- value to the principal Railway Stations in England and Wales.

CAUTION.—Beware of inferior imitations.   All articles bear our name as a guarantee of good workmanship.

BOULTON & PAUL, MANUFACTURERS.

No. 91. LIGHT KENNEL FENCING.

REGISTERED COPYRIGHT.

¼-in. vertical bars.

Cash Prices.
6 ft. high ... ... per yard 5/6
7 ft. high ... ... ,, 6/-

TESTIMONIAL.
From REV. E. L. M. COLVILLE, Leek-Wootton.
Your Dog Kennel Fencing which has been put up is very satisfactory.

TABLETS FOR DOG BENCHES.

Estimates upon application.

BENCHES FOR DOG SHOWS OR KENNELS.

Estimates upon application.

TESTIMONIAL.
From C. R. SHEPHERD, ESQ., Cardiff.
I am exceedingly pleased with the Railing and Benches.

No. 92. NEW KENNEL RAILING.

REGISTERED COPYRIGHT.

¼-in. vertical bars.

Cash Prices.
4 ft. high ... ... per yard 7/6
5 ft. high ... ... ,, 8/6

TESTIMONIAL.
From R. HOLMAN PECK, ESQ., Elmfield.
I have two Runs of yours put up in September, which give every satisfaction.

No. 99. WOOD BATTEN FLOORS,
For Yards of Kennels.

REGISTERED COPYRIGHT.

THIS Flooring is the best for Portable Kennels, as it can at any time be taken up and removed. It is also preferred by some to brick or cement, being drier and warmer for the dogs, and is readily put down.

Price 6d. per foot super, painted.

CAST-IRON DOG TROUGHS.

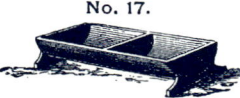

No. 1.   No. 17.

REGISTERED COPYRIGHT.

Cash Prices.
GALVANIZED.	PAINTED.
10 in. diam. ... each 2/-	18 in. long, two divisions each 4/-
12 in. ,, ... ,, 3/-	24 in. ,, three ,, 5/-

No. 98. NEW FOLDING DAY AND NIGHT BENCH.

REGISTERED.

THE advantage gained by this Bench will be seen at a glance. The top floor is made to fold up with the bedding, so that the straw is not trodden upon during the day, and thus kept dry and clean. The lower bench to be used during the day. The loose flap is for making a deep bed for a bitch whelping, or for young puppies.

Cash Prices.
4 ft. by 2 ft. 6 in. ... ... each 15/-
5 ft. by 2 ft. 6 in. ... ... ,, 17/6
6 ft. by 3 ft. ... ... ,, 20/-

Carriage Paid on all Orders above **40/-** *value to the principal Railway Stations in England and Wales.*

CAUTION.—Beware of inferior imitations.   All articles bear our name as a guarantee of good workmanship.

ROSE LANE WORKS, NORWICH.

GALVANIZED SEAMLESS STEEL DOG TROUGHS.

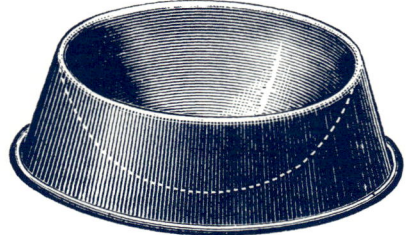

No. 196. ROUND DOG TROUGH
(Ordinary Pattern).
Cash Prices.
Size, 8 in. by 10 in. ... 1/- each.
,, 10 in. by 12 in. ... 1/6 ,,
Cannot be broken. Cannot be upset.

No. 197.
ROUND DOG TROUGH
With Partition.

This is a similar seamless Trough to No. 196, but it has a strong partition and grip for hand.

Cash Price.
Size, 10 in. by 12 in., 2/- each.

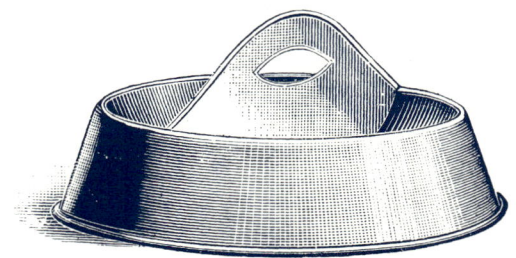

No. 199. OBLONG DOG TROUGH.
Cash Price.
Size, top measure, 8 in. by 5 in., 2/6 each.
Cannot be broken. Cannot be upset.

CAST-IRON DOG TROUGHS.

Cash Prices.

No. 1. Galvanized.
10 in. diam., 2/- each
12 in. diam., 3/- ,,

No. 17. Painted.
18 in. long, two divisions, 4/- each.
24 in. long, three divisions, 5/- each.

No. 17.

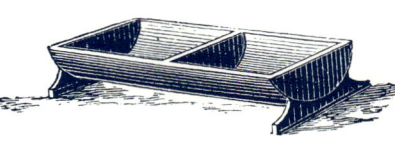

Carriage Paid on all Orders above **40/-** *value to the principal Railway Stations in England and Wales.*

BOULTON & PAUL, MANUFACTURERS,

C.A. MACHINE.

C.A. NEW PATENT "RAPID" DOG BISCUIT AND OYSTER SHELL BREAKER.

This machine is made to suit the varied requirements of dog and poultry keepers. By a simple movement of the regulator, the machine can be adjusted to break either fine or coarse, and is fitted with legs and spout.

Cash Price £1 10 0

C.B. NEW PATENT "RAPID" DOG BISCUIT AND OYSTER SHELL BREAKER.

This machine is very similar to the above, but is made to secure to a wall or partition.

Cash Price £1 7 6

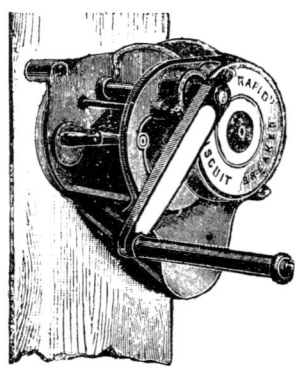

C.B. MACHINE.

RICHARDSON'S PATENT SAFETY POULTRY GRIT CRUSHER.

This machine is specially adapted for crushing old crockery, glass, shell, and other refuse. No danger to the eyes from flying chippings. The machine is arranged to cut fine or coarse, and will last a lifetime.

Cash Price 16/- each.

HUTCH RABBIT KEEPING IN THE OPEN.

Rearing Hutch.

TESTIMONIAL.

The Postern, Tonbridge.

Mrs. WILLIAM VAUGHAN begs to inform Messrs. Boulton and Paul that the two Morant Hutches have arrived, and give great satisfaction.

TESTIMONIAL.

From JOHN BOOTH, Esq., Summerdale.

I have pleasure in enclosing cheque for Rabbit Hutch and Bantam House which appear to be carefully and strongly made, and are also well adapted for the purpose.

THE "MORANT" HUTCH.

A NEW PATTERN GREATLY IMPROVED.

EACH HUTCH STAMPED MORANT'S PATENT.

THE "Morant" Hutch has a Galvanized Wire Netting Bottom which lies flat upon the ground, through which the rabbits graze with perfect ease, and on which the rabbits sit when lifted two or three times a day, thus leaving all impurities behind them, and going on to fresh pasture, exactly as sheep do when their fold is shifted, and, like them, leaving the ground heavily manured and much enriched by their presence. When the object is to separate two bodies, it is not always clear which should move. *We recommend moving the rabbits from the manure, as a better plan than moving the manure from the rabbits.* The Hutches are a pretty object on a lawn or in pleasure grounds, and do not spoil the turf if often moved; in fact, it grows intensely green and thick behind them, and any one who already has nice hutches of the usual pattern, will find one or two in which they can put the young rabbits when weaned a great help in rearing them; not an atom of food is wasted in these Hutches.

Breeding Hutch.

The Hutches can either be constructed handsomely for the garden, or more roughly and cheaper for the farm, and every effort is being made to supply them cheaply, and to contract for large numbers.

Each Hutch is fitted with movable nest, which can be used on the ground with shelf above, or raised with feeding space underneath, and with hayrack and trough.

Cash Prices.

Rearing Hutch, 6 ft. by 3 ft., for twelve rabbits	each	22/6
Breeding Hutch for Does or Single Bucks, 5 ft. by 2 ft.	,,	17/6
Double Breeding Hutch (as illustrated), 6 ft. by 3 ft.	,,	30/-

Two or more Hutches, Carriage Paid. Special Estimates for quantities.

SET OF HUTCHES FOR COUNTRY HOUSE.

Cash Price.

Two Breeding Hutches, Rearing Hutch, Buck's Hutch £3 10 0 per Set.

Children can move them with ease. Require no cleaning.

TESTIMONIALS.

From EDWARD TEW, Esq., Kingswear.

I highly approve of the Morant Rabbit Hutches you sent me the other day, and now order two more; and I will also take a Ferret Hutch such as I saw advertised in *The Field.*

From MAJOR HORROCKS, Muscalls, Kent.

Major Horrocks begs to inform Messrs. Boulton and Paul that he has received the Hutches quite safe, and is much pleased with them.

Carriage Paid on all Orders above 40/- *value to the principal Railway Stations in England and Wales.*

CAUTION.—Beware of inferior imitations. All articles bear our name as a guarantee of good workmanship.

No. 60. OUTDOOR RABBIT HUTCHES.

Constructed upon the most approved principles.

REGISTERED COPYRIGHT.

THE upper Hutches have divisions for breeding purposes, and the lower ones are for single rabbits or young stock. Fitted with zinc trays under floors to top tier only.

Cash Price.
House, as illustrated, 7 ft. 6 in. long, by 5 ft. 6 in. high, £4 10 0

TESTIMONIAL.
From G. SCOVELL, ESQ., Twickenham.
We like the Rabbit Hutches you sent last summer very much, find them clean, and easy to get at.

No. 61. RABBIT HUTCH.

REGISTERED COPYRIGHT.

THIS Hutch is of cheaper construction than No 60, and its price will come within the reach of young fanciers. Made of good seasoned wood, and well finished. Painted outside and whitened inside. Double Hutch, as illustrated, 4 ft. long, 2 ft. 6 in. deep, 4 ft. high.

Cash Prices.
Double Hutch, as illustrated ... £2 2 0
Single Hutch 1 5 0

TESTIMONIAL.
From LADY BEATRIX, Uxbridge.
Lady Beatrix is very much pleased with the Rabbit Hutch.

FEEDING CAGE FOR RABBITS.

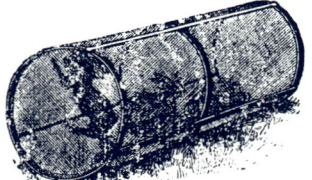

REGISTERED COPYRIGHT.

THE Feeding Cage, as shown, is designed specially for feeding rabbits on lawns; the rabbits are thus enabled to have outdoor exercise without running away, and are an ornamental feature in front of the house. No cleaning is required; the manure falls through on to the grass as the cage is rolled about. They are of especially light construction to facilitate their being easily moved by rabbits, consisting of light wrought-iron hoops, covered with galvanized wire netting, and strengthened by wrought-iron stays as shown. Made in lengths of 3 ft. 4 in. by 20 in. in diameter.

Cash Price 10/- each.

OUTDOOR DOUBLE RABBIT HUTCH ON RAISED LEGS.

BOULTON & PAUL. REG⁰

REGISTERED COPYRIGHT.
Size—4 ft. 8 in. by 2 ft.
Cash Price £1 10 0

TESTIMONIALS.
From C. P. HUTCHINSON, ESQ., Eastbourne.
I have only just unpacked the Hutch, and am very much pleased with its appearance in all respects.

From MISS MABEL LONGSTAFF, Putney Heath.
The Rabbit Hutch arrived safely, and I am very pleased with it.

Carriage Paid on all Orders above 40/- value to the principal Railway Stations in England and Wales.

CAUTION.—Beware of inferior imitations   All articles bear our name as a guarantee of good workmanship.

ROSE LANE WORKS, NORWICH.

Treatise on Poultry Keeping.

Registered Copyright.

To Customers.—We have considered it to be in keeping with the character of this Catalogue to include a few practical hints for the benefit of those who are about to begin Poultry keeping. We feel assured that we shall be able to save our customers much perplexity, disappointment, and annoyance.

Buildings.—The opinion once prevalent that any odd corner, however close, dirty, or generally ill-suited it might be, was good enough for a fowl-house, has been shown to be erroneous, regarded from every point of view. The chief condemnation of such a method of poultry keeping lies in the fact that it is unprofitable, for experience has long since shown that dirt means disease, and that absence of light and air will exercise a very appreciable effect upon the laying powers of any breed of fowl.

Aspect.—We will now suppose that you have purchased your poultry, and provided a *suitable set of appliances* for them. See that the Houses face South or South-West if possible.

Dust Bath.—Remove and frequently renew the surface soil inside your poultry houses, replenish your dust bath with dry sand or road sweepings, in preference to cinder ashes. Likewise a little heap of oyster shells, mortar rubbish, and crushed bone should be placed within the reach of all.

Cleanliness.—Insist on the most scrupulous cleanliness, daily removal of all accumulations, &c. Have all inside partitions frequently brushed over, perches as well, with paraffine; apply it with the toe of an ordinary scrubbing brush. This will make any crevice quite untenable for vermin. Frequently renew straw in nesting boxes; pine sawdust is better, if procurable; and for non-poisonous, effective, and pleasant-odoured disinfectant have your dredger primed with "Sanitas" powder. An important part of the régime is to have the inside of your building distempered or whitewashed frequently. Use as a basis, pink carbolic powder; this, besides the sanitary advantage, will give the inside a warm salmon colour.

Feeding.—Let your fowls be fed sparingly until you have gauged the quantity they will eat *without leaving any*. For birds in confinement vary their diet as much as possible. Barley, maize, and wheat are each too heating for a *long-continued* diet. Maize in particular is calculated to favour the abnormal deposit of fat. Heavy oats are good food for all-round purposes. Buckwheat and hempseed may be given only as a stimulant, sparingly. Do not omit to give the best of all anti-scorbutics—a daily taste of green food. Water must be pure, not from a pond or open butt.

Disease.—A careful following out of these few plain instructions will give you the greatest measure of immunity from the disease brought about by neglect, and the indiscriminate use of foods.

Hatching.—During the three weeks occupied in incubation, the object should be to maintain a humidity rather than ardent dry heat, so that when you are choosing a spot for your hens to sit in, let these considerations guide you. A nest on the ground will, in all probability, be damp enough. Be, however, influenced by the prevailing weather: if hot and dry, lightly sprinkle your eggs with tepid water; and again, cold, piercing winds are very prejudicial to the successful hatching out of a large brood. If the nest be not upon the ground, but raised above it, let a fresh cut turf form the bottom of it. After your eggs have been set a few days, it is well to test the fertility of them or otherwise, by holding them up to a strong light.

Chicken Rearing.—Set your hens in pairs, so that you may, if your eggs turn out very badly, give them all to one hen, and start the other afresh. Should an egg be broken during setting, thoroughly cleanse the débris from the nest and from the other eggs; sponge them with warm water. After hatching out, chickens absolutely require nothing but warmth for twenty-four hours certain, as their abdominal organs are then still in a state of transition. Let the hen have a freshening up under a coop, and a good feed if she will before putting the chicks to her. Their earliest food should be egg boiled hard and chopped fine, with bread crumbs, tiny morsels of meat, and then biscuits soaked in good milk, and grits, and so up to the omnivorous state. Give food little and often, and quite fresh and sweet. Double your attention to them in bad weather. A morsel of raw onion chopped fine is a good preventive of cramp, and makes the youngsters pert as possible.

Conditions of Success.—The conditions necessary to success, then, are cleanliness, regularity, intelligence, and last, but by no means less important, properly-constructed houses, dry, well ventilated, with every facility for keeping scrupulously clean. A dry shed or day shelter having dust bath is also indispensable.

Advice gratis as to the best means of arranging Poultry Yards, and the Fowls suited for the situations, &c.

CAUTION.—Beware of inferior imitations. All articles bear our name as a guarantee of good workmanship.

No. 41a. RANGE OF EIGHT POULTRY HOUSES.

REGISTERED COPYRIGHT.

Main Building 24 ft. by 12 ft.

HAVING central corridor, 3 ft. 9 in. wide, lighted and ventilated from above; four houses on either side, each 6 ft. by 4 ft., fitted with nest boxes and perches; day shelters beneath.

Constructed of strong red deal framing and matchboarding, galvanized corrugated iron roof with capping to ridge and gutters to eaves, strong hinges, handles, and locks to doors, windows glazed with 21-oz. sheet glass. Outside woodwork painted three coats, whitened inside, wood floors to houses. In sections ready for purchasers to erect.

Cash Price, Carriage Paid, £41 0 0

Fencing extra, see page 312.

Special Designs and Estimates free

Gentlemen waited upon and surveys made in any part of the country

---

TESTIMONIALS.

From C. A. LEDWARD, ESQ., Sunning Hill.

The Fowls' House fits together beautifully, and I really believe if I had bought the wood, etc., and made the house myself, it would have cost me more than the price you charge.

From K. E. JAMESON, ESQ., Kelvin Lodge, Balbriggan.

I wish to tell you how highly satisfactory I find the Fowl Houses; warm, dry, neatly put together, and healthy, most useful to any one keeping fowls.

CAUTION.—Beware of inferior imitations. All articles bear our name as a guarantee of good workmanship.

ROSE LANE WORKS, NORWICH 301

No. 41. MODEL POULTRY ESTABLISHMENT.

REGISTERED COPYRIGHT.

REGISTERED COPYRIGHT.

THIS is an arrangement whereby a range of Poultry Houses can be carried out to any scale required. A passage, lighted by a sky-light, runs from end to end of the structure, wide enough to wheel a barrow through, and dividing it entirely, the several compartments for different varieties of poultry being to the right and left hand, and a means of seeing the fowls on their roosts, and of collecting the eggs without going inside the compartment. On each side of the passage, in a somewhat subdued light, are the fattening pens. Beyond are the covered runs or dry shelters and the open yards. During the breeding season any one of the Houses may be used for hatching purposes, either by means of incubators or sitting hens, and afterwards for chickens, or may be surplus stock, etc.

Range of Buildings, consisting of six Roosting Houses, each 8 ft. by 5 ft., and covered shelter, 8 ft. by 5 ft. to each, passage 24 ft. by 3 ft. 9 in. wide, 24 fattening pens.

Cash Price, Carriage Paid, £60 0 0

Fencing extra, see page 312.

THE Plan can also be carried on one side only, if required, as shown in Section B, a wall facing south or south-west forming the back. The buildings are erected in our workshops before being sent away, and have all the joining parts marked so there is no difficulty in fixing them. It will save correspondence if plan of the ground is sent with inquiries

Lean-to Buildings as Section B, consisting of three Roosting Houses, each 8 ft by 5 ft., with passage at back, 3 ft. 9 in. wide, and 12 fattening pens.

Cash Price, Carriage Paid, £24 0 0

Fencing, extra, see page 312.

TESTIMONIAL

From H.R.H. THE LATE DUKE OF ALBANY, K.G., Claremont.
DEAR SIRS,—I am glad to say His Royal Highness is very much pleased with the Poultry Houses erected by you.   J. T. WILLIAMS, H.R.H. Farm Steward.

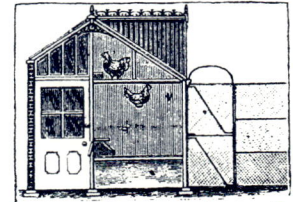

REGISTERED COPYRIGHT.   SECTION B.

Carriage Paid to the principal Railway Stations in England and Wales.

CAUTION.—Beware of inferior imitations.   All articles bear our name as a guarantee of good workmanship.

No. 30. POULTRY ESTABLISHMENT.

Specially designed for farm accommodation.

Provision is made for Chickens, Geese, and Ducks

As supplied to The Right Hon Sir Redvers Buller

Cash Price, Carriage Paid

Size of House, 29 ft. by 8 ft., Shelter at ends, each 12 ft. by 6 ft £48 10 0

In sections for purchaser to erect

CONSTRUCTED of strong red deal framing. Walls covered outside with grooved and tongued matchboarding, weather-boarded roof, sloping floor to upper part of fowls' house, cast-iron gutters and down-pipes to eaves, capping for ridge, gable ventilators. Fitted with perches, nest boxes, dust baths. Fattening house with wood batten floor fitted with pens for 24 fowls. Painted outside, and lime-whited inside.

TESTIMONIAL

From THE VICOMTESSE DESMANET DE BIESME, Cruyshautem, Belgium.

I am pleased to be able to tell you that the Poultry House has reached me in excellent condition, nothing being short. It has been fixed by my bricklayer and carpenter without the least difficulty, and seems to suit my requirements thoroughly.

No. 38. POULTRY ESTABLISHMENT.
ON THE AMERICAN PRINCIPLE.

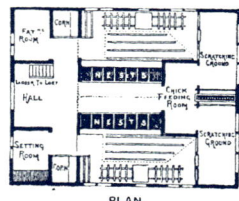

Price upon receipt of particulars of requirements

TESTIMONIAL

From A J. WOODHOUSE, Esq., Helenslea.

I have this day put up the Fowl House and Run, and found everything most satisfactory.

Carriage Paid to the principal Railway Stations in England and Wales.

CAUTION.—Beware of inferior imitations. All articles bear our name as a guarantee of good workmanship.

ROSE LANE WORKS, NORWICH. 302A

No. 74. RANGE OF POULTRY BUILDINGS.

This Range of Buildings consists of Fowls Houses, corn and egg rooms, pigeon loft above in centre block, with side runs, having on each wing either aviaries or pigeon houses. We shall be pleased to give estimates and further plans for Ranges of Buildings carried out in this style, or modified to suit requirements.

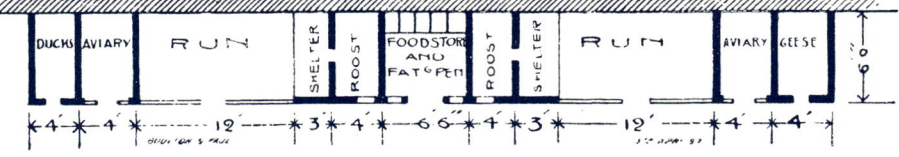

No. 75. ORNAMENTAL RANGE OF FOWL HOUSES.

These Buildings are suitable for a gentleman's residence. Each house comprises roosting place, with nests and perches, covered day shelter, and passage the whole length of building wide enough to wheel a barrow through. The fowls can be seen on the roosts and the eggs collected without going inside the compartment. The passage is lighted by skylight and ventilated on roof as shown.

Special Estimates given on receipt of particulars of requirements.

CAUTION.—Beware of inferior imitations. All articles bear our name as a guarantee of good workmanship.

No. 20. RANGE OF IMPROVED POULTRY YARDS.

REGISTERED COPYRIGHT.

TESTIMONIALS

FOLLY COURT.

MR. HEATHCOTE is much pleased with the Fowl House, as it exactly answers the purpose for which it was needed.

From MRS. SHAWE-STOREY, Andover.

MRS. SHAWE-STOREY takes this opportunity of saying the House has given the greatest possible satisfaction in every way.

THE Roosting and Laying Houses are made of best red deal, grooved and tongued, painted three coats of good paint, and lime-whited inside. Galvanized corrugated iron roof, the roost part lined with wood, which affords good ventilation under the eaves. The floor boards are raised two feet above the ground, which provides a shelter underneath during wet or hot weather. The Houses are fitted with a ventilator and slide to admit of air and light, door with lock for attendant, small door with slide for removal of eggs; floor-boards, nest boxes, perches, and hen-ladder are loose to facilitate cleaning; fitted with cast-iron gutters and down-pipes. The buildings are erected in our workshops before being sent away, and have all the joining parts marked, so there is no difficulty in fixing them.

SIZE OF HOUSES.—5 ft. 6 in. back to front, 4 ft. wide, they require a back wall; 7 ft. high at back, with shed 8 ft. long to each House. The open yards are 12 ft. wide, 18 ft. long, with separate gate to each, enclosed by our Improved Fencing, as described on page 312 of this Catalogue.

The following Prices include Houses, Sheds, and Yards complete, as described above, packed and delivered, Carriage Free, to any principal Railway Station in England.

Cash Prices.

No. 17.	One House, Shed and Run complete, suitable for	12 to 15 fowls	..	..	£8 0 0
No. 18.	Two Houses, Sheds and Runs complete ,,	24 to 30 ,,	..	..	15 10 0
No. 19.	Three Houses, Sheds and Runs complete ,,	36 to 45 ,,	...	..	23 0 0
No. 20.	Four Houses, as illustrated above ,,	48 to 56 ,,	...	..	30 0 0

Wire Netting for covering the tops of the Runs can be had at a small additional cost. This would be necessary when pheasants or game fowls are kept, unless their wings are clipped.

THE small Illustration shows the back of the above Houses boarded, when standing by themselves and not placed against a wall.

There are many advantages in this arrangement, as the fowls can be attended to, and the eggs collected without entering the runs. This adds to the above prices £2 each House

A COVERED passage can be arranged as shown in Section C, and fitted with fattening pens. This is a great advantage where a lady attends to her poultry, as eggs can be collected and all the fowls seen without entering the runs.

REGISTERED COPYRIGHT.
Section C.

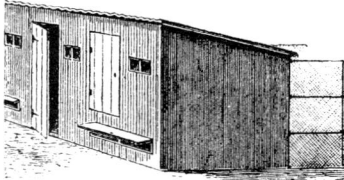

REGISTERED COPYRIGHT.

TESTIMONIALS

From LIEUT.-COLONEL EDWARD LLOYD, Hawkhurst.

The range of Fowl Houses (No. 20) you sent me a few years ago answer the purpose admirably, and stand well.

From N. W. APPERLEY, ESQ., Durham.

I have put up the Lean-to Fowl House myself, and it went up very well and easily, and is a great success.

Carriage Paid to the principal Railway Stations in England and Wales

CAUTION.—Beware of inferior imitations. All articles bear our name as a guarantee of good workmanship.

No. 44. LARGE POULTRY HOUSE.
FOR FOWLS, GEESE, DUCKS, AND PIGEONS, WITH CORN ROOM.

THE plan illustrated has been successfully carried out for accommodating 100 to 150 head of Poultry. It is fitted with a room for corn, etc. The nests are so arranged that the eggs can be taken without entering the Houses and disturbing laying hens. Provision is made at the back for ducks and geese, and also fattening pens. The perches are raised above a sloping floor in such a way that all the dirt falls to the front, and by opening a hinged flap it can be daily removed from the outside. The perches are loose, so that the floors can be swept occasionally. Every facility is made for thoroughly cleansing the inside, and perfect ventilation is given without draughts. The buildings are erected in our workshops before being sent away, and have all the joining parts marked, so there is no difficulty in fixing them. Length 20 ft., 13 ft. deep, 7 ft. high at eaves, 9 ft. to ridge.

REGISTERED COPYRIGHT.

Cash Price, Carriage Paid, £37 10 0
Single House for 50 fowls, without Corn Room or Pigeon Loft, £17 0 0

TESTIMONIALS

From Mrs. STUART, Shrewsbury.

Mrs. Stuart is much pleased with the two Fowl Houses and Sheds, with open yards; they have been put together without any trouble, and are very compact and well made. She thinks poultry will do very well in them, as they are dry and warm.

From FOUNTAIN WALKER, ESQ., Ness Castle, Inverness.

The Poultry Houses that you sent to me last autumn answer very well and look remarkably neat. Their portability is a great advantage.

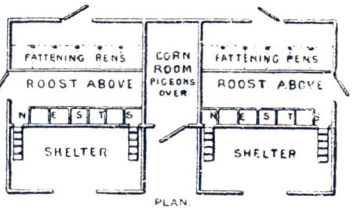

PLAN.
REGISTERED COPYRIGHT.

No. 58. RANGE OF POULTRY HOUSES AND RUNS.
For placing against a Wall.

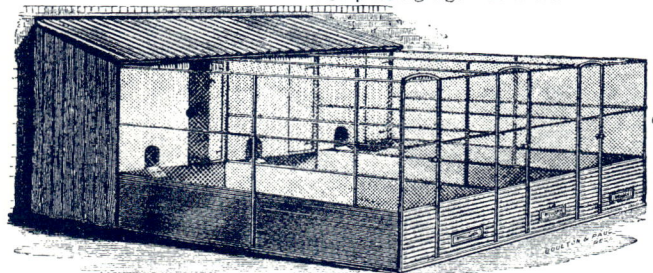

REGISTERED COPYRIGHT.

Size—16 ft. long, 18 ft. wide, 6 ft. high at eaves, 7 ft. at back.

Each House is 4 ft. by 6 ft., and has a Run 12 ft. by 6 ft.

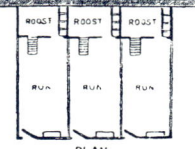

PLAN

Fitted with raised floor, nests, perches, and feeding troughs. In sections, ready for easy erection by purchaser.

Cash Prices, Carriage Paid

One House and Run, £6 0 0 ... Two Houses and Runs, £11 10 0 ... Three Houses and Runs, £16 10 0

Carriage Paid to the principal Railway Stations in England and Wales.

CAUTION.—Beware of inferior imitations. All articles bear our name as a guarantee of good workmanship

No. 46. DOUBLE POULTRY HOUSES WITH SHEDS AND RUNS.

THIS arrangement consists of two Roosting Houses with Corn Room between, from which both fowl houses and pigeon loft are accessible. The whole of the stock can be attended to without entering the fowl houses or yards. The Houses are made of **best red deal**, grooved and tongued boarding, painted three coats of good paint, with Runs under the Houses for shade and shelter; span roof, covered with galvanized corrugated iron, which affords good ventilation; backs boarded. The floor-boards and perches are loose, so that the Houses, &c., may be lime-washed as often as required. A ventilator with slide is also provided to admit air and light. The open yards or grass Runs are 18 ft. long, enclosed with our Improved Poultry Fencing (see page 312). The Building is 28 ft. long, 6 ft. deep, 6 ft. high at the eaves. The Houses are erected in our workshops before being sent away, and all the joining parts marked, so that there is no difficulty in erecting them.

REGISTERED COPYRIGHT.

Cash Price, Complete, £28 0 0, Carriage Paid.
If made with lean-to roof to place against a wall ... £23 0 0

TESTIMONIAL.
From JOHN FARRER, ESQ., Architect, 20 Finsbury Pavement, E.C.
I have had the Fowl Houses (No. 46) erected for my client, and they give every satisfaction; the workmanship and material are good, and altogether there is good value for the money.

No. 59. COMBINATION FOWL HOUSE, AVIARY, AND PHEASANTRY.

REGISTERED COPYRIGHT.

GROUND space, 18 ft. by 14 ft., 6 ft. high at eaves, 7 ft. 6 in. to ridge. Fitted with nests, perches, and feeding troughs complete.

Cash Price, Carriage Paid, £24 10 0

UNSOLICITED TESTIMONIALS.

From G. FERGUSON, ESQ., Denmark Hill.
I have great pleasure in writing to say that the Shed and Fowl House erected by you at the Falcons give every satisfaction.

From WALTER MAKINS, ESQ., Hendon.
I am very satisfied with the Fowl Houses and Runs, and the work altogether

From THOMAS FAIRCLOUGH, ESQ., Croydon.
The Fowl Houses you supplied me with were completed to my satisfaction, and work very well.

No. 60. WIGWAM FOWL HOUSE & RUN.

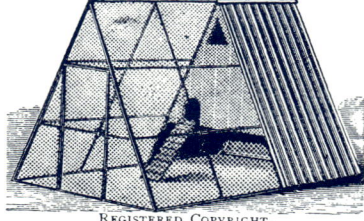

REGISTERED COPYRIGHT.

10 ft. by 6 ft. on ground, 6 ft. high.
Fitted with raised floor, nests, and perches

Cash Price, Carriage Paid	...	...	...	..	£3 15 0
House only	...	...	...	...	2 15 0
On Wheels	..	...	...	extra	0 15 0

TESTIMONIAL.
From S. CHRISTY, ESQ., Stockport.
Mr. S. Christy likes the Fowl Houses, &c., very much. Your people were careful in packing up all the small fittings necessary to make them complete. Our carpenter soon put them together.

Carriage Paid to the principal Railway Stations in England and Wales.

CAUTION.—Beware of inferior imitations.   All articles bear our name as a guarantee of good **workmanship**

BOULTON & PAUL, MANUFACTURERS.

COMBINED FOWL HOUSE, AVIARY, DOVE-COTE, AND DOG KENNEL.

REGISTERED COPYRIGHT.

It consists of Poultry House and Yard, with Roosting and Laying Houses separate, Dog Kennel and Run, Pigeon Loft and Aviary combined; also a Tool House and Corn Loft. It covers a space of about **16 ft.** by **16 ft.** The whole construction takes to pieces easily, thereby rendering it a "tenant's fixture." The design may be altered to suit the taste of the purchaser.

Cash Price, Complete, Carriage Paid, £25 0 0

TESTIMONIAL.
From Miss WILSON, Kendal.

Miss Wilson is more than satisfied with the House and Run sent to her. They are splendid workmanship.

No. 57. COMBINED FOWL HOUSE AND AVIARY.

12 ft. by 16 ft., 6 ft. high at eaves, about 7 ft. 6 in. at top. To stand against the wall. Fitted with nests, perches, and feeding troughs.

Cash Price, Carriage Paid, £14 0 0

TESTIMONIAL
SLANE CASTLE.
Both Lord and Lady Conyngham are very pleased with the Fowl House.

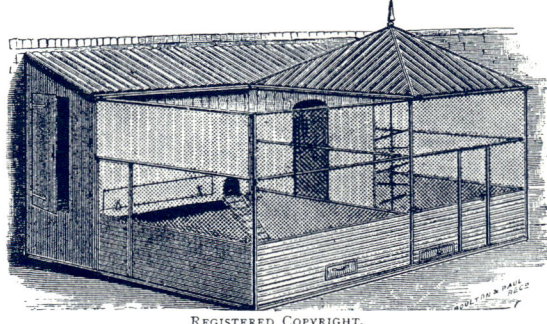

REGISTERED COPYRIGHT.

Carriage Paid to the principal Railway Stations in England and Wales.

CAUTION.—Beware of inferior imitations. All articles bear our name as a guarantee of good workmanship.

BOULTON & PAUL, MANUFACTURERS,

No. 76. LEAN-TO FOWL HOUSE.

This Fowl House answers the same description as the Villa Poultry House, shown on page 307, but is of a more ornamental construction, and would be an artistic addition to garden or grounds.

Cash Prices, Complete, Carriage Paid.

First size, No. 7, 12 ft. long, 4 ft. 6 in. back to front, 5 ft. 9 in. high at back £5 10 0
Second size, No. 8, 12 ft. long, 5 ft. 6 in. back to front, 6 ft. high at back ... 6 10 0
Third size, No. 9, 16 ft. long, 6 ft. 6 in. back to front, 6 ft. 9 in. high at back 8 10 0

Where no wall can be utilized, the back can be boarded for **35/-** extra.

No. 77. RUSTIC FOWL HOUSE.

This House is designed after the manner of Norwegian houses, and is very pretty in effect. It has over-hanging gables and projecting eaves as shown, is fitted with nest boxes and perches, and arranged as a day shelter underneath. This House would have a handsome appearance in the paddock or on the lawn.

Cash Prices, Carriage Paid.

4 ft. square ...	... £3 15	0
5 ft. ,, ...	... 4 10	0
6 ft. ,, ...	... 6 0	0

No. 78. PORTABLE WOOD DUST BATH.

Cash Prices.

5 ft. by 4 ft. ...	£1 2	6
6 ft. by 5 ft. ...	1 12	6

No. 79. PORTABLE WOOD DUST BATH.

Cash Prices.

5 ft. by 4 ft. ...	£1 5	0
6 ft. by 5 ft. ...	1 15	0

CAUTION.—Beware of inferior imitations. All articles bear our name as a guarantee of good workmanship.

ROSE LANE WORKS, NORWICH. 306B

No. 80. CHEAP MOVABLE FOWL HOUSE
For Pastures or Garden.

THIS House is of Novel construction, and is very handy for moving about.

Size, 4 ft. 10 in. long by 3 ft. wide by 4 ft. to eaves, and 5 ft. to ridge.

Cash Price 35/-

TESTIMONIAL.
From W. WHEELWRIGHT, Underwood College.

I beg to inform you that I am particularly well satisfied with the Fowl House you sent me, which arrived yesterday, and has been put up with comparative ease.

No. 81. WIGWAM FOWL HOUSE.

Size, 4 ft. by 6 ft. on ground, 6 ft. high, made of rustic jointed weather-boarding, and fitted with nest and perches, and with raised floor.

Cash Price £2 15 0

TESTIMONIAL.
From D. SMITH, ESQ., Richmond School.

The Fowl House received with thanks. I have not put it up yet, but I can see that, like all your things, it is both good and cheap. Your work is always so well worth the prices you charge.

CAUTION.—Beware of inferior imitations All articles bear our name as a guarantee of good workmanship.

THE VILLA POULTRY HOUSE—"TENANT'S FIXTURE."
SPECIALLY SUITED FOR TOWN GARDENS.

"POULTRY are kept and maintained in an infinitude of places and positions. In towns, especially, gardens abound with them, in numbers far beyond the anticipations of many. Very frequently, we observe, attached to a house for sale or 'to let,' the presumed or real luxury of a 'poultry run,' probably constructed at no little cost, by some former tenant who had a fancy for keeping poultry, and had not scrupled to expend his money on an erection, which upon changing his residence he had been compelled to leave behind for the benefit of his successor. These considerations have induced MESSRS. BOULTON AND PAUL to construct a movable poultry house, which, from its usefulness, deserves notice, the more so as it is in the form of what has become a 'household word' in garden tenancy, viz., a 'tenant's fixture.'"—*Horticultural Record.*

REGISTERED COPYRIGHT.

For placing against a (South or South-west) wall, with galvanized iron roof all over; the roost part being lined with wood. The roost is made of best red deal, grooved and tongued boards, painted three coats of good paint, with Run under the House for shade and shelter; the floor boards are loose, so that the House, &c., may be lime-whited as often as required. Nest boxes are fitted with flap to lock for removal of eggs from the outside. The construction is so simple that they are easily taken to pieces for packing, and are readily fixed by inexperienced persons. This arrangement is of great advantage where space is an object, and is constructed upon the best principles for keeping fowls in confinement, viz., **a warm roost and dry run.**

Cash Prices, Complete, Carriage Paid.

First size, No. 7, 12 ft. long, 4 ft. 6 in. back to front, 5 ft. 9 in. high at back £4 10 0
Second size, No. 8, 12 ft. long, 5 ft. 6 in. back to front, 6 ft. high at back 5 10 0
Third size, No 9, 16 ft. long, 6 ft. 6 in. back to front, 6 ft. 9 in. high at back ... 7 10 0
Where no wall can be utilized, the back can be boarded for 35/- extra.

No. 61.
LEAN-TO FOWL HOUSE.
For placing against a Wall.

SIZE—8 ft. by 4 ft., 6 ft. high at eaves, 7 ft. at back. Painted three coats outside, and lime-whited inside. With raised wood floor, fitted with nests and perches.

Cash Price £3 15 0
In sections, ready for easy erection by purchaser.

REGISTERED COPYRIGHT.

Plan showing Poultry House described above with an outer run made with the Poultry Fencing as on page 312. This can be suited to any situation.

TESTIMONIALS.

From GENERAL FYTCHE, Bournemouth.
The Poultry House has given much satisfaction, and has been put together without any difficulty.

From ALLAN GORDON CAMERON, ESQ., Barcaldine Castle.
The Hen Houses give complete satisfaction. I consider them thoroughly good and moderate in price.

From W. J. DALE, ESQ., Northwich.
The Fowl House arrived quite safely and I am much pleased with it, I put it up this afternoon and shall give you any further orders

Carriage Paid to the principal Railway Stations in England and Wales.

CAUTION.—Beware of inferior imitations   All articles bear our name as a guarantee of good **workmanship**.

No. 47. NEW PORTABLE FOWL HOUSE.

Readily moved from place to place. Suitable for Farm Homesteads.

Size—8 ft. long, 6 ft. wide, 6 ft. high at eaves, 7 ft. 6 in. high to ridge, to accommodate about fifty fowls; fitted with perches, nest boxes, &c. The floor is raised 2 ft. from the ground, so as to use the whole space of the House as a shade or shelter. A flap is provided at the back for removal of eggs, which is fitted with a good lock. Takes to pieces for packing, and is easily put together. It is well made, and only the best materials are used. Perfect ventilation is provided without draught. Painted three coats, and lime-whited inside.

Cash Price, Complete, £9 10 0
On Wheels, 20/- extra.

REGISTERED COPYRIGHT.

TESTIMONIALS.

From A. C. DAWSON, ESQ., Ledgers Park.

I find the Fowl House all that can be desired, and as I am a large poultry keeper, I shall probably, before long, send you another order. I consider it a most valuable kind of Poultry House for summer and autumn use, when it is very important to get the birds out on the land, and away from their winter quarters.

From J. E. V. VERNON, ESQ., Clontarf Castle.

The Poultry House (No. 47 in your Catalogue) was delivered here free of charge for carriage in perfect order. The component parts were very carefully fitted and adjusted, and were put together in complete form by a carpenter in four hours. I have no reason to doubt its perfect success.

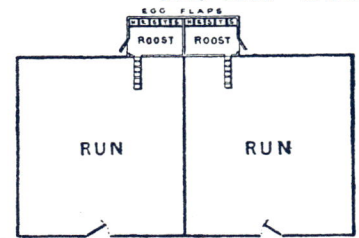

REGISTERED COPYRIGHT.

No. 73. NEW FOWL HOUSE

For moving about pasture or arable land.

Size—5 ft. 6 in. by 3 ft. 6 in., 5 ft. high at eaves, 6 ft. high to ridge.

Cash Price, £3 10 0

TESTIMONIAL.

From Mrs. MACDOWALL, Garthland, N.B.

Mrs. M. is exceedingly pleased with the Incubator House, etc., she has received.

No. 47a. DOUBLE FOWL HOUSE.

Plan showing Runs attached to Double Fowl House for two varieties of Fowls, each Run 18 ft. by 12 ft.

Cash Price, with Runs complete, £16 0 0

TESTIMONIAL.

From E. SPRAWSON, ESQ., Greenwich.

The Fowl House, Run, and Coops are now put together and give great satisfaction. They are made with due consideration to the health and comfort of the fowls.

Carriage Paid to the principal Railway Stations in England and Wales.

CAUTION.—Beware of inferior imitations. All articles bear our name as a guarantee of good workmanship.

BOULTON & PAUL, MANUFACTURERS,

BOULTON & PAUL'S ORIGINAL PORTABLE FOWLS' HOUSES.

These Houses are well made of the best red deals, the frames being **mortised and tenoned**, with tongued and grooved boards, painted three coats of good paint, and whitened inside; corrugated iron galvanized roof **lined with wood and felt**, giving perfect ventilation without draught, day shelter or dust bath under raised floor, egg flap for removal of eggs, nests, perches, etc., complete.

Reduced Cash Prices for 1898, Carriage Paid.

No. 14. For Roosting twelve or more Fowls, 4 ft. square, 5 ft. 9 in. high at ridge £2 15 0

No. 15. For Roosting twenty or more Fowls, 5 ft. square, 6 ft. 3 in. high at ridge 3 10 0

No. 16. For Roosting forty or more Fowls, 6 ft. square, 7 ft. 3 in. high at ridge 4 15 0

If mounted on wheels 10/- extra.

PORTABLE RUN ONLY.

6 feet long with one end.

For No. 14 House, 4 ft. wide, £1 0 0

For No. 15 House, 5 ft. wide, £1 10 0

For No. 16 House, 6 ft. wide, £2 0 0

Extra Ends for using independent of House, 5/- each, by which means a capital feeding coop for young stock is formed.

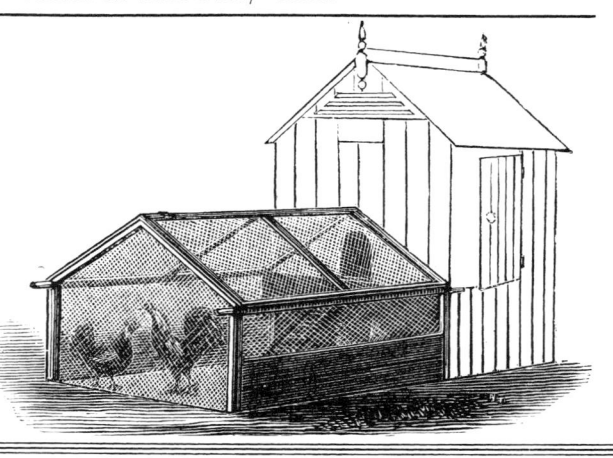

CAUTION.—Beware of inferior imitations. All articles bear our name as a guarantee of good workmanship.

ROSE LANE WORKS, NORWICH.

PORTABLE FOWLS' HOUSE WITH GRASS RUN.

enclosed with the Portable Fencing described on page 312. The eggs can be collected and the house cleaned out without entering the run. **Cash Prices of Runs only. Carriage Paid.**

						£	s.	d.
Run 22 ft. by 18 ft. for No. 14 Houses, with Hurdles, Gate, Angle Pillars, and Trough						3	9	6
,, 29 ft. by 19 ft.	,,	15	,,	,,	,,	4	0	0
,, 36 ft. by 20 ft.	,,	16	,,	,,	,,	4	10	6

For price of Fowls' Houses see preceding page.

CAUTION.—Beware of inferior imitations. All articles bear our name as a guarantee of good workmanship.

310A   BOULTON & PAUL, MANUFACTURERS.

No. 303a.
PORTABLE WOODEN FOWLS' HOUSE.

This House is suited for accommodating a large number of fowls, turkeys, or geese. Red deal framing, covered outside with planed 1-in. grooved and tongued matchboarding, painted one coat outside. Galvanized corrugated iron roof. No gutters. Whitened inside. In sections ready for easy erection by purchaser. Windows glazed with 21-oz. glass. No floor, but fitted with perches and laying boxes.

Cash Prices.

Length.	Width.	Height to Eaves.	Height to Ridge about.	
20 ft.	10 ft.	6 ft. 6 in.	9 ft. 9 in.	£15 4 0
25 ft.	12 ft.	7 ft. 0 in.	11 ft. 0 in.	21 17 0

If roof is lined inside with matchboarding.

Prices.
£16 15 0 ... £24 5 0

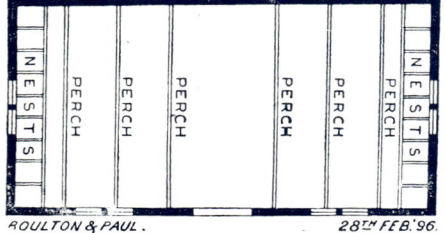

TESTIMONIAL.
From A. E. SPEAD, Esq., Esher.
The Fowl House arrived quite safely yesterday, and was today put up without any trouble. It appears to be the very thing I require.

CAUTION.—Beware of inferior imitations. All articles bear our name as a guarantee of good workmanship.

ROSE LANE WORKS, NORWICH. 310B

CHEAP FOWLS' HOUSES.

Prices Specially Reduced to come within the reach of all classes of Poultry Keepers.

THESE Houses are not morticed and tenoned, but are constructed of match-boarding, well nailed to ledges, and painted one coat. Whitened inside. Strong hinges and lock to door, and including nests and perches.

No. 31. NEW PORTABLE DOUBLE SPAN-ROOF FOWL HOUSE.

Cash Price (with raised wood floor), Carriage Paid. Double House. Each house 6 ft. by 4 ft., **£4 17 6**

No. 32. NEW PORTABLE LEAN-TO FOWL HOUSE.

Cash Price (with raised wood floor), Carriage Paid. Size, 6 ft. by 4 ft., **£2 2 6**

No. 34. CHEAP LEAN-TO FOWL HOUSE

Cash Price, without floor, Size, 5 ft. by 4 ft. by 4 ft. 9 in. to eaves, **£1 5 0**

No. 33. CHEAP SPAN-ROOF FOWL HOUSE.

Cash Price, without floor, Size, 5 ft. by 4 ft by 4 ft. 9 in. to eaves, **£1 10 0**

All Orders amounting to **40/-** *Carriage Paid to the principal Railway Stations in England and Wales.*

CAUTION.—Beware of inferior imitations. All articles bear our name as a guarantee of good workmanship.

No. 28. HYGIENIC COMBINATION POULTRY HOUSE, FATTENING PEN, & SHELTER.

Registered No. 25,606.

This House is designed to meet the wants of those whose grounds are limited, and yet desire to have new-laid eggs, and fatten their own fowls.

The Roosting Compartment is fitted with ladder, wood floor, slide to hen-hole, perches, and nests (which are easily accessible), and lock-up door.

The Shelter under Roosting House is fitted with wire door, and is suitable for a day shelter for fowls, for a dust bath, or a ducks' house.

The Fattening Pen is in two compartments, and has a lath floor. Battens in front are made to slide for the ingress and egress of fowls. Over the front is a hinged flap, which forms a shelter from the sun and rain in the daytime, and can be let down and locked at night. The pen can be used for setting or broody hens. The Wood Feeding Trough is removable, so that it can be easily cleaned when required.

Cash Prices, Carriage Paid.

4 ft. by 4 ft., for roosting 12 fowls and fattening 4 fowls, £3 5 0

5 ft. by 5 ft., for roosting 20 fowls and fattening 6 fowls, £4 5 0

6 ft. by 6 ft., for roosting 30 fowls and fattening 8 fowls, £5 10 0

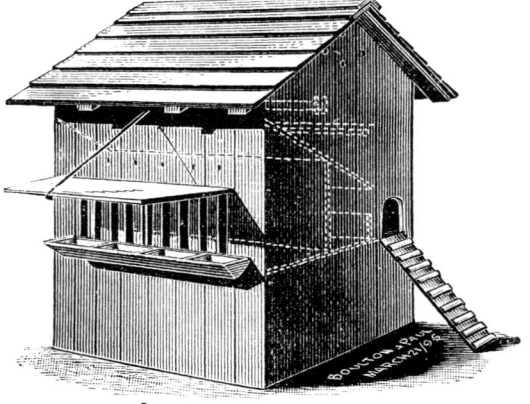

Registered No. 25,606.

CAUTION.—Beware of inferior imitations. All articles bear our name as a guarantee of good workmanship.

ROSE LANE WORKS, NORWICH. 310D

CHEAP FOWLS' HOUSES.

Prices Specially Reduced to come within the reach of all classes of Poultry Keepers.

THESE Houses are not morticed and tenoned, but are constructed of match-boarding, well nailed to ledges, and painted one coat. Whitened inside. Strong hinges and lock to door, and including nests and perches.

No. 29. USEFUL FOWLS' HOUSE.

CONSTRUCTED of match-boarded sides and ends, well nailed to strong ledges, weather-boarded roof. Painted outside one coat of lead colour paint. Inside whitened. Fitted with glazed window, nests, perches, ladder, and lock-up door. Sent out in sections ready for easy erection by purchaser.

Size, 4 ft. 6 in. by 4 ft., 4 ft. 6 in. high to eaves, 5 ft. 9 in. to ridge.

Cash Price, Carriage Paid, £2 0 0

No. 70. MOVABLE POULTRY HOUSE & RUN.

HOUSE, 4 ft. by 3 ft. Run, 8 ft. by 4 ft. Constructed of selected red deal, ledged and well nailed together, painted three coats, and whitened inside. House fitted with nests and perch, skylight in roof. Run covered with 2-in. mesh netting, with feeding trough outside.

Cash Price £2 10 0

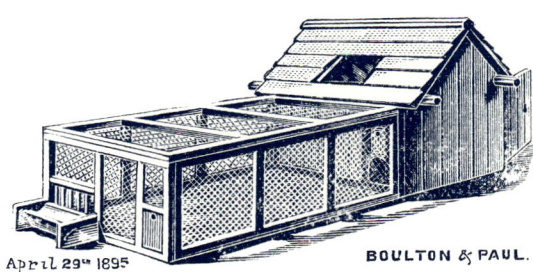

CAUTION.—Beware of inferior imitations. All articles bear our name as a guarantee of good workmanship.

POULTRY KEEPING ON FARMS.
No. 10. MOVABLE POULTRY HOUSE FOR FARMS.

REGISTERED COPYRIGHT.

TESTIMONIAL.
From F. G. RUGG, Broadlake, Kent.
I am much pleased with the Portable Fowl House supplied by you, and wish to have another similar to it.

On Farms generally, too many poultry are kept for home consumption only, but not sufficient for marketing; but if our system were carried out on a sufficiently extensive scale, there would be chickens and eggs enough to pay for sending to a good market twice a week.

Our plan (where it can be carried out) is to have a Movable Fowl House on different parts of the land, shifted about when the plough is at work; and when the cornfields are cleaned, take them into the stubble, with a little soft food to assist them. They keep themselves, and thrive better when thus earning their own living.

Poultry will do well on the poorest land, with natural grass only; if partly covered with brushwood, so much the better; but they must be kept in small flocks, and if our Movable House system is adopted, would well repay the outlay.

Cash Prices, Carriage Paid.
For Fowl House, as illustrated, roughly and strongly constructed for Farm use, on wheels, unpainted;
size 6 ft. by 4 ft., for about Fowls ... £4 0 0
Larger size, 8 ft. by 6 ft., for about Fowls, unpainted 5 15 0
Without Wheels, 10/- and 15/- less.

No. 35. MOVABLE FOWL HOUSE FOR FARM USE.
New Style.

REGISTERED DESIGN, No. 209,452.

All weather-boarded, painted outside one coat, lime-whited inside, strongly constructed. Fitted with nest boxes, perches, and sliding door to open. No floor.

N.B.—The roof is made like No. 37, and not to open in one length as shown above.

Cash Price, Carriage Paid.
Size 10 ft. long, by 3 ft. 6 in. wide, by 3 ft. 10 in
high, for 40 Fowls £4 7 6
On Rails Norwich 4 0 0

No. 37. MOVABLE FOWL HOUSE. (New Style.)
No Floor. This House should be used by every farmer. Can be easily carried by two men.

Cash Price, Carriage Paid.
Size 6 ft. long, 3 ft. 6 in. wide, by 4 ft. 4 in. high;
for 20 Fowls £2 5 0
On Rails Norwich 2 0 0

No. 36. MOVABLE FOWL HOUSE FOR FARM USE.

REGISTERED No. 209,753.

Cash Price, Carriage Paid.
Size 6 ft. 6 in. long, 4 ft. 6 in. wide, by 4 ft. 4 in.
high, for 25 Fowls £4 7 6
On Rails Norwich 4 0 0

REGISTERED No. 209,454.

Carriage Paid to the principal Railway Stations in England and Wales.

CAUTION.—Beware of inferior imitations. All articles bear our name as a guarantee of good workmanship.

BOULTON & PAUL, MANUFACTURERS.

ORIGINAL MANUFACTURERS OF THE
IMPROVED PORTABLE FENCING FOR INCLOSING POULTRY & PHEASANTS.
No. 100.

REGISTERED COPYRIGHT.

THIS is a strong and superior Fencing, made in hurdles 6 ft. long, 6 ft. high, with improved anchor-formed feet, joined with bolts and nuts. The hurdles are painted and covered with strong galvanized netting, the lower portion, 2 ft. from the ground, is 1-in. mesh which is rat proof; the upper part of Fence is 1½-in. mesh netting. This Fencing is easily fixed or removed by any ordinary workman. The gate can be placed in any part of the Fence. A Run or Pen can be formed of any length or shape, without extra charge.

Cash Prices.

6 ft. high, including all necessary bolts and nuts	per yard run 2/7
5 ft. high, including all necessary bolts and nuts	,, 2/3
4 ft. high, including all necessary bolts and nuts	,, 2/-
Doorway complete, 2 ft. wide, including standards and arched stay	each 8/6
Angle-Iron Pillars, for corners, with cast ornaments	,, 1/9
Reversible Troughs, fitted to hurdles, for feeding from outside	5/9
Side Stays	per dozen 9/3

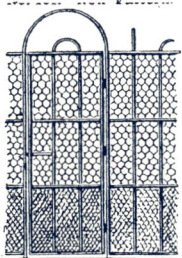

REGISTERED COPYRIGHT.

Cash Prices.
6 ft. high .. per yard 4/6
Gates ... each 10/6

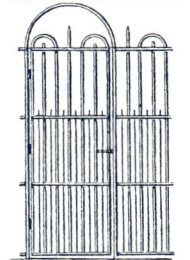

REGISTERED COPYRIGHT.

Cash Prices.
6 ft. high ... per yard 6/-
Gates ... each 15/-

THIS is a more ornamental Fencing than the above. It is made in 6 ft. lengths, and joined together with bolts and nuts.

POULTRY PENS.

RUN only, consisting of five 6 ft. hurdles, one doorway, and two Angle Pillars; 12 ft. long, 8 ft. wide. With reversible feeding trough.

Cash Price, £2 0 0

Cash Prices.

No. 40 Lean-to House 8 ft. wide, 4 ft. deep, to place against a wall, with run 12 ft. long as illustrated, £5 15 0

If boarded back extra £1 10 0

REGISTERED COPYRIGHT.

Estimates given for quantities.

Carriage Paid on all Orders above 40/- *value to the principal Railway Stations in England and Wales.*

CAUTION.—Beware of inferior imitations. All articles bear our name as a guarantee of good workmanship.

ROSE LANE WORKS, NORWICH. 313

ORIGINAL MANUFACTURERS OF THE
IMPROVED MOVABLE FENCING FOR INCLOSING POULTRY & PHEASANTS.
No. 103.

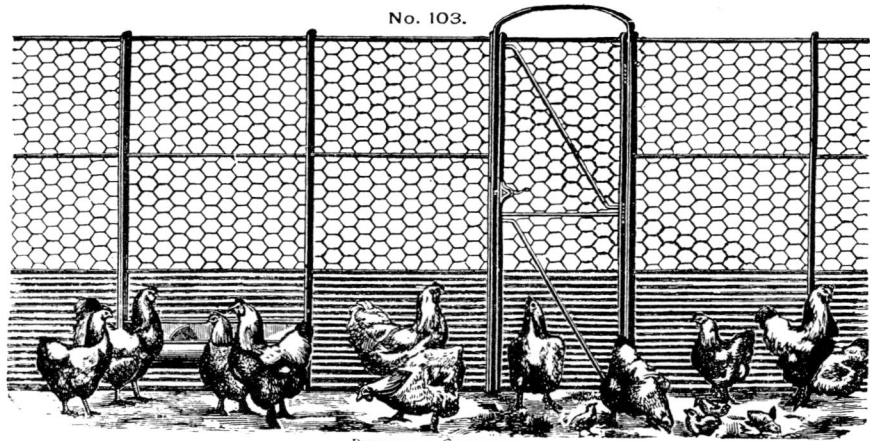

REGISTERED COPYRIGHT.

THE above engraving represents a new and improved Poultry Fence, made in hurdles 6 ft. long and 6 ft. high, with improved anchor-formed feet, and fitted together with bolts and nuts. The Hurdles are painted and covered with strong galvanized wire netting. The upper part of Fence is of 1½-in. mesh netting, the lower portion, 2 ft. from the ground, is covered with *Galvanized Sheet Iron*, specially prepared for the purpose, and is rat-proof, and where the Fence is used to divide Runs, prevents cocks fighting. The birds are also sheltered from the wind. This Fencing is easily fixed or removed by any ordinary workman. The gate can be placed in any part of the Fence. A Run or Pen can be formed of any length or shape, without extra charge.

Cash Prices.

6 ft. high, with galvanized sheet iron 2 ft high	per yard	4/-
Doorway complete, 2 ft. wide, including standards and arched stay	each	9/9
Angle-Iron Pillars, for corners, with cast ornaments	"	1/9
Reversible Troughs, fitted to hurdles, for feeding from outside	"	5/9
Side Stays	per doz.	9/3

No. 103a.

MADE in 4 ft. lengths and 4 ft. high, having the upper portion covered with 1-in. mesh wire netting, and the lower part 2 ft. high of crimped galvanized iron, which affords shelter and protection to young stock.

Cash Price.
Per Hurdle, 4 ft. long, 4/- each.

CHEAP GAME-PROOF & CHICKEN HURDLES.

For inclosing Chickens, Bantams, Brahmas, Cochins, Ducks, or Rabbits.

No. 105.

THE advantages of these are, that the Hurdles are complete in themselves, and are readily put down. They can also be used as pea trainers instead of pea sticks. Made in Hurdles, 4 ft. long, 1-in. mesh.

Cash Prices.
2 ft. 6 in. high, 1/6; 3 ft. high, 2/- each.

No. 180.

Cash Prices.
6 ft. long, 2 ft. 6 in. high, with 2 horizontal bars, as illustrated ... each 2/6
6 ft. long, 3 ft. high, with 3 horizontal bars ... ... each 3/-
Hand Gate, 2 ft. 6 in. wide, and standards ... ... each 7/6

Carriage Paid on all Orders above 40/- *value to the principal Railway Stations in England and Wales.*

CAUTION.—Beware of inferior imitations. All articles bear our name as a guarantee of good workmanship.

ROSE LANE WORKS, NORWICH. 313A

No. 106. GAME PROOF FENCE
for Fowls, Rabbits, &c.

The uprights are of Flat iron, 1¼ in. by 5/16-in., spaced 10 ft. apart, horizontal wires are galvanized, and of No. 10 gauge. Galvanized wire netting 1½ in. mesh, 18 gauge.

Cash Prices.
4 ft. high, with 3 ft. netting, and two wires above ... **8d.** per yard.
5 ft. high, with 4 ft. 6 in. netting ,, ,, ... **9d.** ,,
6 ft. high, with 5 ft. netting ,, ,, ... **10d.** ,,
Terminal straining posts fitted with screws, 4 ft. **10/-** ; 5 ft. **11/-** ; 6 ft. **13/-** each.
Gates, 3 ft. wide. **15/-** each.

No. 107.

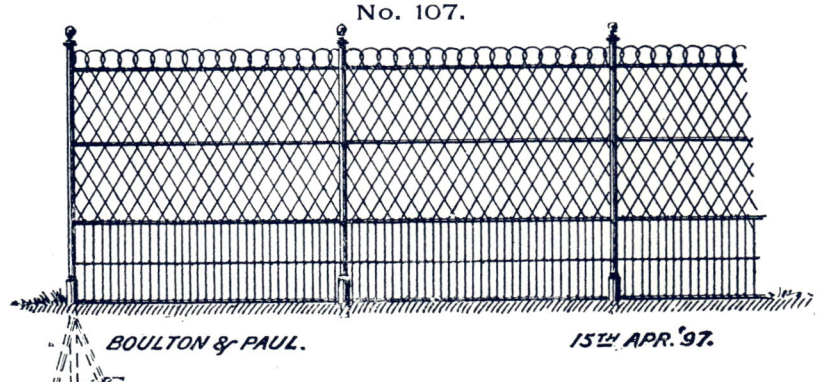

This Illustration represents a highly-ornamental Fencing suitable for enclosing poultry and fancy pheasants, it is also useful as a trainer for plants and forming an effective screen, constructed with Wrought-iron Frames, connected by tubular uprights, and covered with ornamental wire lattice, 2 in. mesh, No. 11 gauge, painted dark green.

Cash Prices.
5 ft. high, per yard **9/6** ; 6 ft. high, per yard **11/6**. Side stays recommended for 6 ft. fencing, **1/6** each.

CAUTION.—Beware of inferior imitations. All articles bear our name as a guarantee of good workmanship.

BOULTON & PAUL, MANUFACTURERS.

REGISTERED COPYRIGHT.

No. 45a. FATTENING PEN.

Fowls will fatten quickly and naturally in these Pens, and be kept warm and healthy.

THE fowls are confined in separate compartments easily accessible by the hinged lids. They stand on a close flooring half its width, with bars along the back, through which the excrement falls into the drawer underneath, which should be filled with earth or ashes, and can be removed *every day* without nuisance, and preserved for manure. The troughs are galvanized, and loose for cleaning.

The birds should be fed about every three hours, the earlier in the morning the first meal is given the better. The following is a good diet:—Equal proportions of ground buckwheat and barley meal mixed together with skim milk to the consistency of crumpy dough, *not pasty*, with milk and water to drink. After feeding, remove the troughs for three hours, and then feed again. Keep the pens in sheds or buildings, moderately warm and not too light.

Cash Price.
For Four Fowls £1 10 0
Two Pens Carriage Paid.

No. 61. NEW FEEDING COOP.

REGISTERED COPYRIGHT.

THE Feeder will supply a want in every poultry yard, it being very important that young stock should be fed apart from the large fowls. The cage should be shifted frequently to fresh ground, and will always keep the food shady and dry. Every poultry rearer should use it, as the saving of chickens alone will repay its cost at once. It can also be used as a coop if desired. The entrances can be partially closed to suit any size chicken.

Cash Price.
2 ft. 6 in. long, 2 ft. 6 in. wide, 2 ft. high ... each £1 0 0
Two Carriage Paid.

No. 45. IMPROVED FATTENING PENS.

REGISTERED COPYRIGHT.

SIX compartments at top for single fowls, two at bottom for six fowls in each, fitted with sliding drawers and troughs complete. 6 ft. long, 4 ft. high.

Cash Price, Carriage Paid.
Complete £4 0 0

No. 61a. PASCALL'S CHICK FEEDING RUN.
New Article.

THIS run has been most successful in keeping chicks of all kinds well fed and we'l protected from other running Poultry. It is made 3 ft. square, by 7½ in. high. The run will be found invaluable by keepers rearing Pheasants under Hens, if placed near the Coops, with a handful of furze or bracken thrown on the top.

REGISTERED COPYRIGHT.

Cash Price 5/6 each.

ATHERSTONE HALL, April 5th, 1894.
MRS. COMPTON is much pleased with the Chick Feeding Runs, they are just what she wanted.

No. 45b. CHEAP FATTENING COOP.
New Article.
As recommended and described in "The Field."

THE coops have a boarded roof sloping backwards, the back and ends are closed, and the bottom made with flat bars with rounded edges, 2 in. wide at the top, and narrower beneath, so as to prevent the droppings sticking to the sides. The front of the coop is made of round bars 3 in. apart, and is provided with feeding trough as shown.

REGISTERED COPYRIGHT.

Cash Price, Coop about 4 ft. long, 2 ft. 6 in. high at front, 2 ft. at back, on legs 2 ft. high, £1 7 6.

Carriage Paid on all Orders above 40/- *value to the principal Railway Stations in England and Wales.*

CAUTION.—Beware of inferior imitations. All articles bear our name as a guarantee of good workmanship.

No. 39. HOUSE ON WHEELS FITTED WITH NESTING BOXES

This House is easily moved from one place to another. 8 ft. long by 6 ft. wide, and will accommodate 96 setting hens.

Cash Price, Carriage Paid.

With 96 nests—two blocks of 48 on each side ... £19 10 0
One block of 48 on one side only 14 10 0

TESTIMONIAL.

From W. J. MACKENZIE, Esq., Glasgow.

The Nest Boxes are strongly and neatly finished, and I think will be much more convenient and useful than Coops which I have used hitherto with rather unsatisfactory results, owing to rats stealing the eggs even from under the hens. These boxes will prevent that.

NEST OF TWELVE LAYING OR HATCHING BOXES.

REGISTERED DESIGN.

All made of red deal, with sliding shutter fronts to each box, ventilated back and front.

Cash Price 30/-

These Nests should be found in every Setting House.

No. 40. THE "COOKSON" RUN FOR SETTING HENS.

To be placed in front of Coops and Setting Boxes. Size, 6 ft. long, by 3 ft. wide, and 2 ft high.

Cash Price 15/- each.

TESTIMONIAL.

From GEORGE J. COOKSON, Esq.

I am expecting much better results from the Hen Runs in the hatching, and I quite expect the increased number of Pheasants hatched will pay me for all the expense in two or three years at least.

SETTING BOXES WITH RUNS.

As described by Major Morant in his book on Poultry Keeping.

No 66

REGISTERED COPYRIGHT.

Each Nest is 15 in. by 18 in. by 18 in. high ; Runs, 2 ft. long, 18 in. high.

Cash Prices.

Five Nests with Runs, complete ... £1 15 0
Two Nests with Runs ,, ... 0 17 6

No. 64. HATCHING & NEST BOX.

REGISTERED COPYRIGHT.

Cash Price.
Lime-whited inside, 3/6 each.

THE "COOKSON" HATCHING BOX.

With Wire Bottom.

REGISTERED COPYRIGHT.

The Hatching Box is well ventilated, and is an improvement on any Hatching Box in the market.

Cash Price 4/- each.

LAYING BOX.

To prevent hens eating their eggs.

No. 67

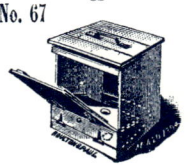

Price 7/6 each.

Can be used as a Setting Box if required.

No. 64a. HATCHING BOX WITH WIRE PEN.

As used on the Elveden Estates.

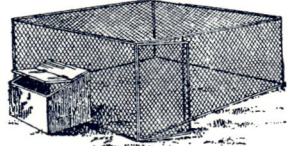

REGISTERED COPYRIGHT.

This is the best system for setting hens on pheasants' or partridges' eggs. The Run is the Cage Trap shown on p. 334 (second size), and can be used as a Trap when not used for this purpose.

Cash Price 19/- each, complete.

Carriage Paid on all Orders above 40/- value to the principal Railway Stations in England and Wales.

CAUTION.—Beware of inferior imitations. All articles bear our name as a guarantee of good workmanship.

BOULTON & PAUL, MANUFACTURERS.

No. 60a. PATENT CHICKEN NURSERY OR BANTAM HOUSE.

No Disease, no Tainted Ground, no Vermin.   Movable, Rat Proof, Thief Proof.

THIS House is invaluable for keeping young stock away from adult fowls after they have left the Coops.

Cash Price.

6 ft. long, 3 ft. wide, 2 ft. 7 in. high   ...   ... £1 5 0 each.
Two Carriage Paid.

TESTIMONIAL.
From
WILLIS CROOKES, ESQ.,
Parson Cross.

The Bantam Houses arrived safely, and my man put them together with the greatest ease. I have had a great number of houses in my time, but these are the best I ever had.

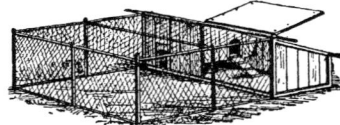

WIRE Run for No. 60a Chicken House, 6 ft. square, made in hurdles, covered with sparrow-proof wire netting.

Price of Run only, with covered top, 15/-

No. 60b. CHICKEN NURSERY & HOUSE FOR GUINEA PIGS OR RABBITS.

Illustration showing Run to Chicken Nursery, No. 60b.

Price of Run only, 10/-

TESTIMONIAL.
From J. W. FREEMAN, ESQ., Syston.
I received the Chicken House all safe, and am very pleased with it. I think it a capital place for what it is intended.

THIS House is similar to No. 60a, but has sliding fronts. The arrangement is a novelty. It has not a wire bottom.

Cash Price.

6 ft. long, 3 ft. wide, 2 ft. 6 in. high, without Run, as Illustrated   ...   .   each £1 5 0
Two Carriage Paid.

No. 24. COCKEREL RUN OR BANTAM HOUSE.

"The new Cockerel Houses and Runs made by MESSRS. BOULTON & PAUL, of Norwich, will supply a real want in many a poultry yard. No one who has kept exhibition fowls, no matter in how small a way, but has been troubled by the fighting of the cockerels, and many a promising bird has had to be sacrificed or sold because of this. We may add that the prices are very reasonable."—*The Fancier's Chronicle.*

Cash Price.
House and Run, 6 ft. 6 in. long, 2 ft. 6 in. wide   ... £2 10 0

TESTIMONIAL.
From H. BELDON, ESQ., Goitstock, Yorkshire.
(*One of the most successful Breeders and Exhibitors of Poultry.*)
I think your new Runs and Coops for Cockerels are very good. I have had similar Coops for years for single cocks; in fact, it was one of the secrets of my success.

REGISTERED COPYRIGHT.

No. 67. WIRE CHICKEN RUNS.

6 ft. long, 18 in. wide, 12 in. high. Covered with 1-in. mesh netting.

Cash Price 1/9 each.
Ends 3d. each.

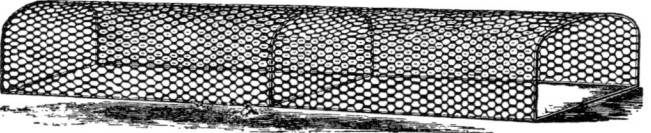

Carriage Paid on all Orders above 40/- value to the principal Railway Stations in England and Wales.

CAUTION—Beware of inferior imitations.   All articles bear our name as a guarantee of good workmanship.

ROSE LANE WORKS, NORWICH.

THE "KEEPER'S" COOPS FOR GAME AND CHICKENS.

Only best Materials and Workmanship employed.

No. 65. COOP AND RUN.

THIS Coop has been enlarged and improved in construction this season.

REGISTERED COPYRIGHT.

Size, 2 ft. square, 18 in. high in front, and 12 in. high at back. Well painted and whitened inside.

Cash Price without Run, 5/6 each.

12 Coops 60/-, Carriage Paid.

Runs, with slide in front, 2/6 each, extra.

No. 65a Pattern.

REGISTERED COPYRIGHT.

1 ft. 10 in. square, 16½ in. high in front, 10½ in. at back.

Without Shutter.

Cash Price 4/6 each.

or 48/- per doz., painted.

No. 65b Pattern.

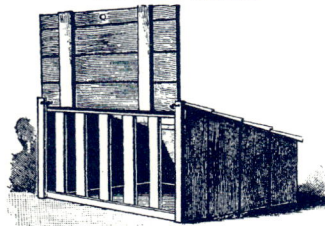

REGISTERED COPYRIGHT.

1 ft. 10 in. square, 18 in. high in front, 12 in. at back. With Sliding Shutter.

Cash Price 7/6.

3/- per doz. less for 12 Coops or more.

THESE Coops are made of best selected red deal well seasoned, painted one coat, and whitened inside, well finished.

TESTIMONIAL.

From R. S. CARR, ESQ. Newcastle.

I am much pleased with the Coops, and have been very successful in rearing this year.

PAYNE-GALLWEY PHEASANT COOP.

As described in *Field*, Jan. 16, 1892.

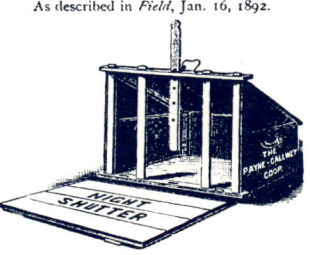

REGISTERED COPYRIGHT.

2 ft. 3 in. wide, 24 in. high in front, 2 ft. deep.

Cash Price 6/6 each.

12 Coops £3 15 0, Carriage Paid.

Wire Netting Bottoms to prevent rats burrowing beneath, 1/- each extra.   Runs for ditto 3/- each.

No. 68. NEW PORTABLE COOP WITH RUN COMBINED.

For Game, Chickens, Rabbits.

CAT PROOF. RAT PROOF. SPARROW PROOF.

REGISTERED, NO. 49,419.

Cash Price.

5 ft. long, 2 ft. wide, 20 in. high ... 15/-.

From A. LLOYD, ESQ., Hern Bank, Petersfield.

I am glad to have the opportunity of saying how very pleased we are with Coop and Run, and consider it well made and most reasonable in price.

No. 69. NEW COOP AND RUN.

The Latest Pattern.

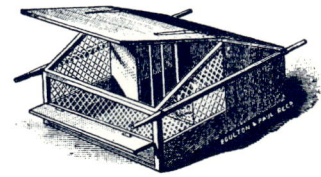

THIS is a well constructed Coop. The Run forms a shelter for chicks until they are strong. Painted three coats.

Cash Price 20/- complete.

Carriage Paid on all Orders above 40/- value to the principal Railway Stations in England and Wales.

CAUTION.—Beware of inferior imitations.   All articles bear our name as a guarantee of good workmanship.

BOULTON & PAUL, MANUFACTURERS.

No. 68. NEW PORTABLE COOP WITH RUN COMBINED.

Chickens and Ducks thrive wonderfully in this Coop.

CAT PROOF. RAT PROOF. SPARROW PROOF.

THE sides and coop are made of wood, well painted; the top and front of run covered with fine wirework. Size, 5 ft. long, 2 ft. wide, 20 in. high.

Price 15/- each.

A larger size, for fifty Young Ducks, or Chickens.

Price 30/-

REGISTERED, NO. 49,419.

IMPROVED POULTRY SHOW PENS.

PIGEON PENS.

IN sets of four. Each pen 15 in. square. To fold flat for packing.

Price 1/- per pen.

CIRCULAR PENS FOR POUTERS.

18 in. diameter. Price 1/6 each.

Showing Pens when folded.

Cash Prices.
1 ft. 9 in. square, 2/3 ; 2 ft. square, 2/9 ;
2 ft. 3 in. square, 3/3 each.

TESTIMONIAL.

From A. T. MILLER, ESQ., Piccadilly.

DEAR SIRS,—The Shelter and Troughs are satisfactory, and I am obliged by your prompt attention.

GALVANIZED FEEDING OR DRINKING CUPS.

For Poultry Pens.

Cash Prices.

Small	...	per dozen	3/-
Medium	...	,,	4/-
Large	...	,,	6/-

Carriage Paid on all Orders above 40/- value to the principal Railway Stations in England and Wales.

CAUTION.—Beware of inferior imitations. All articles bear our name as a guarantee of good workmanship.

BOULTON & PAUL, MANUFACTURERS,

BURALLS' PATENT ECONOMIC SELF-FEEDING POULTRY AND PHEASANT BIN.

Do not starve your Poultry to feed the Sparrows

By throwing food on the ground; in this way you waste as much as your poultry eat.

Everybody knows that Poultry eat less and do better by early and regular feeding. By this ingeniously arranged Bin, Poultry and Pheasants are enabled to feed at any time without personal attention, and need only be visited when the Bin needs replenishing. The saving in time alone, thus effected, soon pays the cost of the Bin.

The food is out of the reach of sparrows and vermin. The feeding places are so made that sparrows cannot stand in them, and only a few grains are in sight at a time. It holds about **six stones**, and any kind of corn can be used. It can be emptied by the fowls to the last grain, and trains fowls, especially game, to stand up well. The most delicate comb cannot be injured.

Mr. CHARLES BOTHWAY, the celebrated breeder of game and winner of numerous cups, writes: "I am very pleased with the Feeding Bin, it answers well, saving both time and food."

Cash Prices.

Large Size, painted **15/-**
Small Size, unpainted **7/6**

INSTRUCTIONS.

LOOSEN the thumbscrew and slide the tins up or down according to the size of grain used. Do not put any loose straw in the case.

Remove any accumulation of dust or dirt from the bottom of the bin before replenishing, otherwise the corn will be prevented from coming freely.

Use only one kind of corn at a time, as Fowls have a preference and will pull out one kind to get at the other; besides, it is better for the Fowls to have a **complete change** of Food.

One is enough for Fifty Fowls.

TANDY'S PATENT CHICKEN REARER.

REGISTERED COPYRIGHT.

Cash Price, Carriage Paid.

Size, 4 ft. by 2 ft. by 2 ft. 6 in. to ridge .. **£2 10 0** each

INSTRUCTIONS FOR USING.

PLACE the Rearer out in the open and move it on fresh ground frequently. The cistern must be three parts filled with hot or cold water. Use the best paraffine oil in lamp, and trim the wick every 24 hours.

A small thermometer should be placed in the warm chamber, which must be kept at about 70 degrees for a start, and decrease as the chicks grow. Put fine peat moss, dust, or sand in the Rearer for the chicks to run on, and clean out often.

Feed the chicks for the first day in the warm chamber, then for a day or two in the cool chamber, afterwards outside the Rearer, and make them come out for their food, do not lock them in, wet or dry, in the daytime.

When the chicks are old enough to do without the heat from lamp, remove water cistern, lamp, partitions, and false floor, but keep the felting over the top as before to check all draught.

Carriage Paid on all Orders above **40/-** *value to the principal Railway Stations in England and Wales.*

CAUTION.—Beware of inferior imitations. All articles bear our name as a guarantee of good workmanship.

BOULTON & PAUL, MANUFACTURERS,

No. 63. NEW FOLDING COOP FOR TURKEYS.

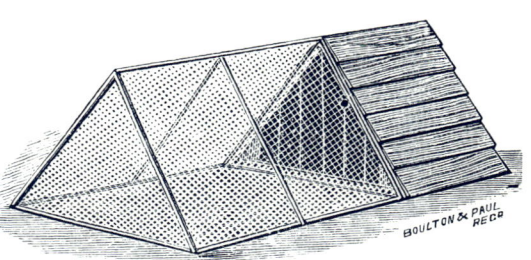

THE advantage of a Coop this shape is, there is less liability of the poults being trodden on by the hen, as the roof sloping to the floor allows of shelter for the poults apart from the hen. The whole packs flat for export, or stowing away when not in use. Painted and lime-washed inside.

Size, 3 ft. high, 3 ft. wide, 2 ft. 6 in. deep.
Cash Price, without floor and run, 10/6 each.
Runs, 6 ft. long, 5/- each extra.

No. 44. FATTENING PEN FOR FOWLS.

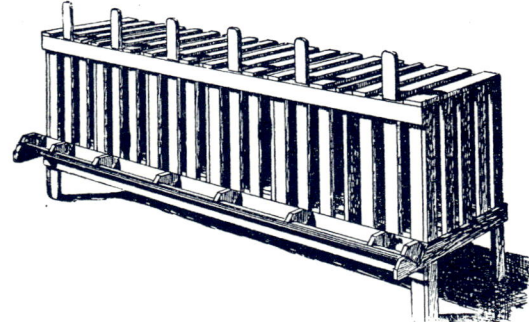

Roughly and strongly made, unpainted.

Size, 6 ft. long, by 1 ft. 6 in. wide, by 2 ft. 4½ in. high.

Cash Price 17/6.

All Orders amounting to 40/- *Carriage Paid to the principal Railway Stations in England and Wales.*

CAUTION.—Beware of inferior imitations. All articles bear our name as a guarantee of good workmanship.

ROSE LANE WORKS, NORWICH. 318B

No. 71. NEW COOP WITH COVERED AND OUTER RUN.

THE Coop is 2 ft. by 2 ft., and has a glazed run, as shown, 2 ft. by 2 ft. ; also outer run covered with wire netting, 4 ft. by 3 ft. Constructed of selected red deal, ledged and well nailed, painted three coats outside, and whitened inside.

Cash Price £1 5 0

No. 72. FUMIGATING BARROW.
For Curing Gapes in Pheasants and Chickens.

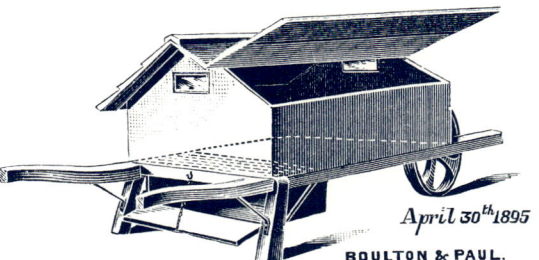

REGISTERED No. 254,893.

THIS Barrow has been used on a large estate, and is recommended as a certain cure for Gapes. A spirit lamp is placed in the box under the floor, shown in the illustration, and a medicated preparation boiled over the lamp, the fumes passing through perforated zinc into the coop, destroy the germs of the disease. The birds should be fumigated about twenty minutes, but care must be taken to watch them through the glass window to prevent suffocation.

    Price complete with spirit lamp, vaporating pan, and sufficient preparation for eight fumigations, £2 5 0, Carriage Paid.

    Fifty to **100** birds can be fumigated at one time.

    The preparation supplied at 5/- per gallon, can included.

    The box can be taken off and the barrow used for removing coops and other purposes.

All Orders amounting to **40**/- *Carriage Paid to the principal Railway Stations in England and Wales.*

CAUTION.—Beware of inferior imitations. All articles bear our name as a guarantee of good workmanship.

24

BOULTON & PAUL, MANUFACTURERS,

No. 81.
FLOATING DUCKS' HOUSE.

No. 82.
RUSTIC DUCKS' HOUSE.

Size, 4 ft. square, with 12 in. ledge all round, 2 ft. to eaves, and 4 ft. to apex.
Cash Price £4 0 0

Size, 6 ft. by 4 ft.
Cash Price £6 10 0

No. 83. DUCKS' HOUSE.

Size, 8 ft. by 6 ft., 6 ft. to eaves by 8 ft. to ridge.
Weather-boarded walls, red deal slab roof, painted.

Well ventilated, having half hatch doors as shown.
Cash Price £7 0 0

PLAIN DUCKS' HOUSES.
Covered with weather-boarding painted.

No. 84. SPAN ROOF.
No. 85. LEAN-TO INDEPENDENT.

Size, 6 ft. by 5 ft., 2 ft. to eaves, 4 ft. to ridge.
Cash Price £2 10 0

Size, 6 ft. by 5 ft., 2 ft. to eaves, 4 ft. to ridge.
Cash Price £2 0 0

CAUTION.—Beware of inferior imitations. All articles bear our name as a guarantee of good workmanship.

ROSE LANE WORKS, NORWICH 319

No. 51. DUCK HOUSE WITH WIRE NETTING ENCLOSURE FOR WATER.

REGISTERED COPYRIGHT.

As made for THE COUNTESS OF LONSDALE.

HOUSE 3 ft. by 2 ft. 6 in., having wood floor, door at back, and sliding shutter for duck hole.

Enclosure 3 ft. by 2 ft. 6 in., with flap for shutting up when required.

Cash Price, Carriage Paid, £3 10 0

No. 52. TRAVELLING COOP FOR EXPORTING BIRDS.

CAN be used as an ordinary Coop or as a Fattening Pen. 2 ft. 6 in. wide by 2 ft., 2 ft. 6 in. high in front, 1 ft. 8 in. at back.

Cash Price.

With batten floor ... 7/6

REGISTERED COPYRIGHT.

53. NEW IMPROVED DUCKS' HOUSE.

THE House is hinged to the floor and arranged to turn over as shown, giving great facilities for a thorough cleaning.

Cash Price, 21/-

REGISTERED COPYRIGHT.

No 66.

REGISTERED COPYRIGHT.

Size—6 ft. by 2 ft. 6 in., 3 ft. high, with partition.

Cash Price, Carriage Paid, £2 10 0

HOUSES FOR DUCKS.

THESE Houses are suitable for placing against a pond or stream for ducks or other water fowl. Each fitted with nests and sliding doors, for shutting ducks in at night, with large door at back. Well painted outside, and lime-whited inside.

From A. E. NELSON, ESQ., Temple.
The Ducks' House, Coops, &c., came to hand yesterday. I am very pleased with them.

From R. CARY, ESQ., Torr Abbey.
The Duck House answers very well.

No. 66a.

Size—4 ft. by 2 ft. 6 in., with floor, ladder, and posts for standing in water.

Cash Price, Carriage Paid, £2 10 0

GALVANIZED DUCK PONDS.

No. 49.

Cash Price.

4 ft. 6 in. long, 2 ft. 4 in. wide, 9 in. deep, 12/6

No. 50.

Cash Price, 3 ft. long, 2 ft. wide, 9 in. deep, 10/6

Carriage Paid on all Orders above **40/-** *value to all the principal Railway Stations in England and Wales.*

CAUTION.—Beware of inferior imitations. All articles bear our name as a guarantee of good workmanship.

BOULTON & PAUL, MANUFACTURERS,

GALVANIZED CORN BINS.

No. 303. EXTRA STRONG GALVANIZED CORRUGATED IRON CORN BIN.

Can be had either Rectangular, Square or Round.

Lettering, when required, charged for extra.

Cash Prices.
RECTANGULAR OR SQUARE.

	With 1 partition.		With 2 partitions.
4-bush.	6-bush.	8-bush.	12-bush.
26/-	30/-	35/-	40/-

No. 14. GALVANIZED IRON CORN BIN.

WITH close-fitting lids and padlocks. All keepers of live stock should use them. They are vermin and damp proof.

Cash Prices.

First size, to hold 1½ bushels	...	10/-
Second size ,, 2 ,,	...	12/6
Third size ,, 3 ,,	...	15/-

Lettering, when required, extra.

REGISTERED COPYRIGHT.

No. 13. GALVANIZED IRON CORN BIN.

With strong Hasp and Hinges.

Cash Prices, without Partition.

Bushels.	Length.	Width.	Depth.	£ s. d.
4	2 ft. 0 in.	1 ft. 6 in.	2 ft. 0 in.	£0 17 6
6	2 ft. 6 in.	1 ft. 8 in.	2 ft. 0 in.	1 1 0
8	2 ft. 9 in.	1 ft. 11 in.	2 ft. 2 in.	1 5 0

Partitions, each 5/- extra.

No. 199. GALVANIZED CORN BIN.

With covered box at top for holding Stable Requisites.
NEW PATTERN.

Capacity.	DIMENSIONS.				PRICE.	
	Length.	Width.	Depth. Front.	Depth. Back.		
Bushels.	ft. in.	ft. in.	ft. in.	ft. in.	£ s. d.	
4	2 6	1 6	1 10	2 5	1 13 0	each
6	2 9	1 7	2 0	2 7½	1 18 9	,,
8	3 0	2 0	2 0	2 9	2 4 6	,,
10	3 3	2 2	2 2	2 11½	2 15 9	,,
12	3 6	2 4	2 4	3 2	3 5 9	,,
16	4 0	2 4	2 5	3 4½	3 18 0	,,
20	4 6	2 8	2 6	3 7	4 10 0	,,

DAVIS' PATENT

PARTITIONS.

4 bush. 4/3.   6 bush. 4/6.   8 bush. 5/-   10 bush. 5/9.   12 bush. 6/6   16 bush. 7/6.   20 bush. 8/6 each extra.

Carriage Paid on all Orders above **40/-** *value to the principal Railway Stations in England and Wales.*

CAUTION.—Beware of inferior imitations.   All articles bear our name as a guarantee of good workmanship.

ROSE LANE WORKS NORWICH

GALVANIZED WROUGHT-IRON POULTRY TROUGHS.

No. 18.

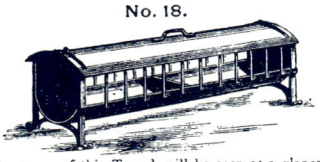

THE advantages of this Trough will be seen at a glance by all practical poultry keepers. Being covered, the food is always dry. The bars prevent fowls getting into the Trough and spoiling the food. By a simple arrangement the top is opened for feeding, and can be entirely removed for cleaning.

Cash Price, 36 in. long, for twelve fowls, 10/-

Repays its cost in six months.

No. 19.

THESE Troughs are light and easily carried. Well adapted for large poultry keepers.

Cash Price, 36 in. long, 5/-

No. 20.

Cash Prices.

36 in. long 3/6
18 in. ,, ... 1/6

TESTIMONIAL.

From MR. HERBERT WOODS, Newport.
DEAR SIRS,—The Fowl Troughs arrived safely to-day, and I am much pleased with them; also the Hen House, which I have put up.

No. 21. POULTRY TROUGH.

Price, 3 ft. long, 6/-

No. 46. POULTRY SHELTER AND DUST BATH.

Made of Galvanized Corrugated Iron.

THE advantage of these will be understood by all practical poultry keepers, for no fowls, whether chickens or adults, can thrive without a good day shelter and dust bath at hand. The shelter should be covered with litter in hot weather.

Cash Prices.

6 ft. span ... ... 2/- per ft. run.
7 ft. ,, ... 2/6 ,,
8 ft. ,, ... 3/- ,,

With necessary bolts and nuts. Ends 7/6 extra.

REGISTERED COPYRIGHT.

PORTABLE DUST BATH FOR POULTRY.

REGISTERED COPYRIGHT.

Cash Price.

3 ft. square, 2 ft. 8 in. high 10/-

TESTIMONIAL.
From
MISS F. SEAWELL, Little Berkhamstead.
Miss F. Seawell has received the Corn Bin and Nest Boxes, and is much pleased with them.

No. 23. EGG CABINET.

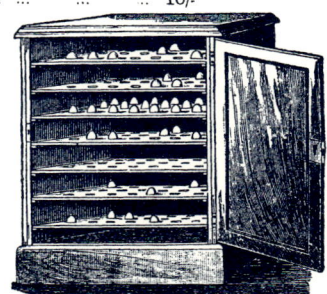

MADE of well-seasoned red deal, varnished. A large number of eggs can be stowed in a small space. The shelves are made to draw out to prevent risk of breakage. Eggs will keep fresh much longer in these Cabinets than in bran or chaff, the air being excluded as much as possible.

Cash Price.

33 in. high, 25 in. wide, 11½ in. deep, to hold about 300 eggs £2 10 0
Smaller size, to hold about 100 eggs 1 7 0
Special sizes made to order.

Carriage Paid on all Orders above 40/- value to the principal Railway Stations in England and Wales.

REGISTERED COPYRIGHT.

CAUTION.—Beware of inferior imitations. All articles bear our name as a guarantee of good workmanship.

BOULTON & PAUL, MANUFACTURERS.

POULTRY TROUGHS AND FOUNTAINS.

NEW WATER FOUNTAIN.

New Drinking Fountain, which we have adopted as a means of supplying pure water to poultry, pigeons, and pheasants. It is constructed on the simplest of hydraulic principles, it is practically unbreakable, and can be scoured out inside.

Cash Prices.

No. 9. 15 in. diameter, for adult fowls. Galvanized each 10/-
No. 10. 12 in. diameter, for chickens. Galvanized each 8/6

REGISTERED COPYRIGHT.

POULTRY & PIGEON FOUNTAIN.

No. 24.

KEEPS the water clean and free from contamination.

Price 3/6 each

POULTRY & PIGEON FEEDING HOPPER.

No. 25.

PREVENTS waste and keeps the food dry, clean, and free from contact with impurities.

Price 3/6 each.

No. 4. CIRCULAR POULTRY TROUGH.

EIGHTEEN inches diameter, with five compartments, the centre one for water.

Cash Price, japanned, each 5/-

No. 5. SEMI-CIRCULAR POULTRY TROUGH.

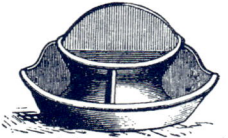

EIGHTEEN inches in diameter, three compartments, the centre one for water. Suitable for placing against a wall.

Cash Price, japanned, each 4/-

No. 6. TRIANGULAR POULTRY TROUGH.

IN two compartments, the upper one for water. Suitable for a corner.

Cash Price, japanned, each 2/6

No. 1. POULTRY TROUGH.

Cash Prices.
JAPANNED.

12 in. long, 2 compartments each 1/9
18 in.   ,,   3   ,,   ,,   2/-
24 in.   ,,   3   ,,   ,,   2/6

No. 2. POULTRY TROUGH.

Cash Prices.
JAPANNED.

9 in. long, 2 compartments each 1/6
24 in. ,,   3   ,,   ,,   3/-
36 in. ,,   3   ,,   ,,   4/-

No. 3. DOUBLE POULTRY TROUGH.

With Improved Movable Guard.
Cash Prices.

30 in. long, 10 in. wide each 7/6
18 in.   ,,   ,,   5/-

No. 7. IMPROVED RABBIT & POULTRY TROUGH.

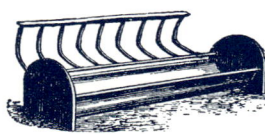

REGISTERED COPYRIGHT.

Cash Prices.

18 in. long, galvanized ... each 3/-
24 in. ,,   ,,   ... ,, 4/-

GALVANIZED FEEDING CAGE.

REGISTERED COPYRIGHT.
WITH LOOSE GALVANIZED-IRON PAN.

Cash Prices.

No. 11. For adult fowls ... each 4/6
No. 12. For chickens   ... ,, 5/6
Trays only   ... ,, 2/-

No. 13. NEW REVERSIBLE POULTRY TROUGH.

For fixing to Wire Hurdles.

REGISTERED COPYRIGHT.

Cash Prices.

24 in. long ... ... each 5/-
36 in. long ... ... ,, 7/6

Carriage Paid on all Orders above **40/-** *value to the principal Railway Stations in England and Wales.*

CAUTION.—Beware of inferior imitations.   All articles bear our name as a guarantee of good workmanship

BOULTON & PAUL, MANUFACTURERS.

REGISTERED COPYRIGHT.

Treatise on Pigeon Keeping.

FEW hobbies can afford more gratification and innocent amusement than the keeping of one or more of our many varieties of the Pigeon. The greatest difficulty, however, is the want of knowing how to manage them. We therefore beg to present the following plain instructions to those who may be desirous to ornament their grounds with a structure and occupants calculated to afford them a most pleasurable contemplation.

Pigeons in Aviaries.—The best varieties of Pigeons for the Aviary are those known as "Toys," viz.—Turbits, Owls, Fantails, Turbiteens, &c. These are most pretty and interesting, and they can be produced in confinement, if sufficient space be given, better than if they are allowed their liberty. Their plumage will come to perfection sooner, and keep cleaner than if allowed to fly about the housetops, and they are not so likely to become a prey to cats.

Breeding.—TO PAIR BIRDS OFF: They should be put separately in pens, and with a wire partition between each cock and hen, the partition being wide enough between the bars for them to put their heads through, at the same time there is sufficient space for either bird to recede far enough to prevent the other bird pecking it on the head. When the birds are matched turn them loose in the aviary, putting a nest pan in every recess. In the bottom of the pan first put a little slaked lime, on the top of that put two handfuls of yellow deal sawdust. This will keep parasites at a respectable distance, a thing you cannot be too particular about, as birds will forsake their young ones from this cause more than anything else.

Feeding.—Give them the best you can purchase, it is the cheapest in the end, it is more satisfactory, and they can clear it up without any waste. Vary their food, not keeping them too long on one thing. The best sorts are: Small Round Maize, Old Grey Peas, Dare, Tares, Buckwheat, and Wheat occasionally. Give them fresh water daily, putting a small piece of sulphate of iron in each fountain (see page 322) once a week, about the size of a pea—this invigorates and gives tone to the birds. Always keep a little heap of red sand, mixed with loam and lime core, such as is cleaned from old bricks, say about half a bushel. This the birds will turn over, more especially if you throw a few carraway seeds amongst it. Give a bath (see page 325) twice a week; place the bath in the middle of the aviary, when done with take it away and throw some dry sand over the wet they have made.

By attending to the above simple directions, you can produce and keep in health some of the most delicate and beautiful specimens of the Pigeon tribe known.

CAUTION.—Beware of inferior imitations.   All articles bear our name as a guarantee of good workmanship.

ROSE LANE WORKS, NORWICH. 325

No. 60. HOUSE FOR GOATS AND OTHER PETS.

TESTIMONIAL.

From O. E. CRESSWELL, Esq., Hereford.

In 1889 you made me a Pigeon House with which I have been well satisfied, and am inclined to try another.

TESTIMONIAL.

From Mrs. PRESTON, Lancaster.

The House is a great success, and very pretty.

REGISTERED COPYRIGHT.

Goat House 9 ft. by 9 ft., wood walls stained, weather-board roof painted, and lined with matchboarding.

Cash Price, Carriage Paid £35 0 0

A cheaper pattern can be made, price on application.

No. 53. NEW ORNAMENTAL DOVE OR PIGEON COTE.

TESTIMONIAL.

From Miss E. T. ROOKE, Dorchester.

Miss E. T. Rooke has received the Pigeon House, and is much pleased with it.

BRACKET PERCHES FOR LOFTS.

Cash Price 8d. per perch complete.

REGISTERED COPYRIGHT.

For placing against a South wall. The whole of the front is made to open.

Cash Price £2 10 0

TESTIMONIAL.

From Mr. CHAPMAN, Congham Lodge.

Mr. Chapman has received the Dove Cote, &c., and is much pleased with them.

SHELF PERCH, FOR LOFTS.

Cash Price 1/6 per perch complete.

No. 15. FEEDING HOPPER FOR PIGEONS OR POULTRY.

REGISTERED COPYRIGHT.
A shake fills it.
Cash Price 10/6 each.

NESTING BOXES.

Cash Price 5/- per Double Nest.

SOLID WOOD TURNED NESTS.

Cash Price 8/ per dozen.

No. 16. BATH FOR PIGEONS AND AVIARIES.

REGISTERED COPYRIGHT.

Cash Prices.

14 in. square, galvanized, 3/6 each.
20 in. ,,   ,,   5/ ,,

Carriage Paid on all Orders above 40/- value to the principal Railway Stations in England and Wales.

CAUTION.—Beware of inferior imitations.   All articles bear our name as a guarantee of good workmanship.

No. 57.
SPAN-ROOF RANGE OF ORNAMENTAL PIGEON COTES.

WOOD HOUSES, stained and varnished outside, curved iron roof with cresting to ridge and eaves as shown, wrought-iron front and divisions to runs, covered with wire trellis, painted, galvanized corrugated iron at the bottom of same. Houses fitted with nest boxes and runs, with bracket perches on each side 16 ft long by 10 ft. wide.

Prices on application.

TESTIMONIAL.
From SIR WALTER FARQUHAR, Dorking
Sir Walter Farquhar is much pleased with the Pigeon Cote, and all the practical arrangements

No. 58. LEAN-TO RANGE OF PIGEON COTES.

CONSTRUCTED similarly to No. 57, but front and divisions are of wood, covered with galvanized wire netting. Roof of galvanized corrugated iron. 16 ft. long by 10 ft. wide.

Prices on application.

TESTIMONIAL.
From CHAS. B. BEDDOE, ESQ., Hampton Park.
The Pigeon House arrived safely, and gives great satisfaction.

Carriage Paid on all Orders above 40/- value to the principal Railway Stations in England and Wales.

CAUTION.—Beware of inferior imitations. All articles bear our name as a guarantee of good workmanship.

No. 61 COMBINATION PIGEON COTE AND FOWL HOUSE.

This novel and most convenient House is constructed of red deal, painted outside, and having iron roof lined with matchboarding.

Size 12 ft. by 6 ft., 9 ft. high. Having accommodation for six pairs of pigeons, and for twenty-four fowls; loft and roost fitted with nesting boxes and perches.

Cash Price, Carriage Paid, £14

TESTIMONIAL
From
T. W. CALVERLEY, Esq.,
Allerthorpe Hall.

I have received the Pheasantry and Pigeon Loft all safe. I am highly satisfied with them in every way, and shall be glad to recommend you amongst my friends, as I find the workmanship good and materials of excellent quality.

REGISTERED COPYRIGHT.

No. 18. FEEDING HOPPER FOR PIGEONS & OTHER BIRDS.

REGISTERED COPYRIGHT

Size 16 in. by 9¾ in

Cash Price 6/-

TRAP FOR SHOOTING PIGEONS & SMALL BIRDS.

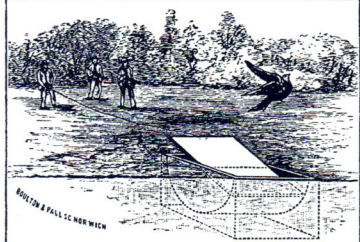

Is so constructed that when the string is pulled the bird is thrown up into the air without injury. Made of wood and lined with zinc.

Price 15/- each.

No. 19. AUTOMATIC FEEDING HOPPER FOR PIGEONS.

REGISTERED COPYRIGHT.

Size 16 in. by 9¾ in.

Cash Price 9/-

No. 52.

REGISTERED COPYRIGHT.

No. 52. PIGEON OR DOVE COTE ON POLE.

FITTED for four pairs of birds. Each hole has double nesting places, with a bridge between, dividing the nests. The pole, which is 14 ft., has a self-fixing base for letting into the ground or screwing to wood blocks. Self-acting corn fountain is provided as shown. Highly finished.

Cash Price, Carriage Paid, £2 10 0

No. 52a. PIGEON OR DOVE COTE ON POLE.

EACH hole has double nesting places, with a bridge between dividing the nests. Pole with self-fixing base.

Cash Prices, Carriage Paid.

For 12 pairs of birds	...	...	£7	0 0
For 16 ,,	...	...	8	10 0
For 20 ,,	...	...	10	0 0
For 24 ,,	...	...	11	7 6

No. 52a.

Carriage Paid on all Orders above 40/- value to the principal Railway Stations in England and Wales

CAUTION.—Beware of inferior imitations. All articles bear our name as a guarantee of good workmanship.

BOULTON & PAUL. MANUFACTURERS.

No. 71. ORNAMENTAL AVIARY FOR PUBLIC GARDENS.

As erected for the South Shields Corporation

Special Estimates for any size free on application.

TESTIMONIAL
From A. SMITH, ESQ., Maidenhead.
Mr. Smith is perfectly satisfied with the Pheasantry that you have just erected

---

No. 70. PIGEON OR DOVE COTE, WITH FLIGHT.

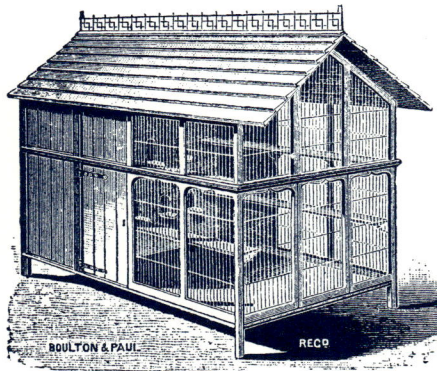

This House is arranged so that all parts can be reached easily for cleaning purposes.

Cash Prices, Carriage Paid.

No. 1 size, 8 ft. long by 6 ft. wide, by 6 ft. high, to accommodate 10 pairs of birds	£10 0 0
No. 2 size, 8 ft. long by 4 ft. wide, by 6 ft. high, to accommodate 7 pairs of birds	8 10 0

No. 72. COMBINATION AVIARIES AND SUMMER HOUSE, WITH CORN ROOM AND ROOSTING HOUSES.

Wood Summer House and Roost, stained and varnished outside, corrugated iron roof, ornamental wrought-iron and wire trellis enclosures to flights, painted, finished with crestings and terminals as shown.

Size about 27 ft. by 15 ft.

Prices on application.

Carriage Paid on all Orders above 40/- value to the principal Railway Stations in England and Wales.

CAUTION.—Beware of inferior imitations.   All articles bear our name as a guarantee of good workmanship.

ROSE LANE WORKS, NORWICH.   329

No. 50. DOVE OR PIGEON COTE.

Painted three coats. Should be placed on a south wall.
Cash Price.
For six pairs of birds £1 10 0
Including iron plates and holdfasts for fixing to wall.
Carriage Paid on Two Cotes.
Pigeon Houses and Lofts constructed upon the latest improved principles.

Registered Copyright.

Special Designs and Estimates given.

NEW PIGEON HOUSE.
No. 54.

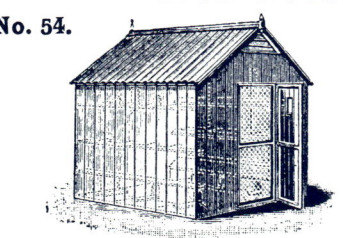
Registered Copyright.

Cash Price, Carriage Paid.

6 ft. long, 6 ft. wide, 7 ft. 6 in. high, fitted with six double nests, and eighteen perches made of wood, with lined iron roof, wood floor, walls painted outside £9 10 0

The pigeons fly out at the end.

No. 49a. DOUBLE COTE

Registered Copyright.

This Cote is constructed upon the same plan as No. 49 but having six compartments.

Cash Prices, Carriage Paid.

4 ft. long, 4 ft. wide, 6 ft. 6 in. high £5 0 0
5 ft. long, 5 ft. wide, 7 ft. high   ..   7 10 0

No. 49.
COTE FOR PETS.

A Nice Present for Children.

This arrangement has been much approved. Each tier is fitted with double nesting-places, with a bridge between dividing the nests.

For specimen birds, or for breeding purposes, this Aviary is well suited.

With three tiers, woodwork painted, wire galvanized.

Cash Prices, Carriage Paid.

4 ft. long, 2 ft. wide, 5 ft. 6 in. high   .   .   £3 0 0
5 ft. long, 2 ft. 6 in. wide, 6 ft. high   .   ..   4 10 0

This will do for Rabbits, Squirrels, Rats, or any other Pets.

Registered Copyright.

TESTIMONIAL.
From A. E. HOLLAND, Esq.
I have received the Pigeon Cote and am much pleased with it, for I think it makes quite a nice ornament in the garden.

Carriage Paid on all Orders above 40/- value to the principal Railway Stations in England and Wales.

CAUTION.—Beware of inferior imitations.   All articles bear our name as a guarantee of good workmanship.

ROSE LANE WORKS, NORWICH. 329A

PIGEON COTES.

TESTIMONIAL.

Miss Osborne has received the Pigeon Cote safely, and is much pleased with it.

No. 84.
BARREL COTE ON POLE.

Fitted for four pairs of birds.

Cash Price £1 15 0

TESTIMONIAL.

From MISS C. LA TOUCHE, Delgany.
The Pigeon House is now quite complete and admired by every one.

No. 83. COTE.
For placing against a wall.

For four pairs of birds.

Cash Price £1 0 0

No. 85. COTE.

For three pairs of birds.

Cash Price 15/-

CAUTION.—Beware of inferior imitations. All articles bear our name as a guarantee of good workmanship.

BOULTON & PAUL, MANUFACTURERS,

No. 80. DOUBLE COTE
for Pigeons and Pets.

Size 8 ft. by 3 ft., by 5 ft. high to eaves. To accommodate 12 pairs of pigeons.

Cash Price £7 10 0

If made single, to accommodate 6 pairs of birds.

Cash Price £5 10 0

Can also be arranged for rabbits.

No. 81. PIGEON COTE AND AVIARY.

WEATHER-BOARDED walls, wood slab roof, painted. Size 8 ft. by 4 ft., by 4 ft. high to eaves.

Cash Price £4 10 0

No. 82. COTE
for Pigeons and Pets.

MATCHBOARDED walls, wood slab roof painted. Size 8 ft. by 4 ft., by 4 ft. high to eaves.

Cash Price £5 0 0

TESTIMONIAL.
From J. HOYLE, ESQ., Wimbledon.
The Pigeon House is erected, and gives every satisfaction.

CAUTION.—Beware of inferior imitations. All articles bear our name as a guarantee of good workmanship.

No. 48. PIGEON HOUSE AND DOVE COTE.

REGISTERED COPYRIGHT.

Suited for Garden or Shrubbery.

LOVERS of Pigeons will find a House of this kind a great source of amusement and delight, as every arrangement is provided not only for the comfort of the birds themselves, but for the convenience of the attendant. By a simple arrangement the birds are set at liberty, or they can be kept inside the House or Aviary at will. The Nesting Boxes and Loft are constructed on the principle figured below, all finished in the best manner. Nesting for about twelve pairs of birds. 12 ft. long, 6 ft. wide, 9 ft. high.

Cash Price, Carriage Paid, £15.

REGISTERED COPYRIGHT.
Interior of Pigeon Loft.

TESTIMONIAL.

From SAMUEL ROGERS, ESQ., Bath.

DEAR SIRS,—I am very pleased with the Pigeon House, it is so well carried out, and all finished in such a satisfactory manner; you deserve the highest commendation. I shall recommend you to all my friends, and you may rely I will do all I can to secure you future orders.

PIGEON LOFT, WITH AREA AND FLIGHT.

REGISTERED COPYRIGHT.

THIS House is designed for the general purposes of the pigeon flyer, and is adapted for erection in a garden.

The area on the top of the Loft House is for flying pigeons at will; the traps can be opened or shut from the outside without disturbing the birds. The inside arrangements are of the most perfect kind for eighteen pairs of birds. See illustration above.

The Flight is formed of iron frames, covered with small mesh netting.

House made of wood, well painted outside, and whitened inside. All finished in best manner.

Size of space covered by House and Flight, 15 ft. by 12 ft.

Cash Price, Carriage Paid, £20.

TESTIMONIAL.

From THE COUNTESS OF LANESBOROUGH, Swithland Hall.

The Countess of Lanesborough begs to say she received the House safely, and is greatly pleased with it.

Carriage Paid on all Orders above 40/- *value to the principal Railway Stations in England and Wales.*

CAUTION.—Beware of inferior imitations. All articles bear our name as a guarantee of good workmanship.

The Pheasantry and Pheasant Rearing.

"The Pheasantries of the Rose Lane Works are perfect gems, and have been constructed with much study of the habits of the various classes of Pheasants."—*St. James' Gazette.*

PHEASANTS in confinement certainly do not appreciate our endeavours to make them more comfortable by giving them a roosting house. We may safely assert that the pheasant suffers no inconvenience from the rigour of our winters. They will generally when penned disregard perches, roofs, and other humane conveniences, and squat on the bare earth.

It is a fact that, excepting gamekeepers, very few persons know anything about the general management of aviary pheasants, or in what respects it differs from that of domestic poultry. We therefore feel assured that a few hints on the subject will be acceptable to those who purpose to embellish their lawns or grounds with a Pheasant Aviary.

We may here remark that (with rare exceptions), none of the Pheasants incubate in Aviaries. The eggs are usually hatched under a game hen; the process occupies about twenty-four days. Like poultry, they require nothing beyond warmth for the first day. A coop with a small enclosed run is used for a few days (see Illustration, page 317), to prevent the escape of the newly-hatched Pheasants, as they are prone to creep under the coop, or through a crevice, then to wander away, and perish miserably. Their first food should be finely-chopped hard-boiled egg, bread, and fresh milk, plain egg custard. After a few days add ants' eggs, mealworms, or fatted gentles. Observe that only one or two of these insect delicacies be given at a time, to qualify a meal rather than compose one. Give them food in the smallest quantities; if not eaten up take it away. Feed them often, very early, and as late as compatible with their settling down under the hen, and being closed in by means of a shutter in front of the Coop every night (see Illustration, page 317.) Let them have as much sun, early and late, as you can manage, by turning the coops round; nor let them lack a kindly shade during meridian heat. You may, after a fortnight, bring the young ones on to a diet of dough, composed of Game Meal, of which there are several good makers, but beware of giving it to them in a sticky, adhesive mass, as this would so disgust them that they would hardly look at the preparation again. Follow the directions generally supplied with the meal, and put the dough or paste before your birds in a state of clean friability, that you may roll it between your palms into pellets, and nothing adhere. Our experience is that young Pheasants should not have the run of water at all times. Give them an opportunity of drinking twice a day, but in the interval remove the supply. Above all, see that the water given is pure spring water, or boiled water; not from a ditch, pond, cistern, or butt, which would be swarming with animalculæ, and would fearfully decimate hand-reared Pheasants. Crushed grain may be given with advantage before the birds are old enough to swallow it whole. Finely-chopped onions or leeks are beneficial, especially if cold winds prevail, or wet weather, when a further addition of scalded and bruised hemp-seed may be added to the meal. Young Pheasants should not be confined to the Aviary before they are pretty well four months old.

TESTIMONIAL.

From HIS HIGHNESS THE MAHARAJAH DHULEEP SINGH, Elveden Hall.

GENTLEMEN,—His Highness directs me to say that the Night House with Run for Game, &c., supplied to him here by you, a month ago, has given him every satisfaction. (*Signed*) F. JEFFERIES.

CAUTION.—Beware of inferior imitations. All articles bear our name as a guarantee of good workmanship.

332   BOULTON & PAUL, MANUFACTURERS.

No. 66. LARGE PHEASANT FEEDER.

TESTIMONIAL.
From
J. LEWIS BONHOTE, ESQ.,
Cambridge.
I must express the great satisfaction the goods have given. I shall have great pleasure in recommending you to any of my friends.

REGISTERED COPYRIGHT.

TESTIMONIAL.
From
ROSS MALLAM, ESQ.,
London.
Have received Poultry Trough, which I think very cheap and altogether satisfactory.

THIS Feeder is so constructed that nothing but a pheasant can feed from it. It is also vermin proof, is very simple, and cannot get out of order. To hold 1½ bushels and feed six birds at a time. Painted green.

Cash Prices, Complete.

With Six Feeders	...	£2 10 0
With Two Feeders	...	1 1 0
A Cheaper Pattern...	...	0 12 0

THE "MECO" PHEASANT FEEDER.

REGISTERED COPYRIGHT.

Cash Price 10/6 each.

ADVANTAGES.

1.—No other birds can obtain the food.
2.—Rats may be caught by an ordinary steel trap without fear of catching the pheasants.
3.—Ten or more pheasants can feed simultaneously.
4.—If the male birds attempt to drive the hens off the food, the hens can feed on the opposite side.
5.—By means of the hoppers a large quantity of corn can be stored, and automatically supplied by the vibrating tray to the birds as required.
6.—The hoppers can be adjusted to suit any size corn.
7.—The hoppers and tray may be easily disengaged from interior of coop, and when fastened on exterior make a complete coop for rearing purposes.
8.—The boards placed outside bars prevent pheasants getting caught in rat trap.
9.—The small hopper attached to hinged board, which fills up half of one side of coop, is intended for depositing food suitable for young fowls. The bars in front, being placed about three inches apart, enable them to readily pass, whilst the older chickens have to remain outside.

Perhaps the greatest advantages in favour of these Feeders is the fact of their being in the shape of a coop, as it is found from experience that hand-reared birds, being accustomed to coops, feed from them without the slightest indication of fear. Wild birds also take to them most readily.
They will hold 3½ stone weight of corn.
These advantages being taken into consideration, with the fact of their being useful to game preservers throughout the entire year, place them in a position far superior to any Feeders as yet introduced to the market.

No. 67. NEW PHEASANT FEEDER.
Cheap Pattern.
Cash Prices.

Single Feeder for 4 birds, made in wood   ... 12/- each.
Single Feeder for 4 birds, made in iron, galvanized   ... 17/6 ,,

REGISTERED COPYRIGHT.

Carriage Paid on all Orders above 40/- value to the principal Railway Stations in England and Wales.

CAUTION.—Beware of inferior imitations. All articles bear our name as a guarantee of good workmanship.

ROSE LANE WORKS, NORWICH   333

No. 63. MOVABLE PENS FOR PHEASANT BREEDING.

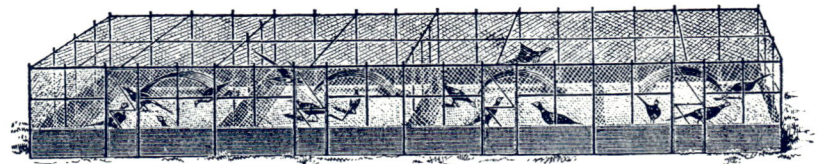

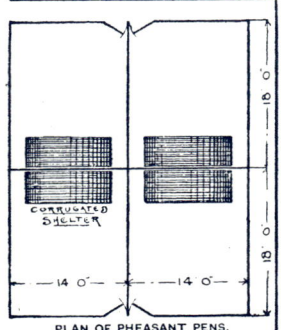

PLAN OF PHEASANT PENS.

Price of Four Pens as plan above, each
18 ft. by 14 ft., £18 0 0

If without string netting top £3 less.
Without Shelters £2 16 less.

THESE Pens are constructed with our Improved Wire Hurdles, and Pens of any size can be readily made by simply bolting the hurdles together. The lower portion, 2 ft. from the ground, is covered with galvanized crimped sheet iron of a special make, which greatly strengthens the hurdle and affords shade and shelter to the birds, and prevents their seeing, and being frightened by passers by. Each Pen has a gate and lock, also corrugated iron shelter. Cord netting is supplied for the top, which prevents birds hurting themselves when flying up. The Pens can be moved to fresh ground each season, if required, without injury to the structure.

Cash Prices, Carriage Paid.

The Price of a Range of Pens is as follows :—

Single Pen, 12 ft. square	£4	0 0
Six Pens	21	0 0
Twelve Pens	40	0 0

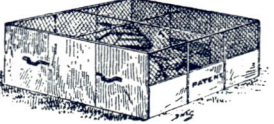

THESE Shelters can be strongly recommended for Pheasantries, as they are very readily put down, are inexpensive, and afford seclusion and shelter ; if covered with brushwood or litter are not unsightly.

Shelters **2/6** per ft. run,
Ends **7/6** each.

PATENT (No. 17,075).

By this arrangement of Movable Pens for Pheasant Breeding, the old plan of fixed pens remaining all the year on the same ground is dispensed with, and a saving of expense is effected.

The dimensions of the Pen are 8 ft. in length by 8 ft. in breadth, made of wood and iron, being entirely closed in upon two sides to afford shelter from the weather, and partly covered in the other sides of the Pen above the pheasants' heads, in order to prevent their seeing, and being frightened at passers by. The rest of the Pen is composed of wire netting.

Low roofs are provided in corners of the Pen for shelter.

Cord netting is placed over the top of the Pen in order to confine the birds, and also to prevent their hurting themselves on flying up.

The bottom of the Pen is formed of galvanized wire netting, through which the pheasants feed.

Handles are arranged so that the Pen *can be moved to fresh ground as often as required*, this provision being very essential ; and also any eggs left upon the ground on the removal of the Pen are thus readily collected without fear of disturbing the birds.

Cash Price, Carriage Paid, £2 10 0 each.

TESTIMONIAL.

From R. J. LLOYD PRICE, ESQ., Rhiwlas.

DEAR SIRS,—I have minutely inspected your **Movable Pheasantry**, and I think it is an exceedingly creditable piece of work, and doubtless, as pheasants are known to do well and lay well in this sort of Pens, they will come, when known, into almost universal use. It is not at all heavy for two men to lift, and the space gives ample accommodation for five hens and a cock. The wire netting does not break the eggs in the act of laying, the birds squat too close ; and it is of incalculable advantage in moving the structure, preventing the handling of the birds, or frightening them by pushing them with the sides along the ground in the act of moving.

Carriage Paid on all Orders above **40/-** value to the principal Railway Stations in England and Wales.

CAUTION.—Beware of inferior imitations. All articles bear our name as a guarantee of good workmanship.

SUDDEN DEATH MOUSE TRAP.

This is the most certain trap ever invented for the destruction of rats and mice, they simply run into it for shelter, and none escape.

Cash Price 2/6 each.

Place trap against the wall, and the mice will run into it for shelter.

Larger size for rats, 4/6 each.

If there is a mouse in the house this trap will catch it.

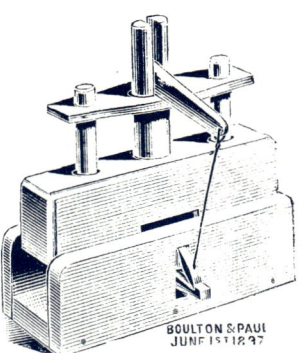

HOUSE FOR CATCHING SPARROWS
AND OTHER BIRDS.

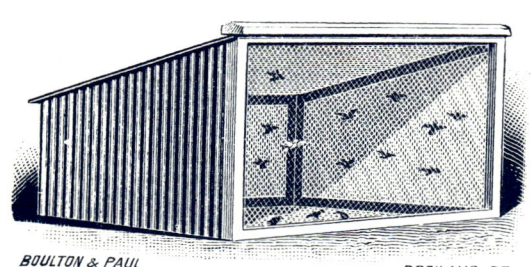

A FOLDING net is arranged so that it can be lifted up and quietly dropped when the birds are in the shed.

Size of shed 8 ft. by 6 ft., 5 ft. high in front.

It can also be used as a Poultry Shelter, or for other purposes.

Cash Price £4 10 0
Carriage Paid.

BAIT CAGE
FOR KEEPING FISH ALIVE.

Special Prices quoted according to sizes required.

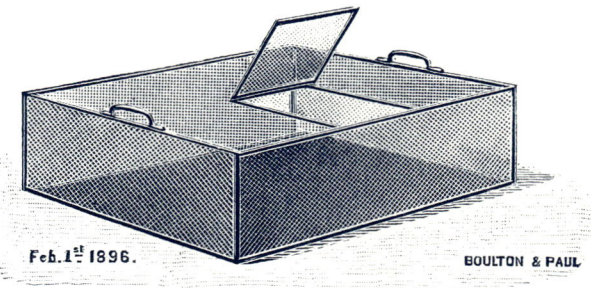

CAUTION.—Beware of inferior imitations. All articles bear our name as a guarantee of good workmanship.

BOULTON & PAUL, MANUFACTURERS,

HOLDFAST PATENT LEVER VERMIN TRAPS.

"SAFE BIND, SAFE FIND."

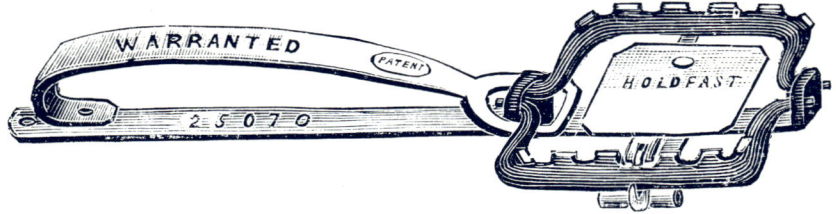

Each Trap warranted, registered, and numbered.

THESE Traps have been proved by demonstration to be the cheapest and most deadly traps ever made. Hundreds of testimonials from the chief keepers in the kingdom.

A Wheel Lever.   B Plain Lever.

  a Tiller Plate.
  b Lever.
  c Wheel securing jaw.
  d Catch securing jaw.

Method of Setting Trap with Diagram sent with all Orders.

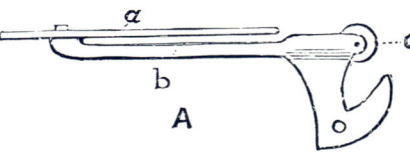

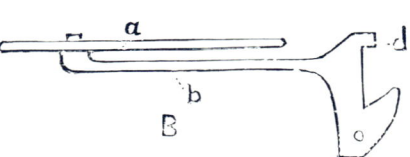

Cash Prices, per doz.

	3 in.	4 in.	4½ in.	Pole Traps.
Plain Lever Traps	14/-	15/-	17/6	24/-
Lever Wheel Traps	17/6	18/6	21/-	30/-
	7 in.			8 in.
Lever Wheel Traps	48/-	For Wolves, &c.		56/-
,, ,, ,, Special	56/-	,,	,,	60/-

TESTIMONIAL.
SANDRINGHAM, NORFOLK.

I am very much pleased with the traps duly received, I find them well made and useful traps, and shall not fail to order patent lever when requiring more traps.

CHAS. H. JACKSON,
Head Keeper to H.R.H. the Prince of Wales.

WIRE CAGE OR TRAP.
FOR CATCHING ANY SORT OF GAME OR VERMIN ALIVE.
(PROVISIONAL PROTECTION GRANTED.)

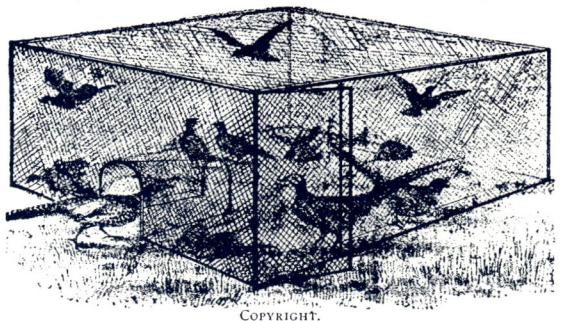

No Injury to Hounds or Foxes.

Requires no Watching is Always Set.

THE Cage is constructed of light iron frames, covered with Galvanized Wire Netting, folds flat for travelling, and is easily put together with bolts and nuts. It is specially suited for taking up Pheasants for the pens. It is always set, requires no watching, and will catch a number of birds at one time without damaging their feathers.

To feed for Game, Sparrows, etc., sprinkle corn—or, better, hemp seed—on the ground inside the trap, and for vermin fix ordinary bait. The ground should be well baited for a few days previous to setting the trap, and, when fixed, the top should be covered with dry fern or fir boughs, to make the trap less conspicuous and more secluded. When the trap is not in use take the top off to allow the birds to fly out.

The birds or vermin enter the pocket or mouth, which is so constructed that they do not appear to be able to find the outlet, a very small percentage escaping.

No. 3. TRAP FOR CATCHING WEASELS, STOATS, & RATS. &c.

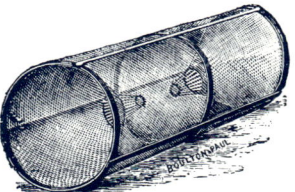

Cash Prices.

No. 1. 4 ft. long, by 3 ft. 6 in. wide, by 2 ft. 6 in. high, covered with 1½-in. mesh netting, for catching Pheasants, Partridges, Rabbits, Hawks, and Wood Pigeons **15/-**

No. 2. 3 ft. 6 in. long, by 2 ft. 6 in. wide, by 2 ft. 6 in. high, covered with ¾-in. mesh netting, for catching Blackbirds, Skylarks, Sparrows, and Cats **15/-**

No. 3. 2 ft. 6 in. long, by 18 in. wide, by 18 in. high, for catching Weasels, Stoats, and Rats **10/-**

This Trap should be set in corner of a wood for Weasels.

TRAP FOR CATCHING EELS.

THESE Traps are made of light wrought-iron hoops, covered with fine mesh galvanized wire netting, and stayed by wrought-iron bars. Each end is fitted with an internal conical pocket of peculiar design, by which the eels enter the trap. Made in lengths of 3 ft. 4 in. by 20 in. in diameter.

Cash Price 12/6 each.

REGISTERED COPYRIGHT.

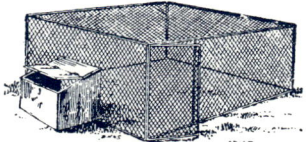
Illustration showing the Cage Trap when used as a run for sitting hen.

From ARTHUR W. PARTRIDGE, ESQ., Watton.
The Wire Eel Traps I had of you are the best I ever used. I have caught large quantities of eels with them.

From J. F. WIGMORE, ESQ., Barton Stacey, Hants.
Mr. F. Adams used his invention (Wire Traps) while he was here as keeper, and was very successful with them, catching a number of Partridges alive. I consider it a most simple way of catching birds without any injury to them.

Carriage Paid on all Orders above 40/- value to the principal Railway Stations in England and Wales.

CAUTION.—Beware of inferior imitations. All articles bear our name as a guarantee of good workmanship.

ROSE LANE WORKS, NORWICH. 335

No. 51. OPEN PHEASANTRY.

REGISTERED COPYRIGHT.

IT is a light and elegant structure, eminently fitted for the purpose (being made in various sizes) of accommodating either a trio of Golden Pheasants or Silver Pheasants, or a family of Common Pheasants, consisting of from four to seven hens and a cock.

It is manufactured in separate frames, so that any one can erect it, or readily shift it when necessary to fresh ground. Besides affording an opportunity for studying the habits of one of the most interesting and beautiful class of birds in existence, it constitutes an ornament, which together with its occupants, must be regarded as a desirable acquisition by every nobleman and gentleman in the country. This form of Aviary can also be supplied in miniature for tropical birds, to stand in a conservatory.

SPECIFICATIONS AND ESTIMATES ON APPLICATION.

It will save correspondence if Plan, with dimensions of ground to be covered, is sent with inquiries.

From MRS. GRAHAM SMITH, Maidenhead.

I am perfectly satisfied and pleased with the Pheasantry that you have just erected at "The Fishery."

No. 52. NEW PORTABLE PHEASANTRIES.

CAN be built to enclose large or small spaces, and adapted to almost any site, varying from 7 ft. to 10 ft. in height. The uprights are of wrought-iron, moulded on the face, and surmounted by cast-iron capital and base to form pillars, spaced from 3 ft. to 4 ft. apart; to these are fixed wrought-iron frames ornamented with scrolls, and covered with ¾-in. mesh galvanized wire netting (sparrow-proof); cast-iron cresting as shown. Top of Aviary is covered with ½-in. mesh wire netting, resting on wrought-iron bearing bars provided for the purpose. Gates with padlocks to each compartment. The framework painted black; pillars and cresting of approved tints.

Cash Price, 7 ft. high, 12/6 per yard run.

Shelters of woodwork supplied, if required, to stand independent of the ironwork, at an extra charge.

Aviary with four compartments, each 10 ft. by 16 ft., covering space 40 ft. by 16 ft., having a back wall.

Cash Price £50 0 0

REGISTERED COPYRIGHT.

TESTIMONIAL.

From W. H. MARTIN, ESQ., Putney.

I beg to inform you the Pheasantry is perfection, at the same time not only useful but ornamental. I should recommend it to any of my friends requiring the same.

Carriage Paid to the principal Railway Stations in England and Wales.

CAUTION.—Beware of inferior imitations. All articles bear our name as a guarantee of good workmanship.

PORTABLE PHEASANTRY, POULTRY, OR PIGEON HOUSE.

REGISTERED COPYRIGHT.

THE Roosting House is made of wood, well painted, with open wire Run, fitted with nest boxes and egg locker. The floor boards and perches are loose, to facilitate lime-whiting. It is easily taken to pieces for packing. The construction is so simple, that they can be readily fixed by inexperienced persons. Can be fitted as a Pigeon House or Aviary.

TESTIMONIAL.

From C. NIGEL STEWART, ESQ.,
Oak Cottage, Hoole, near Chester.
Mr. C. Nigel Stewart wishes to intimate to Messrs. Boulton and Paul that he is in every way satisfied with the Pheasantry, and considers it first-class, and very cheap.

Cash Prices, Complete.

First size,	No. 1.	12 ft. long, 4 ft. wide, 6 ft. high	£5 10 0
Second size,	No. 2.	12 ft. long, 5 ft. wide, 6 ft. 6 in. high	6 10 0
Third size,	No. 3.	16 ft. long, 6 ft. wide, 7 ft. high	8 10 0

No. 13. IMPROVED PORTABLE PHEASANTRY OR POULTRY HOUSE.

THE Illustration represents an Ornamental Poultry House or Pheasantry, suitable for shrubbery or garden. The roof is covered with curved galvanized corrugated-iron sheets, with ornamental cresting. The Night House is made of iron, and *cased inside with wood*, thereby rendering the House warm and dry in all weathers. The Run is covered with sparrow-proof galvanized wire netting. It can be fitted as a Pigeon House or Aviary if required.

REGISTERED COPYRIGHT.

Cash Price.

16 ft. long, 6 ft. wide, 7 ft. high £11 10 0

TESTIMONIAL.

From WILLIAM OELRICH, ESQ., Birkenhead.
The Fowl House has been put together to-day, everything being found to fit exactly, and I am much pleased with its appearance and usefulness.

Carriage Paid to the principal Railway Stations in England and Wales.

CAUTION.—Beware of inferior imitations. All articles bear our name as a guarantee of good workmanship.

Aviaries and their Management.

The Aviary.—Tropical birds will of a necessity be relegated to the Indoor Aviary, placed either in a conservatory, or in a room of sunny aspect, with good ventilation, yet free from draughts. Canaries of most varieties will thrive alike in or out of doors continuously; together with our native birds out of doors, canaries form a pleasing contrast.

Selection.—Those whose tendencies have led them to ornithological study, however slight, will have little difficulty in determining, by the formation of the subject, whether it be of an insectivorous or of a seed-eating variety—carnivorous or graminivorous. These last two orders must of course have a separate

compartment. They include Hawks, Owls, Ravens, Magpies, Jays, Daws, and the Shrike. There may be at any time bickering amongst the inmates of an Aviary, so much so, that the wiser plan may be to remove the pugnacious one.

Feeding.—The food for Finches, Linnets, and hard-billed birds, consists chiefly of seeds of various sorts. The best staple is canary seed; other seed should be given sparingly, or only as a change in diet. They are:—Maw seed, Rape seed, Hemp seed, and Millet. Spray Millet is best, as it gives some occupation to the birds in pecking the seeds out of the husk. Almost every sort of grain is in turn relished by some of the larger of Aviary-kept birds; and should be determinable by reference to their food in a state of freedom. Green food is an important item. It includes:—Groundsel, Plantain, Chickweed, Watercress, &c. Some should be given daily; note, however, not to give it in a wet state. German paste (how to make it anon) may be given to this order of birds, *i.e.*, Canaries, Finches, etc.; and a liberal supply of good, clean, sharp, gritty sand is indispensable. Cuttle fish (not the powder), a chalky stuff, may be purchased at most bird shops; together with sand this serves the purpose of teeth, and accomplishes in the gizzard the trituration or grinding up of birds' food. Water, pure and sweet, and shallow bathing-pans will of course not be overlooked.

Food.—We now come to the soft-billed birds, as the thrush, starling, blackbird, the warblers (including nightingale, blackcap, whitethroat), larks, pipits, wagtails, robins, titmice. These require a different food, containing, as nearly as we can compound it, a substitute for insect food; and when insects are obtainable, as snails, caterpillars, spiders, woodlice, beetles, etc., collect them for the orders of birds enumerated under soft-billed birds. Meal-worms are greatly relished, and they may be reproduced to any extent readily by placing a handful of them in a box or vessel with half a peck of any coarse meal, a few pieces of old sacking in it, and some bones; cover them up and put away in an out-house. Meal-worms are the larval form of a beetle, so that when you discover beetles amongst meal-worms, you will know that the desired process is going on. The staple food for soft-billed birds is termed generally "German Paste," which may be made in the following manner:—Take a pound of meal (three parts peameal and one coarse oatmeal), three ounces of raw sugar or honey, work it into a paste with sufficient salad oil, then add half a pint of well-bruised hemp seed with a dash of maw seed. Vary the food as circumstances require it with the addition of chopped hard-boiled egg, scalded captain's biscuit, ants' eggs fresh or preserved (these last can be bought at all bird dealers'), scraped beef or mutton, or ox heart boiled, dried, and put through a wire sieve; during moult, and when out of sorts, these are the stimulants to give. We lean rather to keeping a morsel of sulphate of iron in the drinking vessels, in place of the rusty nail generally so much relied on.

CAUTION.—Beware of inferior imitations. All articles bear our name as a guarantee of good workmanship.

BOULTON & PAUL, MANUFACTURERS.

No. 55. PORTABLE INDEPENDENT OUT-DOOR AVIARY.

REGISTERED COPYRIGHT.

TESTIMONIALS

From LIEUT.-COL. CHAS. CRIGHTON,
Mullahoden, Naas.

The Out-door Aviary I got from you a few months ago is most successful and satisfactory.

From MRS. GODDARD, Silton Lodge.

MRS. GODDARD sends cheque for account. Had the Aviary put together yesterday, and likes it much.

THIS Aviary can be taken to pieces for removal in a few minutes. It is the best constructed Out-door Aviary that is made. The wood is well-seasoned, varnished, with flight of fine wire netting, galvanized, fixed to light wood framing.

Cash Prices, Carriage Paid.

4 ft. by 4 ft., 4 ft. 3 in. high at eaves, 5 ft. 9 in. high to ridge	£5 10 0
6 ft. by 6 ft., 5 ft. high at eaves, 7 ft. high to ridge	10 0 0

No. 56. OCTAGONAL OUT-DOOR AVIARY.

REGISTERED COPYRIGHT.

TESTIMONIAL.

From BOSWELL G. JALLAND, ESQ.,
Sutton.

GENTLEMEN.—The Aviary supplied by you was erected on my lawn yesterday. It has a very pretty appearance, and gives great satisfaction. It appears to be very strong, and is constructed on the best principle of any Aviary I have seen, a great point in its favour being the ease with which it can be taken to pieces, removed, and re-erected.

MADE entirely of wrought-iron, with zinc roof, and covered small mesh galvanized wire netting, fitted with revolving spiral perches complete. Being in sections, it packs in a very small compass, and is easily removed and put together. The parts are joined with small bolts and nuts, and can be put together by inexperienced hands.

It can be made in miniature for tropical birds, to stand in a conservatory.

The sides should either be matted up or enclosed with glass shutters in cold weather.

Cash Prices, Carriage Paid.

6 ft. diameter	8 ft. 6 in. high	£8 8 0
8 ft. ,,	10 ft. 0 in. ,,	10 10 0
10 ft. ,,	11 ft. 6 in. ,,	15 0 0

Special Designs given for Aviaries to suit the situation.

Carriage Paid to the principal Railway Stations in England and Wales.

CAUTION.—Beware of inferior imitations. All articles bear our name as a guarantee of good workmanship.

No. 53. ORNAMENTAL AVIARIES.

REGISTERED COPYRIGHT.

TESTIMONIALS.

From Mons TIBURCIO CASTANEDA, Madrid.
The two Octagonal Aviaries you sent me here look very pretty and elegant.

From Mrs. GODDARD, Silton Lodge.
Mrs. Goddard sends cheque for account. Had the Aviary put together yesterday, and likes it much.

MADE entirely of wrought-iron, covered with fine wire-work. The columns are wrought-iron, surmounted with ornamental caps and bases. Cast-iron fountains with basins can be supplied with the Aviaries.

We have every facility for carrying out work of this kind.

Designs and Estimates for any situation required.

"The Aviaries manufactured by MESSRS. BOULTON AND PAUL are of various designs, and are fitted up with every accommodation for the birds, at the same time they are made in a tasteful and ornamental manner, while the workmanship is of a very high character."—*Farm Journal.*

No. 54. SEMI-OCTAGONAL OUT-DOOR AVIARY.

REGISTERED COPYRIGHT.

TESTIMONIALS.

From G. MILLER, ESQ., Granville Park.
Much obliged for the Aviary, which has arrived safely, and is a very nice one.

From B. WAKE, ESQ., Abbeyfield.
DEAR SIRS,—The Aviary you have sent appears to be very good work and very well done.

Made entirely of wrought-iron, and covered with fine wire-netting, and zinc roof.

Cash Prices, Carriage Paid. (Without Lean-to Aviary attached.)

	£ s. d.
12 ft. wide, 9 ft. back to front	£13 0 0
14 ft. 6 in. wide, 12 ft. back to front	18 0 0

Carriage Paid to the principal Railway Stations in England and Wales.

CAUTION.—Beware of inferior imitations. All articles bear our name as a guarantee of good workmanship.

BOULTON & PAUL, MANUFACTURERS.

PEAFOWL HOUSE OR AVIARY.

Special Designs for Aviaries prepared to suit any situation.

Sites Surveyed and Estimates prepared free of charge in the event of order.

As erected for PANMURE GORDON, ESQ., Rickmansworth.

FLIGHT enclosure is constructed of wrought-iron work. Roosting house constructed of wood with ornamental cornice and mouldings. Lighted from tinted glass windows and ventilated through openings covered with ornamental perforated zinc. Dome covered with zinc and finished with terminal. Painted three coats any approved tint.

Iron Fountains or other Fittings supplied if desired.

AVIARY OR ORNAMENTAL FOWL HOUSE.

Designed for placing against a Wall.

TESTIMONIAL.
From
B. WAKE, ESQ.,
Abbeyfield,
Sheffield.
DEAR SIRS,—The Aviary you have sent appears to be very good work and very well done.

Constructed as above, but with iron roof and without windows.

We have every facility for carrying out this kind of work. Estimates upon application.
Carriage Paid on all Orders above 40/- *value to the principal Railway Stations in England and Wales.*

CAUTION.—Beware of inferior imitations. All articles bear our name as a guarantee of good workmanship.

No. 73. ORNAMENTAL COMBINATION AVIARIES AND PIGEON COTES.

Wood Cotes and roosts stained and varnished outside, ornamental wrought-iron and wire trellis enclosures to flights, painted, and finished with crestings and terminals as shown.   Size 45 ft. long by 15 ft. wide.

Prices on application.

TESTIMONIAL.
From S. UNDERWOOD, Esq., Sevenoaks.
The Aviary you sent me is now put up, and I am very much pleased with it.

No. 74. COMBINED PHEASANTRIES AND AVIARIES.

As erected for the REV. J. G. COTTON BROWNE, Walkern.

SPECIAL estimates and designs on application for Aviaries or Pheasantries of this description to suit any situation.

Surveys made free of charge.

TESTIMONIAL.
From the REV. J. G. COTTON BROWNE, Walkern.
I like the Aviaries very much ; your men have done their work well.

Carriage Paid on all Orders above 40/- *value to the principal Railway Stations in England and Wales.*

CAUTION.—Beware of inferior imitations.   All articles bear our name as a guarantee of good workmanship.

ROSE LANE WORKS, NORWICH.   343

No. 78. COMBINATION COTE FOR PET BIRDS, RABBITS, GUINEA PIGS, &c.

MATCHBOARD sides, back and divisions, weather-board roof, painted outside, whitened inside, wood frame and wire lattice front to flights, wrought-iron bars to rabbit hutch, floor raised one foot above the ground.

Cash Price,
Carriage Paid, £12

TESTIMONIAL.

From
T. W. CALVERLEY, ESQ.,
Allerthorpe Hall.
I am highly satisfied with the Pheasantry and Pigeon Loft in every way, and shall be glad to recommend you amongst my friends.

No. 79. WOOD AND WIRE LATTICE AVIARY AND PIGEON COTE.

No. 77. CONSERVATORY OR IN-DOOR AVIARIES OF VARIOUS DESIGNS.

8 ft. diameter, 16 ft. high. Stained or painted in two tints.
Cash Price, Carriage Paid, £16 10 0

TESTIMONIAL.
From SAML. ROGERS, ESQ., Bath.
I am very well pleased with the Pigeon House—it is so well carried out and all finished in such a satisfactory manner.

TESTIMONIALS.

From J. H. CANTON, ESQ., Stafford.
We received the Aviary safely this morning, and are very much pleased with it.

From MRS. GODDARD, Silton Lodge.
Mrs. Goddard had the Aviary put together yesterday, and likes it much.

From S. UNDERWOOD, ESQ.
The Aviary you sent me is now put up, and I am very much pleased with it.

ELEVATION.

Constructed of wrought-iron and lattice wire, and finished according to price.

Carriage Paid on all Orders above 40/- value to the principal Railway Stations in England and Wales.

CAUTION.—Beware of inferior imitations.   All articles bear our name as a guarantee of good workmanship.

ROSE LANE WORKS, NORWICH. 343A

AVIARIES.

No. 17. AVIARY
for placing inside Conservatory.

CONSTRUCTED with Wrought-iron Frames covered with Wire Lattice, having stained and varnished wood base.

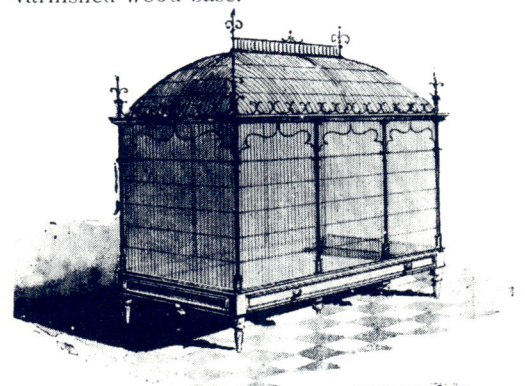

No. 19. OUTDOOR AVIARY
with Cosy Sleeping Compartment.

CONSTRUCTED of red deal, stained and varnished, and well finished. Flight enclosed with Wire Lattice.

No. 18. OCTAGONAL AVIARY.

THIS Aviary is similar in construction to No. 17.

Special quotations for these Aviaries given on receipt of particulars as to requirements.

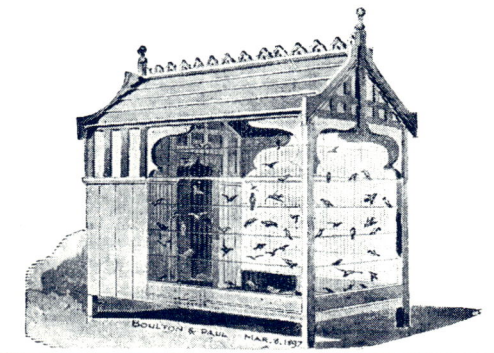

CAUTION.—Beware of inferior imitations. All articles bear our name as a guarantee of good workmanship.

No. 75. ORNAMENTAL TRIPLE AVIARY.

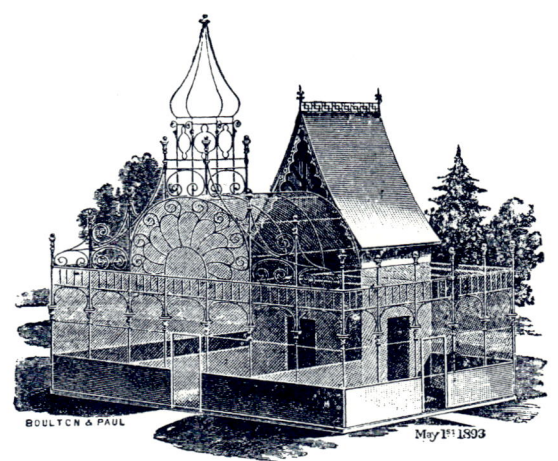

This Aviary is divided into three flights, with three roosts within one house.

The House constructed of wood, stained and varnished, with painted weather-board roof, wrought-iron and wire lattice flights.

Estimates on application.

TESTIMONIAL.

From HENRY TATE, Esq., Streatham

I am glad to say I am very satisfied with the work you have done.

No. 76. RANGE OF ORNAMENTAL PIGEON COTES AND FLIGHTS.

As erected by us for ROBERT PRESTON, Esq., Ellel.

Wood House with shelter beneath, painted weather-board roof, wrought-iron and wire lattice flights.

Prices on application.

TESTIMONIAL.

From OSCAR TROEGAR, Esq., Kirn.

The Aviary arrived quite safe, and I am very pleased with it.

REVIEW.

"The Aviaries manufactured by Messrs. BOULTON AND PAUL are of various designs, and are fitted up with every accommodation for the birds; at the same time they are made in a tasteful and ornamental manner, while the workmanship is of a very high character."—*Farm Journal.*

Carriage Paid on all Orders above 40/- value to the principal Railway Stations in England and Wales.

CAUTION.—Beware of inferior imitations. All articles bear our name as a guarantee of good workmanship.

ROSE LANE WORKS, NORWICH. 344A

ORNAMENTAL AVIARY AS ERECTED FOR H.R.H. THE DUKE OF YORK AT SANDRINGHAM.

Estimates with full particulars free on application.

CAUTION.—Beware of inferior imitations. All articles bear our name as a guarantee of good workmanship.

AVIARIES.

Specially designed for Interiors of Conservatories. Constructed of wrought-iron and lattice wire, and finished according to price.

No. 10.

ELEVATION.

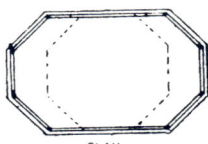

PLAN.

AVIARY for Centre Space.

No. 13.

ELEVATION.

AVIARY for placing in Centre Space or against a Wall.

MONKEY HOUSE OR AVIARY.

REGISTERED COPYRIGHT.

CONSTRUCTED of red deal, having shaped gable and turned finials, well finished, stained, and varnished, and having lattice wire front, and glazed doors for closing at night, or when weather is cold, well ventilated. Size—3 ft. long, 2 ft. wide, 5 ft. 6 in. high.

Cash Price, Carriage Paid, from £6 10 0.

Estimates and Catalogues free on application.

No. 11.

ELEVATION.

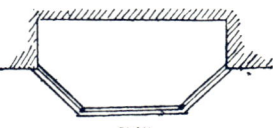

PLAN.

AVIARY for placing in Recess.

No. 12.

ELEVATION.

AVIARY for placing in Centre Space or against a Wall.

No. 14.

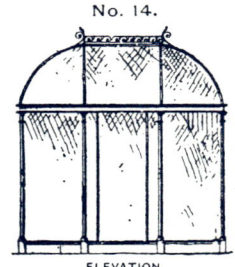

ELEVATION.

AVIARY for placing in Centre Space or against a Wall.

CAUTION.—Beware of inferior imitations. All articles bear our name as a guarantee of good workmanship.

No. 125. EAGLE HOUSE AND PHEASANTRY.

As erected by us for E. Sydney Woodwiss, Esq., Upminster.

Prices on application.

TESTIMONIAL.
From E. SYDNEY WOODWISS, Esq.
Your men have erected the Bear House, Aviary, Kennel Railing, &c., to my satisfaction.

No. 126. AVIARY FOR TROPICAL BIRDS.

As erected by us in the Nottingham Arboretum.

Special designs for Aviaries prepared to suit any situation. Sites surveyed and estimates prepared free of charge.

Prices on application.

TESTIMONIAL.
From W. H. WHITCHURCH, Esq., Nottingham.
I am greatly satisfied with the Aviary you made me, and think it will answer my purpose in every respect.

Carriage Paid to the principal Railway Stations in England and Wales.

CAUTION.—Beware of inferior imitations. All articles bear our name as a guarantee of good workmanship.

ISBN 0-921335-21-0